MODERNISATION, MECHANISATION AND INDUSTRIALISATION OF CONCRETE STRUCTURES

MODERNISATION, MECHANISATION AND INDUSTRIALISATION OF CONCRETE STRUCTURES

Edited by

Kim S. Elliott
Precast Consultant, Derbyshire, UK

Zuhairi Abd. Hamid
*Construction Research Institute of Malaysia (CREAM),
Kuala Lumpur, Malaysia*

WILEY Blackwell

Library of Congress Cataloging-in-Publication Data:

Names: Elliott, Kim S., editor. | Hamid, Zuhairi Abd., editor.
Title: Modernisation, mechanisation and industrialisation of concrete
    structures / edited by Kim S. Elliott, Zuhairi Abd. Hamid.
Description: Chichester, UK ; Hoboken, NJ : John Wiley & Sons, 2017. |
    Includes bibliographical references and index.
Identifiers: LCCN 2016044825| ISBN 9781118876497 (cloth) | ISBN 9781118876510
    (epub)
Subjects: LCSH: Concrete construction industry.
Classification: LCC HD9622.A2 M63 2017 | DDC 338.4/76241834 – dc23 LC record available at
https://lccn.loc.gov/2016044825

A catalogue record for this book is available from the British Library.

Cover Design: Wiley
Cover Image: Background image: Smileyjoanne/Gettyimages
              Top landscape image: wsfurlan/Gettyimages
              Left bottom image: Highly automated plant in 2013/RIB SAA Software Engineering GmbH
              Centre bottom image: bikeriderlondon/Shutterstock
              Right bottom image: Canadastock/Shutterstock

Set in 10/12pt MinionPro by SPi Global, Chennai, India
Printed and bound in Malaysia by Vivar Printing Sdn Bhd

10 9 8 7 6 5 4 3 2 1

Contents

About the Editors

Kim Stephen Elliott, Precast Consultant, Derbyshire, UK

Kim Stephen Elliott is a consultant to the precast industry in the UK and Malaysia. He was Senior Lecturer in the School of Civil Engineering at Nottingham University, UK from 1987–2010 and was formerly at Trent Concrete Structures Ltd. UK. He is a member of fib Commission 6 on Prefabrication where he has made contributions to six manuals and technical bulletins, and is the author of Multi-Storey Precast Concrete Frame Structures (1996, 2013) and Precast Concrete Structures (2002, 2016) and co-authored The Concrete Centre's Economic Concrete Frames Manual (2009). He was Chairman of the European research project COST C1 on Semi-Rigid Connection in Precast Structures (1992–1999). He has lectured on precast concrete structures 45 times in 16 countries worldwide (including Malaysia, Singapore, Korea, Brazil, South Africa, Barbados Austria, Poland, Portugal, Spain, Scandinavia and Australia) and at 30 UK universities.

Zuhairi Abd Hamid, Construction Research Institute of Malaysia (CREAM), Kuala Lumpur, Malaysia

Zuhairi Abd. Hamid has more than 32 years of experience in the construction industry. His expertise lies in structural dynamics, industrialised building systems, strategic IT in construction and Facilities Management. He is active in engineering education and research and has been appointed by universities in various capacities; from Adjunct Professor, Research Fellow and member of the industry Advisory Panel. He is the Regional Director of South East Asia and Guest Member for the UN support International Council for Research and Innovation in Building and Construction (CIB). Currently, as the Executive Director at CREAM, he actively engages in construction research for industrial publications in Malaysia.

Notes on Contributors

Ahmad Hazim Abdul Rahim, Construction Research Institute of Malaysia (CREAM), Kuala Lumpur, Malaysia

Ahmad Hazim Abdul Rahim is currently the Manager at Construction Research Institute of Malaysia. He has been involved in research and development (R&D) fields in construction and civil engineering for more than 15 years. He currently heads the structural engineering laboratory at CREAM, an ISO/IEC 17025 accredited laboratory and in charge of the technical and quality aspect of the day to day operational of the laboratory. He holds a B. Eng (Hons) in Civil Engineering and completed his Masters in Engineering Science in 2007. His area of interests is conformity assessment of structural component, full-scale structural test and construction materials testing and evaluation. He has published various publications ranging from books to peer-reviewed journals, technical papers, proceedings and articles.

Foo Chee Hung, Construction Research Institute of Malaysia (CREAM), Kuala Lumpur, Malaysia

Foo Chee Hung is a researcher from the Construction Research Institute of Malaysia (CREAM) – a research arm under the Construction Industry Development Board Malaysia (CIDB). He is currently the Head of Consultancy and Technical Opinion Unit.

He obtained both his first (Environmental Engineering) and master (SHE Engineering) degree at the University of Malaya. He then further pursued his PhD (Urban Engineering) in the University of Tokyo. His research interest is in sustainability, affordable housing, green building, building quality assessment, Industrialized Building System (IBS), and urban ecosystem.

He is a member of the Institution of Engineers Malaysia (IEM), and the GreenRE manager.

Gan Hock Beng, G&A Architect

Gan Hock Beng is the Founder of G&A Architect. He is currently engaged in a number of residential and commercial projects in Georgetown, Penang. The projects he has worked on include landmarks like Times Square, Moonlight Bay, University Place, The View, etc. He has been the invited speaker at conferences organized by the Singapore Ministry of Housing, and the South Korea Ministry of Housing. He was given the award of "Most Innovative Design" in a competition organized by the Ministry of Housing Malaysia.

Susanne Schachinger, Precast Software Engineering, Wals-Siezenheim, Austria

Susanne Schachinger is an International Sales Representative at Precast Software Engineering. As a co-author she brings in expertise from her daily work with precast companies in various countries. University studies in Graz (Austria), Volgograd (Russia) and Prague (Czech Republic) led her to the consumer goods industry before she joined Precast Software Engineering in 2011. She represents the company at IPHA (International Prestressed Hollowcore Association).

Thomas Leopoldseder, Precast Software Engineering, Wals-Siezenheim, Austria

Thomas Leopoldseder works as an international BIM Consultant and Product Manager of TIM (Technical Information Manager) at Precast Software Engineering in Austria (Part of Nemetschek Group) which is developing high-end CAD and BIM solutions for the precast industry. After studying at the Vienna University of Economics and Business, he first worked as an IT consultant and then at various levels (CFO, General Manager) in the precast industry. He is now one of the leading experts in BIM solutions concerning the precast industry.

Robert Neubauer, SAA Software Engineering GmbH, Austria

Neubauer has been a Managing Partner at SAA Software Engineering in Vienna, Austria since 1999, leading the development of CAM and Control-Software for the precast concrete manufacturing industry. In 1986 he was engaged in automation in the construction industry, realizing the first control- and master-computer-software for the first automated precast concrete plant. After graduating in mechanical engineering at the Technical University of Vienna in 1993 he was previously at Ainedter Industry Automation and at Sommer Automatisierungstechnik (Austria).

During the past 30 years, Mr Neubauer has been working on automation in the prefabrication industry for construction, developing and conducting development for new solutions, collaborating with different vendors for plants and machinery and leading committees. At the end of 2015, SAA merged with RIB Software AG/Germany, and as Managing Partner, he is accompanying RIB SAA Software Engineering GmbH into a BIM-5D integrated future for smart production of construction systems.

Gerhard Girmscheid, ETH, Swiss Federal Institute of Technology, Zurich, Switzerland

Gerhard Girmscheid, studied construction engineering in Darmstadt (Germany), occupying management posts at German and American construction enterprises, involving assignments abroad that included major construction sites in Egypt and Thailand, as well as the fourth tunnel tube under the River Elbe in Hamburg. Since 1996, he has been Professor of Construction Business Management and Construction Process Engineering at ETH Zurich (Switzerland). He was recently awarded emeritus status.

In research and teaching, Professor Girmscheid focuses on construction processes, and strategic and operational construction enterprise management. His SysBau® research targets improved, more efficient, and new sustainable life cycle-oriented construction processes and portfolios aimed at strengthening the innovative and competitive abilities of construction industry providers. His research activities have produced numerous dissertations and research reports, together with more than 100 peer-reviewed specialist publications. He has written several books on construction enterprise and process management.

He sits on the Board of Directors of general contracting and property company Priora AG, and prefabrication specialists Müller-Stein AG. He also manages CTT-Consulting in Lenzburg (Switzerland), advising companies and training staff on improving bidding and execution processes, managing claims, and implementation.

Julia Selberherr, ETH, Swiss Federal Institute of Technology, Zurich, Switzerland

Julia Selberherr received her Civil Engineering diploma from the Vienna University of Technology in 2009 as well as her Business Management diploma from the Vienna University of Economics and Business in 2010. She then conducted a research project focused on the provision of sustainable life-cycle offers in the building industry at the Institute of Construction and Infrastructure Management at the Swiss Federal Institute of Technology. Her research is dedicated to the optimization of operational and strategic processes across a building's life cycle through the integrated cooperation of the key stakeholder using customizable industrial production technologies. She has contributed several international journal papers and conference publications establishing innovative approaches to a life-cycle service provision in the building industry. Dr. Selberherr completed the project with the development of a new business model in her PhD thesis in 2014. As a renowned expert in the field of organization and process design for sustainable building, she is currently working as a senior consultant in the real estate industry in Zurich.

Preface

The modernisation and industrialisation of concrete structures, through the means of prefabrication of concrete elements together with the computerization of design, detailing and scheduling, is taking an awful long time to come to fruition. The once aspired paperless journey from the architect's concept to the factory floor and beyond is gradually closing in. Critics may cite the post WW2 boom in the construction of high-rise apartment buildings in part of northern and eastern Europe as 70-year-old industrialisation, but it was nothing more than concrete construction on such a large scale that it was thought to be "industrialisation" - the linear and manual processes of design, detailing, scheduling and manufacture were no more advanced than early twentieth-century construction.

There was little automation in the concrete industry until the combined technologies of long line pretensioning of steel wires and the extrusion of semi-dry concrete lead to such elements as prefabricated hollow core floor slabs in the 1950s. Except for a number of step-change advancements such as (i) the hydraulic extruder, (ii) the "carousel" method of casting and moving beds, and (iii) higher performance/strength concrete, hollow core units are still made in much the same way. Changes came in the 1970s after the Japanese taught the Europeans and Americans how they made cars – with forward/sideways/up/down production of the individual components leading to the whole. We now see such automotive methods used in the carousel table-top production of concrete wall panels and façade units, and together with CAD/CAM, Auto-CAD systems, TIM scheduling, and the automated supply of drawings and component schedules to the factories, the age of modernisation, mechanisation and industrialization (MMI) of concrete structures has finally arrived.

Architects and consulting engineers are still wary of the term "building systems", with images of shoe-box designs typified by the 1960s national building frames, carelessly exploiting modularization and standardisation possible in precast concrete. Today fully bespoke and individually tailored precast concrete elements can be designed and erected into many diverse forms to cover the huge spectrum of building architecture – all of which are industrialised by MMI. The term IBS (Industrialised Building Systems) can now be used with architectural and engineering freedom, for example, Sydney Opera House's torus-shaped prestressed beams and tiled facades. During a precast concrete workshop in Singapore in the mid 2000s an architect asked (something like) "what are the major features that distinguish precast concrete buildings from cast *insitu*". The reply, given by one of the authors of this book, was that "precast is used when the client or architect sees concrete as something special – both structurally of aesthetically, and maximises the operations that you can only carry out in the controlled environment of a factory", and so on.

The move to increased automation in the factory has coincided with the automation of spatial design – the use of three-dimensional co-ordinate orthogonal geometry, well known to school boys, to build 3D models from rectilinear 2D building plans and elevations, now known as BIM (Building Information Modelling) and the accompanying software for the design and detailing of precast (and steel, timber, etc.) structures. Professor C. J. Anumba of Pennsylvania State University addressed a Seminar & Workshop on the Developments and Future Directions in BIM (Kuala Lumpur, 2012) thus *Developments in BIM have resulted in significant industry interest and uptake. Most new building projects are dependent on BIM for resolving coordination, schedule, integration, estimating and other functions. Advances in information and communications technologies are continuing to open up new opportunities and applications. As such, more needs to be done to fully exploit the potential of these technologies and to meet the requirements of increasingly complex projects.*

Against this background of MMI and BIM the aims and objectives of this book were, as conceived by Dr Zuhairi, from CREAM (Construction Research Institute of Malaysia), to provide a concise text to show how the modernisation, mechanisation and industrialisation of prefabricated concrete structures can be achieved through the knowledge of best practice, information modelling, and the procedures and management of factory engineered concrete products and systems. The main objectives were to:

i.   show how previous R&D and present design and manufacturing techniques can be best exploited for the construction of modern precast concrete structures,

ii.   show how the IBS ethos can control the supply chain from the client to sub-contractors, and can best utilise BIM methods and design/detailing software,

iii. introduce the best concepts of automation and robotics in concrete production, and

iv.   exploit the industrialisation of off-site production and on-site processes, including low cost housing in south east Asia.

The authors were selected from the UK, Germany, Switzerland, Austria and Malaysia, each having expertise and a (fairly) long history in items (i) to (iv). Of significance was Mr Robert Neubauer of SAA Software Engineering, a production/structural engineer able to harmonise the requirements of prefabrication in design with automated production; Mr Thomas Leopoldseder and Ms Suzanna Schachinger, of Precast Software, with abilities to exploit BIM and related software to the full advantage of precast solutions; Prof Gerhard Girmsheid and Dr Julia Selberherr, of ETH (Swiss Federal Institute of Technology, Zurich) specialising in the respective roles of industrialisation of off-site and on-site construction; CREAM consultants Dr Zuhairi, Mr Gan Hock Beng, Foo Chee Hung and Ahmad Hazim Abdul Rahim responsible for the technical advancements of IBS for low-cost housing; and Dr Kim S Elliott, precast consultant, summarising the modernisation and optimization of precast and prestressed elements and structures.

This book is divided into three key themes, as reflected in its title:

Part 1: MODERNISATION

Chapter 1: Historical and Chronological Development of Precast Concrete Structures
Chapter 2: Industrial Building Systems (IBS) Project Implementation
Chapter 3: Best Practice and Lessons Learned in IBS Design, Detailing and Construction

Part 2: MECHANISATION AND AUTOMATION

Chapter 4: Research and Development Towards the Optimisation of Precast Concrete Structures
Chapter 5: Building Information Modelling (BIM) and Software for the Design and Detailing of Precast Structures
Chapter 6: Mechanisation, Automation and Robotics in Concrete Production

Part 3: INDUSTRIALISATION

Chapter 7: Lean Construction, Part 1 – Industrialisation of On-Site Production Processes
Chapter 8: Lean Construction, Part 2 – Planning and Execution of Construction Processes
Chapter 9: New Cooperative Business Model - Industrialisation of Off-Site Production
Chapter 10: Retrospective View and Future Initiatives in IBS and MMI
Chapter 11: Affordable and Quality Housing Through Mechanization, Modernization and Mass Customization

A number of chapters address the issues of modern housing. Concrete has great potential to offer building and housing construction works towards improving the function, value, and whole life performance, especially in the era where quick, efficient, and inexpensive construction and delivery are becoming the necessity and desires of the societies. Precast concrete construction is a technology that possesses the potential to eliminate building site inconveniences, reducing the lapsed time and cost of construction, and contributing to an end product that conforms to the required standards and codes.

However, buildings and houses produced with such technology have a rigid structure, an interlocking plan, and predetermined functions, where very few of them are sufficiently open plan to enable retrofitting and reconfigurations to be made quickly, economically, and repeatedly. Moreover, various negative perceptions, opinions, and images spring to mind when considering the concept of prefabrication and standardisation in housing, as a result of a number of buildings constructed in the past making use of prefabrication were judged to be of poor quality. This book will provide insight to builders of the potential for building and housing design system that makes use of the prefabrication construction to produce a variety of housing design options that meet possible user requirements not yet identified at the design stage, while retaining principal uniformity to facilitate the execution of simple but accurate construction with a minimal initial cost.

It is believed that only by having combined design and construction systems that take advantage of mass production and mass customisation, the efficient design of offices, parking structures, shopping and residential buildings, coupled with housing affordability and liveability can be achieved. A home that can be altered with minimum effort and expense at a time of change in the lives of its owners is a home that evolves with the lifecycles of its household rather than becoming rigidly obsolete in the conventional manner. As such, the affordable housing needs to be designed in such a way that it is economically and easily adjustable, as well as adheres to the context of contemporary technology, climate adaptation, and cultural responses.

Part 1

Modernisation of Precast Concrete Structures

Chapter 1

Historical and Chronological Development of Precast Concrete Structures

Kim S. Elliott

Precast Consultant, Derbyshire, UK

An overview of the four major phases in the twentieth-century history of precast concrete construction: developing years; mass production and standardisation; lightweight structures and longer spans; thermal mass design, shows how the beneficial issues in each period has lead to the present-day movement towards modernisation, mechanisation and industrialisation (MMI) and the interface with industrialised building systems (IBS). Timelines of market share, building height, span/depth, thermal efficiency, and hybrid and mixed precast construction are drawn through the phases from 1920 to 2010. The benefits of composite and continuous construction for prestressed concrete beams and slabs have decreased the mass of the floor construction by about 30% over the past 25 years. The conclusion shows how MMI serves and suits the demand for prefabrication of concrete-framed structures.

1.1 The five periods of development and optimisation

From a historical background, the prefabrication of concrete and the development of precast concrete structures for residential, commercial and industrial buildings have passed through four major periods:

1. The Developing Years (1920–1940) including the technological breakthrough of prestressed concrete (psc) and the further advancement of reinforced concrete (rc) in terms of improvements in the strength of materials, the optimisation of design and durability and resilience of the resulting elements. Figure 1.1 shows the first use of precast concrete, called *ferro-cement* at the time, in a multi-storey building.

Modernisation, Mechanisation and Industrialisation of Concrete Structures, First Edition.
Edited by Kim S. Elliott and Zuhairi Abd. Hamid.
© 2017 John Wiley & Sons Ltd. Published 2017 by John Wiley & Sons Ltd.

Figure 1.1   Weaver's Mill, Swansea. The first precast concrete skeletal frame in the United Kingdom, constructed in 1897–1898 (Courtesy Swansea City Archives).

2. The Mass Production and Standardisation Period (1945–c.1970) involved rebuilding residential post-war Europe, as well as developing south east Asia, using mainly wall panel construction (Figure 1.2), and semi-automated floor slabs such as prestressed long-line extruded or slip-formed hollow core units (hcu), eventually leading to the development of modularised "national building frames", for example, in Figure 1.3.

3. The Lightweight and Long-span Period (1970–2000), driven by the need to produce leaner structures with greater span-to-depth ratios by using composite, continuous and integrated designs in hybrid (precast with *insitu* concrete) and mixed materials (e.g., precast with steel, timber and masonry). Figure 1.4 shows total prefabrication of a steel frame supporting prestressed hcu having a span-depth ratio of about 40, and floor area-to-structural depth ratio of nearly 250 m^2/m.

4. The Thermal Mass Period (2000 to date) responding to the demand for the sustainable and environmentally advantageous used of factory engineered concrete and off-site construction philosophies, energy storage, improving admittance of the building fabric and lowering transmittance (U-values) requirements. Figures 1.5 and 1.6 show the use of so-called "FES", active fabric energy storage in the precast concrete elements.

There is now a new era, although some would argue this is already established in many countries, taking in the beneficial aspects of the latter day periods towards the increasingly popular trend for automated manufacture and off-site prefabrication:

Figure 1.2   Wall panel and hollow core floor slabs used in residential buildings of the 1950s being demolished in 2002.

Figure 1.3   Example of the National Building Frame, comprising modular spandrel beams, columns and slabs.

Figure 1.4 The Big Apple retail and car park near Helsinki, Finland. Sixteeen m long × 400 mm deep prestressed hollow core floors are supported on prefabricated inverted-tee steel beams, minimizing the structural depth.

Figure 1.5 Fabric energy storage at the Jubilee Library, Brighton (courtesy of Bennetts Associates Architects, London).

5. The Automated Period involving the modernisation, mechanisation and industrialisation (MMI) for the design, detailing and manufacture of concrete structures. On top of this we may add in building information management (BIM), the co-ordinated control of the building services, the structure, scheduling and construction, giving us the full spectrum of MMI.

This chapter aims to show how the (first) four major periods in the evolution, development and optimisation of precast structures has shaped the course of architectural and structural design, culminating with the types of buildings shown in Figures 1.6 and 1.7. This fifth major period will be introduced as the focus of this book develops in the subsequent chapters.

The objectives of this chapter are to analyse and criticise some of the key aspects of each period, and to show how the present has benefited from the advancements made against a background of mistakes made in the past. There are no better examples than the wall panel and framed structures shown in Figures 1.2 and 1.3, already reaching the end of their service lives after around 40 years. Taking examples from the Europe, North and South America (mainly the USA and Brazil) and south east Asia (mainly Singapore and Malaysia), lessons have been learned from the excesses of post-WW2 Europe to the streamlined buildings of twenty-first century design.

The focus of this book is on the three main types of precast concrete structures, and their compounded elements, such as columns, walls, beams, rafters, slabs and staircases. These are:

1. Skeletal frames – a beam and column framework supporting slabs and stairs, which may, depending on height, need to be stabilised by walls and/or cores. The classical skeletal frame is shown in Figure 1.7, comprising long-span psc hcu with a span-depth ratio of about 35 (for office loading), supported by psc internal beams acting compositely with the floor slab to minimise the downstand, or L-spandrel beams at the perimeter. Columns, made from high-strength concrete (up to $f_{ck} = 100$ N/mm^2 compressive cylinder strength at 28 days) are the minimum possible width to avoid slenderness, because the framework is stabilised by only a few precast concrete shear walls or cores. The specific volume of precast concrete (mass to the whole) is less than 4%. Figure 1.8 shows how it is possible to take advantage of frame action by using the partial stiffness, known as "semi-rigid" of the beam-to-column connection, in this case for a 10-storey building with a large number of internal columns and connections.

2. Wall frames – a wall and slab structure, inherently soundproofed and stabilised by walls in one or both directions. A popular use is shown in Figures 1.9 and 1.10 for student accommodation, but also for hotels, and some hospital and college buildings. The bathroom "pod" is often totally precast. The specific volume in these buildings is around 7%.

3. Portal frames – columns and rafters making large open spaces for industrial warehouses or factories, with or without additional cross-bracing (typically steel rods of small sections) and secondary beams for internal offices. Figure 1.11 shows long-span prestressed I-shape rafters, simply supported onto stabilising deep columns, which act as vertical cantilevers from the foundation. Other options include sway frames with moment resisting eaves connections.

Figure 1.6   Modular design of the predalles enabled rapid manufacture at the Jubilee Library, Brighton (courtesy of Bennetts Associates Architects, London).

Figure 1.7   Precast skeletal frame comprising of long-span lightweight floors, beams with hidden connections, and narrow columns braced using precast shear walls.

Figure 1.8   Precast skeletal sway frame using semi-rigid beam columns connections to provide stability up to 10 storeys. (Note that the imposed wind pressure is quite low in countries where this form of stability is used.) (University of Recife, courtesy of T&A, Recife, Brazil).

Figure 1.9   Precast wall frame under construction for student or teacher accommodation, in Kuala Lumpur, Malaysia.

Figure 1.10   Wall frame for residential and commercial use in Sydney, Australia.

Figure 1.11   Precast portal frame (courtesy David Fernandez-Ordoñez, Escuela Técnica Superior de Ingeniería Civil, Madrid).

There is no doubt that the skeletal frame has been the most challenging, offering architects, design engineers and services engineers the greatest freedom, thereby attracting the most interest in research and development (R&D) worldwide, based mainly in the USA, Northern Europe, Canada, UK, Scandinavia and Brazil (listed approximately according to output). This has lead to a number of key text books (Glover, 1964; Haas, 1983; Federation International de la Precontrainte (FIP), 1986; Morris, 1987; Sheppard & Phillips, 1989; New Zealand Concrete Society, 1991; FIP, 1994; Bljuger, 1988; Bruggeling & Huyghe, 1991; Elliott, 1996; El-Debs, 2000; Elliott, 2002; Bachmann & Steinle, 2011; El-Arab, 2012, Elliott & Jolly, 2013; Elliott, 2016), lecture packages (e.g., *fib* Master classes and workshops by Van Acker, Elliott, and Vambersky (1990 to date, see reference list) and computer-based teaching aids. Whilst the advancements in wall frames have changed little since the 1960s, except for the benefit of the thermal mass era, and the design of portal frames hardly at all, except for higher strength concrete and pretensioning longer rafter spans, skeletal frames have evolved considerably through several changes in the "modularised", "lightweight" and "thermal" eras.

This was accelerated in the 1990s by the conscious decision to use "mixed" construction (Goodchild, 1995; Fédération Internationale du Béton (*fib*, 2002)): the term *conscious* here meaning the *deliberate* rather than the *accidental* employment of different building mediums acting structurally and compositely with precast concrete, or simply architecturally. These included structural steelwork, *insitu* reinforced and post-tensioned concrete, timber (in particular glue-laminated timber), masonry, structural glazing and occasionally polymers (plastics, pultrusion sections). The publication "Precast Concrete in Mixed Construction" (*fib*, 2002) noted that

> "mixed construction is being used in more than 50 per cent of new multi-storey buildings in the western world. The increased use of precast concrete over the past ten to twenty years is due largely to the move towards greater offsite prefabrication of structural and non-structural components, and the sound economics, quality and reliability of doing so. However, some of the limitations found in precast concrete inevitably lead to it being used with other materials in a cost effective manner, e.g. to provide structural continuity and/or robustness using small quantities of cast *insitu* reinforced concrete (rc), or to form long span steel or timber roofs."

Structurally, combinations may work together or independently, but together they can be preferable to single material. The examples in Figures 1.4, 1.12–1.15 show that mixed construction was used essentially to meet the architectural requirements and for the speed of construction, both of which translated into substantial overall savings.

Mixed construction should, by definition, yield best value solutions. Prefabrication is not limited to concrete elements alone; the general term for this is "off-site fabrication" (OSF) and can equally include reinforcement cages and formwork for *insitu* concrete as well as precast concrete, steelwork or timber prefabricates. OSF strategies may be different from traditional on-site procedures, and should be agreed as early into the project as possible. All possibilities should be explored and precede the finalisation of OSF strategy. The greatest benefit from mixed construction is at the concept stage where possibilities are many and the cost-of-change is minimal. Gibb (1999) shows the benefits of making early key decisions where off-site fabrication used, as

Figure 1.12 Hybrid construction using precast spandrel beams and hollow core slabs, supported by *insitu* concrete columns in the VNO Building, Netherlands (courtesy of Corsmit Consulting Engineers, Netherlands).

Figure 1.13 Structural steel beams with welded shear studs enable composite action with prestressed concrete hollow core floor slabs.

Figure 1.14   Mixed construction using precast concrete and timber, both indigenous materials to the local construction industry in Norway (courtesy of Spenncon, Norway).

Figure 1.15   Load-bearing prefabricated masonry columns, supporting arched precast concrete floor slabs, together with a steel roof, combine to create a mixed structure at Inland Revenue, Nottingham, UK.

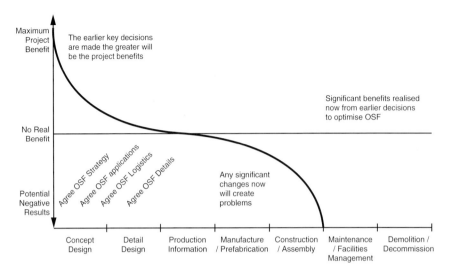

Figure 1.16   Timing of key off-site fabrication (OSF) decisions to maximise project benefits (Gibb, 1999).

shown in Figure 1.16. At the detailed design stage, a large number of options remain available but by now the cost-of-change increases rapidly. The designer who conceives the building must be comfortable with all the viable options. The process of choosing between the options should be part of a "value engineering" exercise, where the entire design, construction and client team should examine whole-life costs. These procedures will be examined in Chapter 2 under the implementation of projects.

The *fib* publication (*fib*, 2002) notes that one of the most common combinations is of precast concrete with cast *insitu* concrete. Speed of construction and the quality of precast combine with the economy and robustness of *insitu* concrete to give high-quality, aesthetically pleasing structures, quickly and economically. This is especially true in seismic countries where the means of connecting precast elements is to form monolithic branches of reinforced concrete, which shift the failure zone away from the column and connection. In some countries, this construction is called "*hybrid concrete construction*". Structurally, these elements may be made to act together (compositely) or independently (non-compositely). It is sometimes difficult to say when an element or structure is composite or not, so boundaries are vague. The two parent materials may not even have any physical contact with one another but their use in one structure provides the benefits of both materials to specific projects. The benefits may not just be in the structural sense; advantage can be gained by combining architectural and structural functions, and structural and building services functions.

Another example is of structural steelwork and precast concrete. Their respective industries, which once competed with each other in the construction market, have now joined forces to provide the industry with new dimensions, seeking synergy in respect to their individual and mutual strengths or weaknesses. Designers of steel structures have found economy in using long-span prestressed concrete floors, having a span-to-depth ratio of up to 50 and spans of up to 20 m, rather than being limited to short-span metal decking.

The emergence of structural timber, particularly glue-laminated timber, has provided new economic options for long-span construction, for example, Figure 1.14. When combined with precast concrete, timber provides good aesthetic and structural solutions as well as using local resources of timber and concrete. Innovative designs are emerging especially at the connections which are, by nature, visible in the final structure. The architect is allowing the structural engineer the freedom to express complex structural details.

Precast concrete flooring has had a long association with structural masonry, particularly in low-rise housing. However, it is only recently that the two industries have formed alliances and provided technical design data, especially at bearings. Precast flooring now dominates the market share in load-bearing masonry buildings.

Mixed construction methods vary considerably with the type of construction and building function. These reflect local trends, environmental and physical conditions, relative material and labour costs, and local expertise. Although the permutations for mixed construction are numerous, Table 1.1, adapted from *fib*, 2002, is a guide to their most common forms and combinations.

The design, manufacture and construction of skeletal frames has also been enhanced by the introduction of high-strength concrete with compressive strengths in the range $90 - 120 \text{ N/mm}^2$ compared with $40 - 60 \text{ N/mm}^2$ for the most part of the twentieth century, leading to the construction of 16- to 36-storey skeletal frames in Belgium, as shown in Figure 1.17 and 1.18, where 600 mm diameter precast columns were factory-manufactured using grade C95 ($f_{ck} = 95 \text{ N/mm}^2$) in the lower floors, and C60 elsewhere (Van Acker, 2006). A huge impact on precast production methods has been the use of self-compacting concrete (SCC), which as the name suggests will flow and compact without external vibration. After the problems of low early strength gain were resolved around 2002, this has been widespread in all types of frames, flooring, stairs and terraces. Figure 1.19 shows SCC used to cast the seating terraces for stadia, undeniably the highest quality of precast concrete. The precast industry has exploited the advancements made in concrete technology more than in any other part of the construction industry.

By comparison, changes to the design of wall-framed buildings has been largely due to the increased spans available in lightweight floors, such as hollow core units, which are today manufactured using routine methods in depths of between 150 and 500 mm (1000 mm deep has recently been achieved) and spans around 12–18 m for office loading. This has enabled the construction of single-bay apartments between two external walls, known as the "envelope system" as shown in Figure 1.20, offering the architect total freedom of space unhindered by internal columns or walls. The stability of the single-bay wall frame for open plan offices is provided by a deep horizontal floor diaphragm, typically spanning 50 to 70 m between the gable end walls. A further development is the so-called "split skin" wall frame, in which a modular load-bearing wall panel, typically 3.6 m wide × storey height is over-clad with an irregular façade, creating the image of variable storey height, as shown in Figure 1.21. The architectural storey height is varied relative to a fixed structural height. The service life of the façade may be designed for say 25 years, and replaced after this time, whereas the load-bearing structure may have a design life of up to 100 years. This has been a popular option in some Middle East countries since the 1990s, allowing the re-cladding, often with glass fibre concrete (GRC), and re-use of buildings according to the client's preferences.

Table 1.1   Building types using precast concrete in mixed construction (*fib,* 2002).

Building type	Mixed precast construction methods	Comments
Commercial offices	Insitu concrete frame with precast flooring and facades. Steelwork frame with precast shear walls, flooring and facades. Precast frame with steel raker ("Mansard" type) or steel roof truss. Precast frame with pitched timber truss.	All combinations possible with insitu concrete under-ground or ground floor podium.
Retail and shopping	Insitu concrete or steel frame with precast flooring and facades. Precast load bearing wall with cast insitu floors. Masonry load bearing walls with precast floors.	Ditto
Educational buildings	Steel frame with precast floor, with steel or timber roof Load bearing masonry with precast floors	Maximum clear spans to allow for changes in use
Parking Garages	Insitu concrete or steelwork frame with precast flooring and cast insitu topping Precast frame with glue laminated timber or steel roof	Long-span double-tee floors up to 20 m
Industrial and warehouses	Steel frame with long span precast wall units. Precast columns with steel roof truss. Steel frame with precast floors (office areas).	Hollow cored or sandwich walls give thermal insulation. Long-span lightweight roof.
High-rise residential	Precast load bearing walls with cast insitu floors. Masonry load bearing walls with precast floors.	Composite floor plank often used because of complex floor plan layout.
Domestic, low-medium density	Masonry load bearing walls with precast floors. Precast walls with timber floors.	Beam-and-block precast and hollow core dominates.
Stadia	Steel frame including raker beams, with precast terraces. Cast insitu frames with precast terraces. Precast columns with steel raker beams and precast terraces.	All combinations possible with steel or pretensioned precast roof.

Long-span hcu have also usurped the double-tee slab market for single spans in car parks, typically 400 to 500 mm deep with a 75 to 100 mm structural *insitu* topping. Figure 1.22 shows long-span prestressed hcu's supplying two rows of parking and the driveway, based on a 7.2 m grid. The high emissivity of the soffit of the hcu, even unpainted, is thanks to the smooth steel bed on which the units are cast. This has helped to reduce the demand for lighting creating a pleasing ambience.

From the MMI perspective, these variations for the three main types of precast frames present equal challenges, but there are clearly greater potential benefits for mechanisation, not only in manufacture, but also in the design, detailing and shop floor schedules in the entire BIM domain for skeletal and wall frames. Unfortunately,

Figure 1.17 Thirty-six-storey precast skeletal tower buildings in Belgium (North Galaxy, Brussels, courtesy of Ergon, Belgium).

Figure 1.18 Sixteen-storey and an elliptical shape to this precast skeletal frame in Belgium (Central Plaza, Brussels, courtesy of Ergon, Belgium).

Figure 1.19   Self-compacting concrete used for stadium terraces, have clean, sharp lines of precision (Bison Manufacturing Ltd., UK).

Figure 1.20   The "envelope system" of load-bearing facades and long-span single-bay floor slabs (courtesy of Van Acker, Belgium)

Figure 1.21   "Split-skin" frame concepts involve different structural and architectural storey heights (courtesy of Strängbetong, Sweden).

Figure 1.22   16 m span hollow core floor units × 7.2 m span spine beams creates six parking spaces and the driveway per column.

the precast concrete industry has been rather slow in adopting the mechanisation that BIM affords, for example, the paper-less construction site that was promised by some of the larger Scandinavian precasters in the mid-1990s is not yet a reality in most other countries. Electronic updating of construction drawings, and the strong links between component schedules, shop drawings and erection drawings, has let down the purposeful well meaning of advanced software (such as "Teckla" offers). This has not been the case in the parallel field of structural steelwork? Whilst mechanisation in the design, scheduling and manufacture of precast elements, such as slip-formed and extruded hollow core floor or wall units, has advanced to total CAD-CAM interfaces in many of the larger producers, the full MMI gambit still eludes us.

Table 1.2 gives a historical summary of the key developments in the precast industry. Developments in materials and technology are presented separately as they cut across all components and frame types, often taking the lead role in the exploitation of technology behind which structural frame designers play catch-up, typically by 5–15 years. For example SCC emerged from a Japanese laboratory in about 1990, and was finally adopted in some wet-cast precast production between 1995 and 2011. High-strength concrete, say 120 N/mm^2 compressive strength was used in bridge beams in Spain in the mid-1990s but not in structural frames until around 2005. A significant feature in Table 1.2 is the number of gaps where no major developments have taken place – for example, the number of changes in facades compared with prestressed hcu. MMI offers the opportunity to accelerate many of these features to facilitate more than just a face change to precast concrete structures.

Throughout these periods the height of multi-storey buildings has steadily increased, thanks to changing technology, architectural demand in some countries and cities, and the increase in the compressive strength of concrete. This is shown in Figure 1.23, since 1950, in terms of compressive cube strength and the corresponding axial load capacity of a 300 × 300 mm size r.c. column. The data show a three-fold increase in capacity, from about 2000 kN to nearly 6000 kN, due in part also to the increase in the yield strength of high-tensile reinforcement from 410 to 500 N/mm^2 (in fact cold worked mild steel grade 250 N/mm^2 was used in many precast columns until 1960s). This equates to a reduction in the size of the column, for the same axial load, of 48%, that is, in the year 2010 to carry an ultimate axial load of 2000 kN, the column section would need to be only 145 mm (down from 300 mm). Of course slenderness effects have nullified these advantages, but the message is still clear. Reduction in partial safety factors for dead and live loads will further enhance this trend, for example, introduction of Eurocode EN 1991 (EC, 2002a) has reduced the ultimate load demand by about 10%.

As a consequence, building heights have increased. The total building height, or number of storeys for either totally precast frames (dashed line) or skeletal or wall frames braced using insitu cores, is shown in Figure 1.24. The increase in skeletal frames has more or less followed the trend shown for concrete strength and column capacity in Figure 1.23, showing that this is a key parameter in the development of tall precast buildings. The sudden jump from year 2000 to today is reflected in the increase in readily achievable high strength + high performance concrete developed in the cement and concrete laboratories of Scandinavia, France and the USA. The rises, and falls, in wall frames have been more associated with social trends and the effect of major collapses such as Ronan Point in 1968, than technological advancements. Given that the height of a wall frame could easily reach 200 m or 60 to 70 storeys,

Table 1.2 Historical developments in precast concrete technology.

Period	Materials	Technology	Floors	Skeletal frames	Wall frames	Facades
Developing 1920–1940	Higher strength C50.	Ferro cement products. Reinforced. Prestressed	Ribbed slabs. X-beams and blocks. Wet cast hollow core.	Limited height, small regular spans, some psc X-beam and block.	Limited height with insitu or X-beam floors.	Precast panels.
Mass production 1950–1970	Calcium chlorides. HAC. Low-slump. RHPC White OPC	No-fines concrete. Sandwich panels. Steel and polypro fibres, thin wall formwork. Half-slab, semi-automated cages.	Solid or ribbed slabs. Double-tee and single-tee. Extruded & slip-formed hcu (≤250)	Standardised. Modular grids. National Building Frames. Patented connections systems. Steel rafters and Mansard roofs	Wall and cross-wall frames, 12–25 storeys. Mass production. Bathroom pods. Ronan Point collapse.	Exposed aggregates. Sandwich panels. Load bearing wall panels.
Lightweight and long-span 1970–2000	Lightweight aggs. Colouring pigments. SCC. Steel fibre high bond. HSC to 100 N/mm². Reactive powder up to 250 N/mm².	Stability ties for progressive collapse. Brick slips. Composite and continuous design. Semi-rigid connections. Generic algorithms for production.	Prestress beam and polystyrene blocks. 400 mm deep hcu. Deeper hcu for specials (730). Diaphragm action tests. Bubble-deck. Carousel casting plant. Cast-in lifting hooks in hcu.	Hybrid frames. Integrated frames. Longer composite spans. Mixed construction. Semi-rigid and PRESSS research, NIST connector.	Knee-joint open frames (wall-hcu). Single-span twin wall. Split skin.	Coloured, painted. Polished. Blast resistance. Large area panels.
Thermal and acoustic 2000–2010	Blended cements. PFA and GGBS. HSC 120 N/mm². Rapid hardening SCC	Thermal and acoustic design.	500 deep hcu. Thermally insulated ground floor beam-and-block floors and floor slabs	22 storey seismic frame (NIST joints)	Twin-wall (precast-insitu infill). Up to 54 storey with insitu cores (Netherlands)	Thermal mass and insulated.
MMI in the future						

HSC = High-strength concrete, HAC = high alumina cement, OPC = ordinary Portland cement, RHPC = rapid hardening Portland cement, PFA pulversized fuel ash, GGBS = ground granulated blastfurnace slag, SCC = self compacting concrete, hcu = hollow core floor unit.

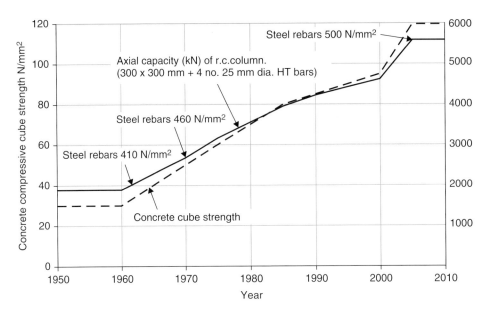

Figure 1.23 Approximate compressive strength (left ordinate) used in the precast industry since 1950s, and the corresponding axial load capacity of a 300 × 300 mm reinforced concrete column (right ordinate).

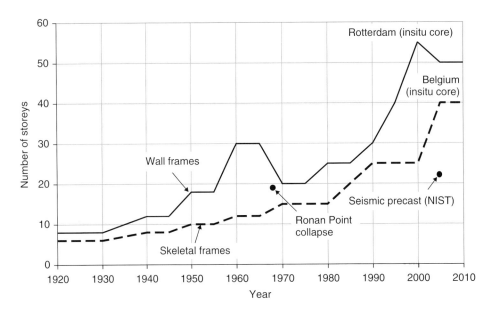

Figure 1.24 Typical and approximate maximum heights for precast concrete skeletal and wall frames since 1920s.

Figure 1.25   Strijkijzer building on completion (courtesy of J Vambersky, Netherlands).

it is only a matter of architectural preference that wall frames are still limited to 40 to 50 storeys in a few isolated countries of the world. The 44 storey Strijkijzer ("Skyriser") wall panel-frame in The Hague, Netherlands (Figure 1.25) has certainly heralded a new generation of totally precast wall panel-frame buildings, thanks to the novel deep nib shear connection shown in Figure 1.26. This enables high shear load, together with vertical tie bars for the design against progressive collapse, to be contained within the projecting nib between the walls, turning a vertical interface shear problem into a bearing and compressive strut solution.

Figure 1.26   Deep nib shear key used to stabilise the Strijkijzer building.

The increase in floor capacity, in terms of span-to-depth is shown in Figure 1.27. This has been due mainly to improvements in the technology of slip-formed or extruded prestressed hollow- core units (hcu), the optimization of their lighter weight cross-section, together with continuous, integral and composite design. Precast reinforced concrete (r.c.) has been left behind, and is perhaps the chosen medium where span-depth is not critical, or the spans and depths available do not warrant the extra effort of pretensioning, for example, many floors governed by thermal mass may as well be in r.c., such as ribbed floors, double or triple-tee units. The recent reduction in span-depth ratio is due to the same reasoning, that of thermal mass increasing the slab thickness or removing some of the voids. Of course there are still many projects where the floor span-depth ratio is still around 40. More detail is given in Sections 1.3.3 and 1.3.4.

The increase in beams' span-depth ratio is largely a function of design rather than materials. Given that the flexural capacity of beams is governed mostly by the strength and quantity of reinforcement, the strength of concrete used in most beams is made constant by each producer as 40 to 50 N/mm². There are some situations where a prestressed or partially post-tensioned beam may benefit from strengths of around 90 N/mm², but the data shown in Figure 1.27 do not include these. Increases in span-depth are due to composite and continuous design, and the use of wide inverted-tee (or drop beams) beams with shallow downstands, pioneered in Italy but common elsewhere. Figure 1.28 shows a 1200 mm wide precast beam awaiting composite action with hollow core slabs, and continuity reinforcement at the supports. This form of construction yields a bay area = beam span × floor span,

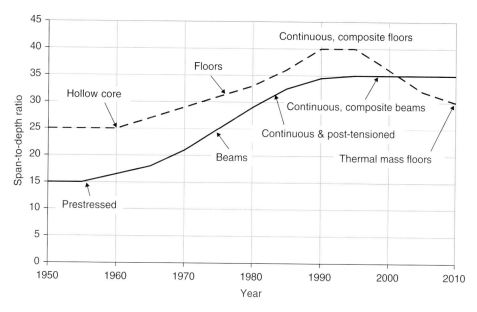

Figure 1.27 Approximate span-to-depth ratios with some key designs for precast concrete floors and beams since 1950s.

Figure 1.28 Wide inverted-tee beams with top rebars to resist the compression on the temporary stage. After hollow core floor units are concreted into the top of the beam, the beam is composite and continuous at the supports, leading to a span=depth ratio of about 35. (courtesy of APE, Spa, Italy).

Figure 1.29 Large span-depth ratio in beams achieved using semi-rigid beam-column connections at internal columns. The reduction in mid-span bending may be around 200 kNm, enabling a reduction of 150–200 mm in the depth of the beam.

of about 70 to 100 m^2 for a total structural depth of 400 to 500 mm. More detail is given in Section 1.3.4.

An alternative approach to increasing beam span-depth ratios is to utilise the semi-rigid negative restraint at the ends of precast r.c. or p.s.c. beams, as shown in Figure 1.29. Applying a fixity factor of about 0.75 (pinned = 0, fully rigid = 1) to the connection, enhances the rotational stiffness at the beam-column connection sufficient for the connection to carry in the region of 200-kNm for beams between 400 and 600 mm depth. Reductions in mid-span bending moments and deflections enable the beam depth to be reduced by 150 to 200 mm typically (see Chapter 3 for more detail).

Trends in the strength of concrete, column, beam and slab capacity, together with reductions in structural zones and column sizes have all combined to allow consultants to hone and optimise the design of precast r.c. and p.s.c. frames, heralding the lightweight and long-span period between about 1970 and 2000.

1.2   Developing years and the standardisation period

The 1930s saw expansions by companies such as Bison (UK), Partek (Finland, Belgium) and Spancrete (USA), with establishments often positioned close to aggregate reserves near rivers. Elliott and Jolly (2013) provide a summary of the developing years:

"The reason why precast concrete came into being in the first place varies from country to country. One of the main reasons was that availability of structural timber became more limited. Some countries, notably the Soviet Union, Scandinavia and Northern Continental Europe, who together possess more than one-third of the world's timber resources but experience long and cold winters, regarded its development as a major part of their indigenous national economy. Structural steelwork was not a major competitor at the time outside of the United States, since it was batch processed and thus relatively more expensive. During the next 25 years developments in precast frame systems, prestressed concrete long span rafters (up to 70 feet), and precast cladding, increased the precasters' market share to around 15 per cent in the industrial, commercial and domestic sectors. Influential articles in such journals as the "Engineering News Record" encouraged some companies to begin producing prestressed floor slabs, and in order to provide a comprehensive service by which to market the floors these companies diversified into frames. In 1960 the number of precast companies manufacturing major structural components in Britain was around thirty. Today it is about eight."

Early structural systems were not burdened with the need to minimise structural zones, which were often around 800–1000 mm in offices, giving rise to span-depth ratios of less than 10. This was known as the "heavy" period, as shown in the hand-book "Structural Precast Concrete" (Glover, 1964). As the relative cost of facades and building services to the static costs of structural components began to increase in the 1970s, designers' hands were forced by the architects into reducing structural zones or increasing spans without increasing unit costs. A saving of say 150 mm per floor could translate into 1.5 m over 10 storeys, saving let's say 200 m^2 of external cladding/façade panels and associated weather sealants, etc. at \$80/m^2 = \$16k. This saving could be made by utilising the rotational stiffness of the beam-to-column connection, or making the beams continuous at supports, essentially reducing the depth of beams for no extra charge. Thus the "lighter" precast period heralded increasing structural efficiency as shown in Figure 1.27 in the 20 years after 1970.

Attempts to standardize precast building systems lead to the development of so-called national building frames (NBF) and quite often standardised designs known as Public Building Frames (PBF). The Malaysian Public Works Department's (JKR) school building shown in Figure 1.30 has already had a life of 60 years. NBF's were designed to provide " ... a flexible and economical system of standardized concrete framing for buildings up to six storeys in height. It comprises a small number of different precast components produced from a few standard moulds" (National Building Frame Manufacturers' Association, undated).

The NBF structural models were simple and economical: simply supported long-span prestressed concrete slabs, usually double-tee units, up to 12.6 m long × 500 mm deep with a 100 mm topping were half recessed in to beams of equal depth, spanning 7.2 m. By controlling the main variables, such as loading (3 + 1 kN/m^2 superimposed was used throughout), concrete strength and reinforcement quantities, limiting spans were computed against structural floor depths. Columns were based on 200 or 300 mm dimensions for unbraced buildings of 3 or 4 storeys, but larger in 50 mm increments for lower levels of braced buildings up to 10 storeys. Bracing was

Figure 1.30   Public frame school building in Malaysia first designed in 1950s is still constructed today in both cast *insitu* and precast concrete.

by gable shear walls acting with the top flange and insitu topping of the double-tee floor diaphragm action over a distance of not more than 50 m. Storey heights of 3.0 or 3.3 m gave clear heights of 2.4 and 2.7 m. Figures 1.31 and 1.32 show one such building and examples of the details of these frames from a UK reference (National Building Frame Manufacturers' Association, undated), which states "The system is designed in accordance with the principles introduced by the Ministry of Public Building and Works in its dimensional co-ordination publications, and is controlled by the National Building Frame Manufacturers – an association of independent companies throughout the country, who are specialist manufacturers and erectors of frame components." In other words a "closed" system for what we now term RISP which stands for "Registered IBS System Providers". Following the demise of NBFs and PBFs, precast frame design evolved towards more of a client-based concept. Standard frame systems gave way to the incorporation of standardized components into bespoke solutions or the creative use of standard products. The result, shown in Figure 1.33, established the new generation of versatile precast concrete frame concepts of the present day.

In making a comparison of the developments in precast buildings in Europe and in North America, Nilson (1987) states "Over the past 30 years, developments of prestressed concrete in Europe and in the United States have taken place along quite different lines. In Europe, where the ratio of labour cost to material cost has been relatively low, innovative one-of-a-kind projects were economically feasible. … In the U.S. the demand for skilled on-site building labour often exceeded the supply, economic conditions favoured the greatest possible standardisation of construction … "

Figure 1.31   National Building Frame in UK. Highbury Technical College (now Highbury College) Portsmouth, opened in September 1963 (courtesy of National Building Frame Manufacturers Association, London (now defunct)).

The precast industry is still struggling to overcome the misconceptions of modular precast concrete buildings. This is not surprising as many texts (Bruggeling & Huyghe, 1991) refer to " ... *the design of a precast concrete structure on a modular grid. The grid should preferably have a basic module of 0.6m* ... ". The phrase "modular co-ordination" was used to describe the interdependent arrangement of dimensions, based on a primary value accepted as a module. This dimension was 30 cm horizontally and 10 cm vertically. Moreover the storey height in precast concrete apartment

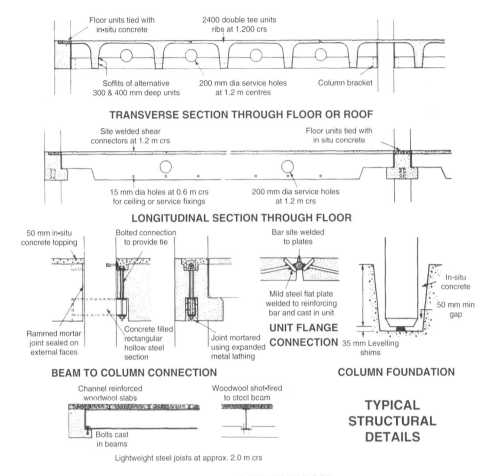

Figure 1.32   National Building Frame components and details (courtesy of National Building Frame Manufacturers Association, London (now defunct)).

buildings was fixed at 280 cm with the horizontal grid dimension on a 30-cm incremental scale between 270 cm and 540 cm. Strict observance of these rules facilitated the optimum assembly of prefabricated structures.

There is a clear distinction between "modular coordination" and "standardization". Modularisation offers zero flexibility off the modular grid. The end product is evident in the comparison of the two buildings adjacent to Vauxhall Bridge in London, and shown in Figure 1.34. Interior architectural freedom is possible only in the adoption of module quantities and configuration, and one cannot escape the geometrical dominance and lack of individuality of the older building on the left of the photograph.

Industrial modularized buildings were introduced in the 1950s during the mass construction period following the Second World War. The problems in the architectural and social environment brought a re-emergence of traditional methods, and the closer control on design and factory production. This has inevitably led to a new

Figure 1.33 Arlington Court. Sand-blasted precast concrete spandrel beams and columns all derived from standard products (Reed Baines Photographer, Nottingham, UK, now defunct).

philosophy in what is called the "modulated hierarchical building systems" (Sarja, 1992), which aims at the sub-division of a building into:

- functional systems, that is, space utilisation both vertically and horizontally, personnel coordination, adaptability to changing needs
- technical systems, that is, the structural design, the façade, mechanical and electrical, waste disposal, and air conditioning systems.

Figure 1.34   Examples of past and present use of exposed precast concrete. Vauxhall Bridge, London.

Figure 1.35   Exposed concrete columns, beams and roof units in reconstructed stone at Paddington Station, UK.

Precast modular frame manufacturers have been able to synthesize these requirements through continuous development of improved products and creative use of limited ranges of precast concrete products, for example, in the exposed concrete columns, beams and vaulted roof units shown in Figure 1.35.

Standardisation is quite different from modularisation. It refers to the manner in which a set of predetermined components are used and connected. The buildings shown in Figures 1.17 and 1.18 constructed using more or less the same family of standardised components. By adjusting beam depths, column lengths, wall positions, and so on, the same components in any of these buildings could have been used to

Figure 1.36 High-rise modular buildings in Wembley, London. (Courtesy of FutureForm Modular Ltd, West Sussex, UK.)

Table 1.3   Examples of levels of off-site manufacture.

	Levels of Off-Site Manufacture			
	1. Manufactured components	2. Elemental or planar systems	3. Modular and mixed construction systems	4. Complete building systems
Examples of construction technologies	Precast concrete slabs Timber roof trusses Composite cladding panels	Structural precast concrete, steel or timber frames Light steel framing Structurally insulated panels	Prefabricated plant rooms Modular lifts and stairs Modules at podium level Bathroom pods in framed buildings	Fully modular buildings
Proportion*	10–15%	15–25%	30–50%	60–70%
Reduced time**	10–15%	20–30%	30–40%	50–60%

*Proportion of off-site manufacture in value terms.
**Reduction in construction time relative to level 0 = site intensive construction using traditional material with little off-site work except for items such as doors, windows, and so on.

make a completely different structure, as discussed in Chapter 2, Section 2.2. This is not possible with the modular system.

Modular construction is being established in an increasing number of building projects, particularly volumetric construction in offices, schools, supermarkets and residential buildings. Prefabrication of concrete, steel, timber and plastics has been at the forefront of this emerging market. The latest technology is summarised by (Lawson *et al.*, 2014) in a large number of case studies (over 40) mainly from the UK against a background of planning, development and construction models for the factory production of modules. Figure 1.36, completed in the past three years, is a good example of this work. Lawson agrees with Bruggeling's approach to precast frame design and notes "dimensional and special planning is crucial to the success of modular construction. To maximise building use and flexibility, modular units may be combined with planar elements or structural frames in hybrid construction". Elliott's approach (Elliott & Jolly, 2013) is the reverse, to incorporate volumetric elements such as lift shafts, bathroom pods, stair cores to skeletal or wall frames. The Housing Development Board in Singapore has applied the same logic to extending apartments with volumetric bathrooms or other rooms.

The efficient use of off-site prefabrication is summarised (Lawson *et al.*, 2014) in Table 1.3 in terms of their level 1–4 of OSM (off-site manufacture). The percentage use of OSM leads to an almost corresponding reduction in construction time, with more improvements for bad weather, congested sites or with repetition of elements, although the longer lead-in time for manufacture may compensate this advantage.

1.3   Optimisation and the lightweight period

1.3.1   *Minimising beam and slab depths and structural zones*

The years 1970–1990 saw a major shift in the design of precast concrete buildings from the massive and less efficient floor and beam components to being shallower, lighter and more structurally demanding, none more so than prestressed floor slabs. Figure 1.27 shows an increase in the span-to-depth ratio for floor slabs of around 25:1 to 40:1. Depending on the span and magnitude of imposed loading these increases would be from 25:1 to 35:1 for higher loads and shorter spans, and 30:1 to 40:1 for lighter loaded longer spans.

Improved manufacturing techniques, thanks largely to the combined efficiency of long-line pretensioning and new extrusion and slip-forming machines, produced robust cross-sections with thinner flanges and webs, together with:

- increase in concrete strengths from C35 to C50 (compressive cylinder strength $f_{ck} = 35$ to 50 N/mm^2)
- cement additives and admixtures such as air entrainment
- increased level of prestress in strands and wires of up to 75% ultimate
- greater ultimate strength of tendons from about 1600 to 1860 N/mm^2

enabled 10 to 15% lighter units with 25 to 35% greater bending capacity.

Reductions in self weight due to the use of lightweight aggregates in concrete (LAC) were never possible with prestressed units due to deficiencies in shear-tension capacity and brittle failure modes, and the need for high early strength at detensioning (30 N/mm^2 at 16 hours). However, LAC with densities from 1600 to 1800 kg/m^3 (30% reduction compared with normal aggregate concrete) could be used in wet-cast solid slab or some wet-cast hollow core units that are not critical in shear, and of course in structural toppings or finishing screeds.

To take an example, consider a 1200 mm wide × 250 mm deep prestressed hcu manufactured by slip-forming techniques with 11 no. 170 mm × 50 mm oval-shaped hollow cores. The self weight of the finished floor (including site placed infill in the joints between hcu) is 4.33 kN/m^2. The design parameters are:

- concrete cylinder/cube compressive strength $f_{ck}/f_{cu} = $ C32/40
- pretensioning is 10 no. 12.5 mm diameter helical strands, stressed at 0.65×1600 N/mm^2
- internal exposure without flexural cracking
- permissible tension in soffit $f_t = 0.45\sqrt{f_{cu}} = 2.85$ N/mm^2
- characteristic imposed dead, partitions and live loading is 2.0, 1.0 and 4.0 kN/m^2, respectively.

The serviceability moment of resistance for a maximum span L = 8.98 m is $M_{sR} = 137$ kNm. Given that the second moment of area $I = 1330 \times 10^6$ mm^4, long-term deflection of the floor after the application of finishes is $\delta = 10$ mm. The maximum dynamic deflection is $\Delta = 18$ mm mm and the natural frequency is $f = 4.3$ Hz. Comparing this to the currently optimised cross-section and materials using 190 × 60 mm cores, C50/60 concrete, super strands stressed at 0.70×1860 N/mm^2 and $f_t = 3.49$ N/mm^2, the self weight reduces to 3.83 kN/m^2 (12% saving). To satisfy permissible tensile stresses during transfer of prestress, and to achieve L = 8.98 m only 7 no. 12.5 mm strands are required to achieve L = 9.14 m, a huge and important saving of 30% steel. By using compound section properties based on transformed area of tendons, the stiffness term reduces by 6% to $I = 1255 \times 10^6$ mm^4 such that $\delta = 9$ mm, $\Delta = 15$ mm (slight reductions due to

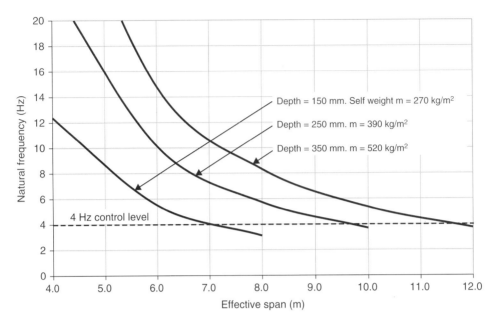

Figure 1.37   Relationships between natural frequency, floor mass and span for prestressed hollow core floor units.

14% increase in Young's modulus) and $f = 4.5$ Hz is almost the same. If the limiting transfer stress is ignored or overcome, for example, by debonding strands at the ends of units (not an easy operation in machine cast hcu) or by delaying detensioning or raising the detensioning strength to 40 N/mm², pretensioning the current hcu using the same number of 10×12.5 mm strands increases the resistance by 30% to $M_{sR} = 177$ kNm and the span by 16% to L = 10.46 m. However, at this span $\delta = 18$ mm (close to the usual limit of span/500 or 20 mm for masonry or brittle partitions), $\Delta = 28$ mm and $f = 3.4$ Hz (not acceptable for many applications).

Unfortunately, increased spans and reduced mass comes at the expense of greater deflection and lower natural frequency. Figure 1.37 shows the effect of changes in span, depth and self weight (or surface mass m kg/m²) on the natural frequency (first fundamental mode) f of a non-composite prestressed hcu, using 0.05 for the dynamic damping factor in the presence of partitions and walls. The data show by reducing m by 33% (520 to 390 kg/m²) the reduction factor for f is 1.45, and so on for other values of m. An important point is where $f \approx 4$ Hz, the threshold frequency where architects and computer floor users become concerned about harmonics and peak acceleration exceed about 0.5g (see ISO 2631-1, 1997). The structural capacity of the slabs is almost reached when $f < 4$ Hz. (For the record, the type of hcu used in the data in Figure 1.37 are 1200 mm wide, contain 11 oval cores, use C50/60 concrete with $E_c = 32$ kN/mm² and are optimally prestressed with respect to M_{sR} for characteristic imposed dead and live loads of 2.0 and 5.0 kN/m².)

The data in Figure 1.37 assume rigid supports and do not include deflections of the supporting beams, which may be in the order of one-third to two-thirds of the mid-span slab deflections, resulting in even lower values of f depending on the aspect ratio of slab-to-beam spans. The optimal geometry for precast slabs and beams is typically where the span of prestressed slabs is 1 to 1.5 times the span of the beam, the reason being that concentrated loads on beam makes bending moments and deflections more sensitive to longer spans than is the case for slabs. In addition, floor slabs should be recessed within the depth of the beam, as is the case of ledger beams, spandrels and inverted tee beams, as shown in Figures 1.22, 1.28 and 1.29. Because of the recess, the solution to minimizing the mass and structural zone of skeletal structures is based on minimum beam depth for the least number of components, that is, some function of floor area divided by structural depth. This means we require a solution for the correct orientation of beams with respect to the floor space.

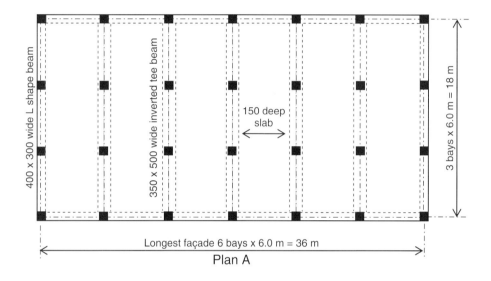

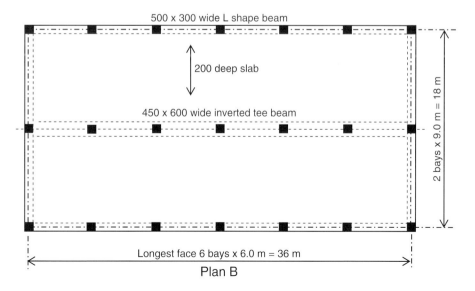

Figure 1.38   Alternative floor plans based on the *orientation rule* of floor and beam span directions.

1.3.2   Orientation rule

This approach leads to one the most useful conceptual design rules for skeletal frames, known as the *orientation rule*: that the main beams should run parallel, or along the face of the longest building dimension, and the (mainly prestressed) floor slabs should span perpendicular to the beams and hence tie in to the building façade. Of course, this rule cannot always be exercised in irregular or small courtyard buildings, but where it is possible in rectangular buildings with aspect ratio (length-to-depth) of more than 2 times the reduction in structural zone, that is, slab depth (including allowance for prestress camber of about span/300 and any structural topping typically 50 to 100 mm) plus beam downstand (drop beam depth) is around 100 mm for spans of 6 to 8 m, and about 150 mm for greater spans. This orientation pattern often conflicts with the designer of cast *insitu* buildings, where continuous beams span between internal columns perpendicular to the façade, or even in both directions, with short shallow slabs continuously between the beams either as one-way or two-way spanning. This does matter in *formwork and fill* casting methods as the same operations apply to either orientation, but it has a much bigger impact on precast and prestressed construction.

In converting beam and slab layouts from a cast *insitu* solution to a precast solution, for example, shown in Figure 1.38, the cast *insitu* designer will unwittingly produce Plan A, compared to Plan B. Plan A has 33 beams, 28 columns and 90 slabs (1.2 m wide), albeit shorter in the length than Plan B but still require the same transportation and erection, totaling 151 components. Plan B has 22 beams, 21 columns and 60 slabs totaling 103 components, 48 fewer. For 10 storeys that is 480 fewer operations! (assuming single-storey columns rather than three- storey columns, but even that would have 451 fewer components). Furthermore, there would be 220 fewer beam-column connections, a much more time-consuming activity than erecting the beams.

For the structural schemes, consider office loading of 5 kN/m^2 live load and 2 kN/m^2 imposed dead load (finishes, services, etc.). The structural requirements for Plan A and Plan B are given in Table 1.4.

1.3.3   Composite and continuous floor slabs

Precast concrete floor slabs may often be designed with a cast *insitu* structural topping (sometime referred to a screed, but this author prefers the term screed for non-structural sand/cement finishes) for reasons of:

- stability, for example, the topping contains a steel mesh for means of catenary action in the event of partial collapse
- to enable the topping to act as the horizontal floor diaphragm
- point loads and water tightness in car parks
- horizontal steel mesh and loose rebars ties to prevent frame dilation
- enhancing structural capacity, mainly bending strength and stiffness, *namely,* reduced deflections due to live loads
- continuity of negative bending moments in multi-bay or cantilevered slabs.

Table 1.4 Structural requirement for the alternative floor plans shown in Figure 1.38.

	Plan A	Plan B	Differences
Clear span of floor slab (m)	5.50	8.55	
Depth of prestressed floor hcu (mm)	150	200	Plan B deeper floor but does not affect structural zone
Self weight of floor slab (kN/m²)	2.7	3.3	Plan B 22% greater mass but 50% fewer units
Total floor service load (kN/m²)	9.70	10.30	
Total floor ultimate load (kN/m²) ($\gamma_f = 1.25$ (dead) and 1.5 (live))	13.38	14.12	Within 6% of each other
Clear span of beam (m)	5.70	5.70	Same, $6.000 - 0.300$ m
Depth × breadth of inverted-tee beams (mm)	350 × 500	450 × 600	Plan B increase of 100 mm per storey
Ultimate load per internal column per storey (kN)	470	740	Plan B requires 400 × 300
Column size (assume five-storey) (mm)	300 × 300	400 × 300	Plan B 33% greater size but 33% fewer components
Floor bay area (m²)	36.0	54.0	Plan B provides 50% greater area between column obstructions
Floor bay/structural depth ratio (m²/m)	103	120	Plan B has 1.17 superior ratio in spite of deeper beams
Total weight of precast per storey	2425	2675	
Proportion of self weight due to floor slab	68%	77%	Plan B is greater due to deeper slab
Number of components per storey	151	103	Plan B 46% fewer

The topping may, but not always, increase the structural capacity of the floor slab, depending on (i) span, as the large bending moment due to self weight in longer spans may nullify the reduction in imposed bending stress, and (ii) thickness of the cast *insitu* topping relative to the depth of the precast r.c. or p.s.c. unit, acting as the sub-base. Designers should not forget to add the pre-camber in prestressed units (taken as span/300 for higher prestressed units, or span/750 for the lowest prestress) to the thickness of the topping *t* as specified at mid-span, for example, if t = 75 mm on a 8 m span highly prestressed unit, the thickness at the support = 75 + 8000/300 = 102 mm. If both a structural topping and a finishing screed are used, the composite floor slab will be structurally inferior in resisting service stresses than its non-composite unit for the reason that the r.c. topping working monolithically with the precast unit is not as structurally efficient as the unit alone, particularly if prestressed. The solution is to power float the topping to a finished floor, as shown in Figure 1.39, in order to dispense with the screed. Composite r.c. slabs are not affected by the enhancement in elastic section properties because these slabs are critical at the ultimate limit state, and therefore capacity is proportional to total depth alone. Shear capacity will be enhanced, but not by as much as the increase in thickness due to the location of the shear failure lies close to the narrowest part of the webs in both hollow core and double-tee units.

Figure 1.39 Power floating a structural topping on hollow core slabs in order to minimize further dead weight due to additional finishes.

The design of composite prestressed and r.c slabs is well documented, for example, Chapter 6 in Elliott and Jolly (2013) and in particular by the FIP document on interface shear between the insitu topping and precast unit (FIP, 1982). In brief, the maximum stage 1 bending moment M_{s1} due to self weight of the precast unit plus wet *insitu* topping, averaged for thickness at mid-span and the support, is resisted by the precast section alone, causing a tension stress (sign negative) is $\sigma_{b1} = -M_{s1}/Z_{b1}$ in the bottom of the floor unit, where Z_{b1} is the section modulus of the precast unit alone. When the topping has hardened, the monolithic section modulus increases to Z_{b2}. The bottom stress due to stage 2 imposed dead and live loads moment M_{s2} is $\sigma_{b2} = -M_{s2}/Z_b2$. If the floor unit is prestressed, after all long-term losses, the compression in the bottom fibre is $+f_{bc}$ and the total stress in the bottom $= +f_{bc} + \sigma_{b1} + \sigma_{b2} \geq -f_t$, where f_t is the limiting tension allowed by national codes, for example, $0.45\sqrt{f_{cu}}$ (British BS8110), $0.3f_{ck}^{2/3}$ (Eurocode EC2) or $0.62\sqrt{f_{ck}}$ (American ACI-318), typically 3.2 to 4.0 N/mm^2. Knowing f_t, f_{bc}, Z_{b1} and Z_{b2} the allowable imposed load is back-calculated from $M_{sR2} = (f_{bc} + f_t - M_{s1}/Z_{b1}) Z_{b2}$. Total $M_{sR} = M_{sR2} + M_{s1}$. Limiting compression should also be checked at the top fibre, but this is rarely critical, except in some shallow double-tee slabs. At ultimate limit state, the area of tendons required to resist M_{u1} is subtracted from the total area provided in the bottom of the unit, and then the ultimate moment of resistance M_{uR2} is calculated from the remainder, based on the strength of the *insitu* topping, such that the total strength is $M_{uR2} + M_{u1}$.

Table 1.5 gives examples for 200 and 400 mm deep composite hollow core slabs with 50 and 100 mm thickness topping, respectively, and compares allowable spans with those for non-composite units when carrying an imposed load of 5 kN/m^2. The result for the 400 mm deep unit plus 100 mm topping demonstrates the point made earlier about the stage 1 moment, due to the self weight of 14.68 m span wet topping

Table 1.5 Comparison of allowable spans under fixed loads for continuous (propped), continuous-composite slabs, and composite and non-composite simply supported floor units.

Precast unit depth (mm)	Thickness of topping t (mm)	Span type	Imposed dead & live load (kN/m²)	Self weight including topping (kN/m²)	M_{s1} (kNm)	σ_{s1} (N/mm²)	M_{sR2} (kNm)	Eff. Span (m)
200	Nil	Simply	5.0	3.27	108.8	–	–	9.36
	50	supported		4.60	61.8	−8.56	67.2	9.47
		Continue*		4.64	72.5	−10.05	51.8	10.73
		Propped**		4.64	56.0	−7.77	75.5	11.05
400	Nil	Simply	5.0	4.93	352.9	–	–	15.39
	100	supported		7.34	241.4	−10.33	161.7	14.68
		Continue*		7.34	272.2	−11.66	116.0	16.06
		Propped**		7.34	199.4	−8.54	220.7	16.74

Examples based on EC2 with f_{ck} = 50 N/mm² (precast) and f_{ck} (*insitu* topping) = 25 N/mm².
Let f_{bc} after long-term losses = +11.0 N/mm² and f_t = −4.07 N/mm² for internal exposure in all cases.
Precast units are hollow core with 6 no. 140 mm wide cores.
*Continuous design has 3 no. equal spans, for example, 200 deep slab has 3 × 10.73 m = 32.2 m building length. Stage 1 simply supported length bearing onto 500 mm wide beams = L − 400 mm.
**Propped at two positions at ¹/₃ and ²/₃ span.

nullifying the prestress, leading to 5% lower span compared to 15.39 m for the totally prestressed non-composite unit.

The design for continuity is not very common when using prestressed floor slabs, requiring specific conditions of reasonably uniform spans (within 30% of each other is usual) and multiple bays of at least three or four to make the additional work worthwhile. Conditions for ensuring the continuity of compression in the bottom of the slabs at the supports, both in service and long-term ultimate conditions allowing for creep, shrinkage and thermal gradients, must be maintained. This usually rules out double-tee or other ribbed slabs unless the ends are made solid over the full width, as shown in Figure 1.40. However, the effect can be dramatic in increasing the load carrying capacity, as shown in Table 1.5, particularly under heavy loads where the second-stage negative continuity reduces main span moments and deflections. Moment redistribution is not used in prestressed floor units. Continuous-composite action can help to reduce the depth of the floor slab and hence its mass, which can form a major portion of the self weight in a framed structure (as indicated in the examples in Table 1.4 as 68% to 77% of total mass).

The enhancement is due to negative bending moments M_{s3} at the internal supports due to imposed loads only, which may be 1/2 to 7/8 the mid-span moment M_s, (depending on the ratio of the imposed-to-dead loads) thereby considerably reducing mid-span bending and deflections. For example, in the 200 mm deep hcu + 50 mm topping slab in Table 1.5, M_{s3} = −69.8 kNm and M_s = 124.3 kNm. The increase in allowable span of a three-bay continuous composite slab is 1.14 (for 200 + 50 slab) and 1.10 (for 400 + 100 slab) compared to simply supported. However, it must be remembered that to achieve continuity, top reinforcement is required in the structural

Figure 1.40 Double-tee floor unit with solid ends designed for continuous spans with a structural topping.

topping, which 200 + 50 slab is H16 bars at 180 mm spacing, curtailed at 1.7 m from supports. The major advantage of continuity is a reduction in slab depth and hence self weight, for example, the 400 mm deep simply supported unit could be reduced to 350 mm if designed in continuous spans, making 9% saving of self weight.

Propping floor slabs during construction of the topping considerably reduces the stage 1 moments, by inducing a negative bending moment when the wet *insitu* topping is poured. Table 1.5 shows increases of around 4% in allowable span where twin props at one-third and two-thirds span have been used. Once again, there is a penalty to pay for the increased capacity or span by virtue of the additional labour involved in propping, but the extra 680 mm span may be critical in some design situations. Overall, from simply supported to continuous-with-props, the proportionate increase in composite spans is 1.14 and 1.17 for the 200 + 50 and 400 + 100 mm deep slabs, respectively, for the one example of 5 kN/m^2 imposed load.

Figure 1.41 shows the same features for the full range in increments of imposed load from 1 to 10 kN/m^2. The striking feature is the enhancement in span for the greater load carrying capacities, that is, when the imposed load approaches 10 kN/m^2 three-stage continuous-composite construction is clearly beneficial due to the larger negative moments at supports, and so on for greater loads (even though patch loading patterns means that the imposed live load in adjacent spans is taken as zero). The main benefit in continuous construction at lower loads is the reduction in deflections, where the negative curvature at the support can have the effect of reducing the long-term deflection by 50%, for example, the deflection in the 400 + 100 mm slab spanning 18.4 m under imposed load of 1.0 kN/m^2 was 56 mm compared to the simply supported deflection of 84 mm for the same span.

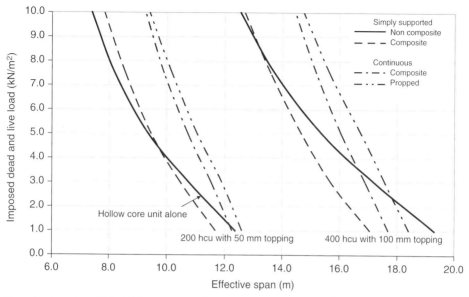

Examples based on EC2 with f_{ck} = 50 N/mm² (precast), $f_{ck}(t)$ = 30 N/mm² at transfer, and f_{ck} (*insitu* topping) = 25 N/mm². Let f_{bc} after long-term losses = +11.0 N/mm² and f_t = −4.07 N/mm² for internal exposure in all cases. Self weight of slabs given in Table 1.4.

Figure 1.41 Imposed uniformly distributed load versus span for continuous (propped) and continuous-composite slabs, and composite and non-composite simply supported hollow core floor slabs.

1.3.4 *Composite and continuous internal beams*

The most common design restriction for internal beams is total depth, or more significantly the beam span/beam depth ratio, and the floor bay/beam depth ratio, where the competition is with post-tensioned flat slabs or slim-floors, for example, flush steel plate beams in Figure 1.4, leading precast designers to consider shallower beams. As with floor slabs, prestressed beams offer the greatest potential for shallower depths, particularly if the floor slab is completely recessed within the depth of the beam, leading to inverted-tee beams, otherwise known as double-boot or ledger beams. Given that the depth of the floor slab is usually the same, whatever type of beam is used, the key issue is to minimise both the total depth and the depth of the downstand, or boot. As the boot becomes shallower, its behaviour begins to resemble that of a shallow nib, where strut and tie design methods are hampered by the minimum size of the links that can be placed in the boot. The links must provide sufficient resistance to prevent spalling and cracking at the top corner of the boot, as well as providing the correct structural mechanism directly beneath the slab bearing. This usually calls for small diameter high tensile rebars, with small bend radii, at closely spaced centres, for example, H10 at 150 mm spacing. Mild steel rebars are easier to bend by hand on site if larger diameter bars are required. A unique solution is to anchor a full-length steel angle section, such as 150 × 100 × 4 mm in the top of the boot, which is designed to carry the floor reaction and work independently of the concrete in the boot.

The key design input includes concrete strength, where grade C50/60 is specified to give a high tensile and cracking resistance in the soffit, steel tendon strength and degree of prestress, composite action between beams and floor slabs, and/or continuity of bending through or to the immediate sides of columns, as shown in Figure 1.28. Prestressed beams may be designed in two or three stages:

Stage 1:   prestressed beam alone resisting bending stresses due to the self weight of itself plus that of the slab and small quantities of *insitu* concrete infill;

Stage 2a: prestressed beam acting compositely with the floor slabs to resist imposed dead and live loads; or

Stage 2b: the composite beam is post-tensioned (p-t) to resist imposed live loads.

The latter benefit from long straight bays of roughly equal bay size, that is, both floor and beam spans, in order to optimise p-t and to not exceed the combined compression due to prestress plus p-t near to the supports, particularly where the beam carries small imposed live loads. The p-t tendons are deflected in the same region as the development length of the tendons, typically 800–1000 mm from the ends of the beam. To add further capacity at the serviceability limit state of stress, the beam may be propped at mid-span or at one-third and two-thirds span points during construction of the floor slab.

Figure 1.42 shows the benefit of composite and continuous design, leading to longer spans or greater load capacity, for a 600 mm wide × 400 mm deep prestressed inverted-tee beam supporting 6 m span × 200 mm deep prestressed hcu of self weight $3.27 \, kN/m^2$, for (i) basic beam, (ii) composite beam with 200 mm deep slab, (iii) composite beam continuous over three equal spans, and (iv) as (iii) propped at mid-span during construction of the floor slab. The total beam service load (in kN/m) excludes the self weight of the floor slab and beam = 24.4 kN/m. In this example the effective breadth of the compression top flange is b = 1650 mm. The beam contains 22 no. 12.5 mm diameter helical strands stressed to $70\% \times 1770 \, N/mm^2$ at a centroid height of 136 mm. Under such conditions, the critical service stress will be the total flexural tension in the top of the beam at the support, for which additional static rebars will be required. The beam is designed to Eurocode EC2 (EC, 2002b).

Taking one example from Figure 1.42, for example, 8.0 m span continuous-and-propped beam, Figure 1.43 shows the design service (dashed) and ultimate moments and resistances for an imposed load of 40.2 kN/m, totalling 59.8 kN/m inclusive of floor slab dead load. This means that if the floor load is $3.27 + 5.0 = 8.27 \, kN/m^2$, the effective span of the floor slab $\leq 59.8/8.27 = 7.23$ m, for which the 200 mm hcu is capable of (as shown in Table 1.5). Final iteration of this equality leads to floor span = 7.03 m. If the slab bearing = 75 mm, the spacing of the beam is $7.03 \, m + (0.6 - 0.075) \, m = 7.56$ m. If column size = 300 mm and the beam bearing point = 100 mm from the face of the column, the column centres = 8.0 + 0.5 = 8.5 m, resulting in bay area/depth ratio = $8.5 \times 7.56/0.4 = 161 \, mm^2/m$, a huge value for a downstand depth of only 200 mm.

In fact, a 150 mm deep hcu will span 7.03 m, reducing the slab load by $0.64 \, kN/m^2$ and enabling the beam to span 8.35 m, further increasing bay area/depth ratio = $167 \, mm^2/m$. Under similar conditions, the non-composite basic beam span = 5.35 m and bay area/depth ratio = $111 \, mm^2/m$, a reduction of 50%.

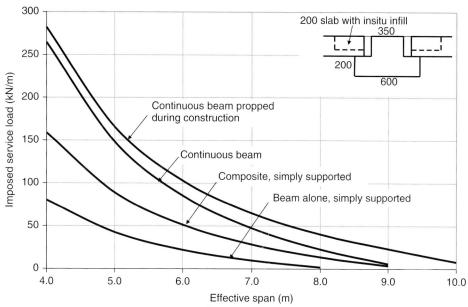

Examples based on EC2 with f_{ck} = 50 N/mm^2 (precast) and f_{ck} (*insitu* infill) = 25 N/mm^2.
Let f_{bc} after long-term losses = +14.8 N/mm^2 and f_t = −4.07 N/mm^2 for internal exposure in all cases.
Self weight of beam and slab = 24.4 kN/m.

Figure 1.42   Imposed uniformly distributed beam load versus span for continuous (propped) and continuous-composite beams, and composite and non-composite simply supported inverted-tee beams.

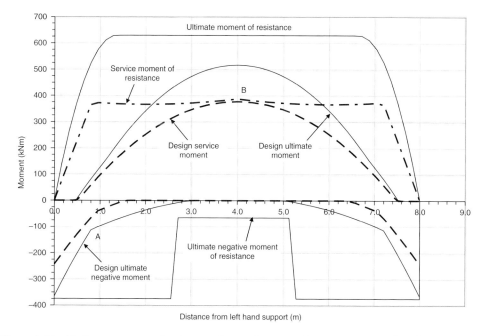

Figure 1.43   Design moments and resistances for continuous and composite inverted-tee beam (in Fig. 1.41), propped at mid-span. The kink at A is due to the different loading patterns for maximum and zero live load is alternate spans. The effect of the prop at mid-span is visible at B.

These examples have shown the potential for minimisation of the depth of beams in typical floor layouts, offering the architect greater headroom of greater spans whilst utilising the same structural components. Together with the optimisation of component cross-sections, for example, hcu and double-tee slabs, and improvements in material strength and pretensioning techniques, this was the main feature of the lightweight period in development of slim-line precast frame design.

1.4   The thermal mass period

1.4.1   Background to fabric energy storage in precast framed and wall structures

In the mid-1990s the precast concrete industry turned its attention towards fabric energy storage (FES) as a means of promoting precast concrete buildings as an energy efficient solution. This was not only in the completed envelope, by the use of higher-specification thermally insulated wall panels and complete façade systems, but also in the lower energy demands in the manufacture of the same – self-compacting concrete (SCC) being a material that factory-cast concrete could embrace with ease to drive down both energy and labour demands (see Section 2.1.3). The key to low energy design is the interaction between the mass of the structural elements and the external and internal environment. Whilst medium to high values of thermal resistance and admittance can be provided by the mass of the concrete itself, it is the added insulation in precast sandwich panels and soffit-insulated ground floor slabs that offer the greatest benefit, minimizing whole life costs at little additional expense. Ducting air in the opened cores of prestressed hollow core floor slabs, to form a continuous conduit of cooling or warming air, as shown in Figure 1.44, is a simple value-added task for the producer of the slabs to make. There may be some problems in making connections across the joints, but overall systems known as TermoDeck® have been very successful in hot climates. A well quoted example (Boardass & Leaman, 2012) is a four-storey office building that has a total energy consumption per annum of $90 - 100 \text{ kWh/m}^2$; by comparison a typical domestic house built in the 1960s today consumes $150 - 180 \text{ kWh/m}^2$, although the exposed perimeter per m^2 floor area of a domestic house is much greater than the offices. Night-time storage of cool air is used to ventilate buildings during the day, a cycle known as fabric energy storage (FES). Over the past 10–15 years many concrete building awards have embraced thermal (concrete mass), acoustic (concrete mass, blocking flanking passages, impact transmission) and dynamic (mass damping) criteria, all of which are associated to some extent by FES. A simple example is the night time storage of cool air being released gradually throughout the day through, say 30 mm of dense concrete, typically at the rate of $0.5 - 0.6 \text{ kW/m}^2$ over a temperature differential of 10°C.

The publication "Utilisation of Thermal Mass in Non-residential Buildings" (De Saulles, 2006) focuses on the role of FES in concrete structures, highlighting the use of profiled and exposed concrete soffits and the embodiment of CO_2 in the production and maintenance of concrete, both of which are features of factory-engineered concrete. De Saulles notes "*FES can do much to simplify building design and operation, however, it also brings with it specific design issues that are not present in more*

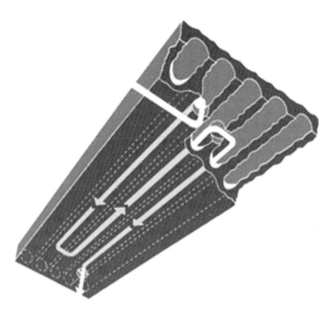

Figure 1.44   Passage of air through the extruded cores in precast hollow core floor slabs. (Courtesy of TermoDeck®.)

traditional office design. These issues mostly arise from the use of exposed concrete soffits, which has implications for acoustics, lighting, routing services and the general design process." These words are reflected by some of the key issues given in this book's Chapters 7 to 9 by Girmscheid and Selberherr in relation to the expertise of "system suppliers", such as the acoustic specialist working with the prefabricator, and the role of the "system integrator" who has the breadth of know-how towards a sustainable, optimized building. The 39,000 m² prefabricated concrete and masonry office building shown in Figures 1.15 and 1.45 (also in Figures 1.5 and 1.35) demonstrates this point. Michael Hopkins, architect for the buildings, notes *"At night the inherent thermal mass of the concrete is exploited and purged with fresh air to pre cool the structure. At the corners of the buildings, the air within the glass block stair towers* (see Figure 1.15) *warms and rises on sunny days, giving extra drive to the ventilation system. Fabric umbrellas on the tops of the towers act as large dampers, lifting to exhaust hot air and closing, on cool days, to conserve heat"* (Hopkins, 2014). At the time of construction (1994) it was said that if this building was scaled to the size of a domestic house it would have an annual energy bill of £60/$100!

The CIBSE Report (CIBSE, 2005) predicts that by 2050 a typical naturally ventilated office designed in the 1960s is likely to overheat to 28°C for around 15% of its occupied life compared to just 2% to 3% for a building of high thermal mass based on the principles of FES. In the UK, the Building regulations (HMSO, 2006) Part L2A sets this target as little as 20 hours per annum, or 1% of the occupied period for most office environments. Furthermore, the combined and solar gains (due to people, lighting, equipment, etc.) per unit floor area, averaged over daily occupancy, should not exceed 35 W/m² when the building is also subjected to solar irradiances. This calls for synergy between the design of the façade as well as the roof and floor slabs. Precast concrete façade panels of typical thickness 225 mm (including about 75 mm of

Figure 1.45   External elevation of prefabricated offices at Inland Revenue, Nottingham, UK, including precast columns in masonry, precast head stones, precast vaulted floor units, together with a steel roof.

insulation with conductivity $\lambda \approx 0.025$ W/m°K) will achieve a U-value (including low emissivity and sheltered surface resistances) of about 0.26 W/m²°K which is less than 0.3 W/m²°K required for most national building regulations.

1.4.2   Admittance and cooling capacity

The basic principle for the operating performance of FES in buildings of high thermal mass is a slow reaction to changes in ambient conditions, and the ability to reduce peak temperature (Braham *et al.*, 2001), namely, a cave versus a tent. The UK Building Research Establishment (BRE) Digest 454, Part 1 (Braham *et al.*, 2001) shows that high thermal mass delays peak internal temperature by around six hours for a mid-afternoon period as shown in Figure 1.46. After the peak temperature is achieved the FES cycle is reversed making night-time ventilation an effective means of removing accumulated heat from the concrete elements, lowering its temperature for the coming day.

The response of the thermal mass of structural elements (either singularly or composite building fabric including floor of wall finishes) is quantified by "admittance" – the ability of an element to exchange heat with the environment when subjected to simple sinusoidal variation of temperature (e.g., Figure 1.46) over a given period (24 hours). Admittance is measured in W/m²°K, where K = difference between mean daily temperature and the temperature at a point in time. The upper limit for naturally ventilated spaces is 8.3 W/m²°K. Table 1.6 lists values for common building fabrics, including precast concrete sandwich panels rated at 5.48 W/m²°K.

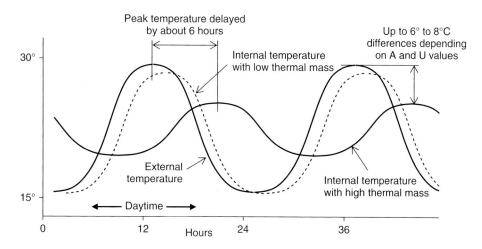

Figure 1.46   Stabilising effect of thermal mass on internal building temperature (after De Saulles, 2006).

Table 1.6   Admittance values for walls and facades (based on CIBSE Guide A – Environmental Design (CIBSE, 2006)).

Wall construction	Admittance W/m^2°K
Precast concrete sandwich-wall panel	5.48
Brick and dense concrete block cavity wall	5.75
Brick and block cavity wall	2.95
Internal block wall	2.09
Timber frame brick wall	0.86

The Concrete Centre's publication (De Saulles, 2006) gives cooling capacity data for a range of FES systems, including

- flat and profiled soffits in cast *insitu* and precast (e.g., double tee units) concrete
- raised floor added to flat and profiled soffits
- precast hollow core slabs with mechanical ventilation in the conduit-cores
- water cooled slabs
- chilled beams with exposed soffits.

The data, summarized here in Table 1.7, show the benefits of using precast concrete floors which have the highest cooling capacity without the need for water cooling.

Precast concrete floor slabs have been used in a number of significant FES-driven projects in Europe, the USA and the Middle East, where both hollow core-conduit slabs and bespoke ribbed or vaulted slabs were specified in the face of competition from water-cooled flat slabs. The benefits of precast prestressed slabs with cooling features are:

- long-span/shallow-depth (12 to 20 m in depth of between 300 and 450 mm are possible)

- large unit sizes (e.g. 16.0×3.6 m in Figure 1.45)
- factory repetition (on 15 different units)
- low embodied energy (SCC without vibration, minimal cross-section and steel strand area)
- reduced energy sapping site work is replaced by efficient machine production (e.g., hollow core slab)
- clean, smooth, ex-steel mould finishes to receive paint directly or be exposed concrete decorative finish.

The latter point is clearly obvious where hollow core slabs are used in car parks spanning 16 m without interruption of drop beams, reducing lighting requirements by 25% to 50%. Bespoke precast elements may also be manufactured using white Portland cement, further reducing emissivity and lighting requirements. A disadvantage (in some countries) is the need for high capacity craneage and transportation, for example the floor units in Figure 1.45 weighed about 12 tonnes and required a 60^T capacity crawler crane. Sometimes major disadvantages with hollow core floor units are the limiting modular width (1200 mm mainly but 600 and 2400 mm are possible) without the need for expensive cut-widths, and difficulties with balconies, curved floor areas, etc. The camber (due to prestress, typically span/200–400) has to be considered in the conduit-core ventilation system for two reasons; (i) conduit-cores crossing at beam grid-lines may misalign by at 20–30 mm (ii) differential camber between adjacent units may be 10–15 mm (depending on span).

Table 1.7 Cooling capacity for a range of FES and ventilated floor systems (based on De Saulles, 2006).

Description	Application	Cooling capacity W/m^2	Benefits
Flat or profiled floor slabs with natural ventilation	Offices (with low internal gains), schools, universities	15–20 (flat) 20–25 (profiled)	No fan energy; minimal maintenance
Slab with raised floor with mechanical ventilation	Offices, public and commercial buildings	20–30 (flat) 25–35 (profiled)	Mixed mode ventilation system with windows; allows convective heat transfer with top of slab
Hollow conduit-core slab with mechanical ventilation*	Offices, schools, hotels, universities, theatres	40 (basic) 50 (+ supplementary cooling) 60 (+ switch flow)**	Large floor spans; air introduced at any level; well established technology
Water cooled slabs (plastic pipework)***	Offices, schools, hotels, universities	64 (flat) 80 (profiled)	High cooling capacity; water from boreholes, wells; good temperature control, precast option
Concrete slabs with chilled beams suspended below	Refurbished and new offices, as above	15 – 30 (FES†)	High cooling capacity; high chilled water tem-perature may allow free cooling; good temperature control

*Open-ended cores with controlled flaps may be used
**Flow may easily be reversed in these systems
***Pipe floor or Thermocast are precast concrete options
†Chilled beams provide additional $100 - 160$ W/m^2

Other precast options for water cooled slabs include half-slab (e.g., Omnia, fili-gree, permanent formwork) comprising a precast soffit plank acting as permanent formwork (i.e. reinforced with exposed lattice girders or prestressed with roughened top surface) of 60–100 mm thickness, on which polybutylene pipes are set into an *insitu* structural topping of 75–150 mm depth (depending on span and loading). The solution offers:

- speed of construction
- high quality soffits (ex-steel mould) as mentioned above
- longer spans up to about 8 m with propping and continuity over beams
- and overcomes the awkward shape restrictions and positions of service voids of hollow core units.

Polystyrene (or similar) void formers (typically 2 no. 400×100 mm cross-section per 1.2 m width) may be placed on the top of the precast plank. For a solid slab thickness of 225 mm, such insulating voids will reduce self weight by about 30% and improves the passive thermal resistance R_T by a factor of about 2.2.

The thermal transmittances of floors, roofs and walls, the so called "U-value" = $1/R_T$, gives useful comparative values of the insulation properties of different floor constructions, and are used in building regulations in some countries for the limitation of heat losses through and floors and walls.

1.4.3   Thermal resistance and U-values for precast ground and suspended floors

The thermal transmittance between internal and external environments, "U-value" (units $W/m^2 °K$) may be calculated for determining heat loss through floors and walls, of singular or composite construction. It is carried out according to ISO international standards:

- floor thermal resistance according to ISO 6946 (ISO 6946, 2007)
- floor thermal transmittance to ISO 13370 (ISO 13370, 2007)
- thermal conductivity of materials to ISO/DIS 10456 (ISO/DIS 10456, 2007).

It is necessary to calculate the thermal resistance R_f of each floor element, including surface resistances R_{se} and R_{se}, in the construction (see Section 1.4.3.4). According to Figure 1.47, three situations are analysed:

- Internal-soil, where the slab bears directly onto ground and h = zero.
- Internal-external, at suspended ground floor where h = height from internal ground level to soffit of slab.
- Internal-internal, at upper floor level where h = storey height.

A worked example for each case is given in Section 1.4.3.5. Note that thermal bridging is not considered in this chapter, as it is a matter for bespoke building design and may be different for each project. The general arrangement for the composite

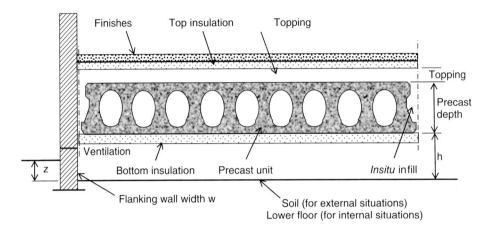

Figure 1.47   General arrangement of precast concrete floor slabs and walls in the calculation of U values for construction.

Table 1.8   Thermal conductivity of ground and suspended materials.

	Thermal conductivity λ (W/m°K)
Clay or silt	1.5
Sand or gravel	2.0
Homogenous rock	3.5
Concrete	1.7
Insulation	Manufacturer's data

construction is shown in Figure 1.47. The thermal properties of the ground and materials and surface resistances are given in Tables 1.7 and 1.8.

It is first necessary to determine the characteristic dimension of the floor area B', defined as the ratio of the area A of the room floor bounded by insulating walls to the exposed perimeter P, as:

$$B' = A/0.5P$$

In the case of basements, B' is calculated from the area and perimeter of the floor of the basement, not including the walls of the basement, and the heat flow from the basement includes an additional term related to the perimeter and the depth of the basement floor below ground level. P is the exposed perimeter of the floor: the total length of external wall dividing the heated building from the external environment or from an unheated space outside the insulated fabric.

Thermal resistance is represented by its equivalent thickness, which is the thickness of ground that has the same thermal resistance, that is, d_t = equivalent thickness for floors; and d_w is the equivalent thickness for walls (of own thickness w) of basements below ground level. The steady-state ground heat transfer coefficients are related to the ratio of equivalent thickness to the characteristic floor dimension, for example, d_t/B'.

1.4.3.1   Slab-on-ground floor (known as "internal-soil")

Referring to Figure 1.48, the equivalent thickness for floor $d_t = w + \lambda_s (R_f + R_i)$ where λ_s is for the ground conditions given in Table 1.8. R_f = thermal resistance of the floor slab including surface resistance R_{si}. However, dense concrete slabs should be neglected, leaving only the resistance of the air gap in the cores. R_i = thermal resistance of any all-over insulation layers above the floor slab, and that of any floor covering.

Then if $d_t < B'$ $U = \frac{2\lambda}{\pi B' + d_t} \, ln \left(\frac{\pi B'}{d_t} + 1 \right) W/m^2°K$

Otherwise $d_t \geq B'$ $U = \frac{\lambda}{0.457 \, B' + d_t} \, W/m^2°K$

1.4.3.2   Suspended ground floor (known as "internal-external")

Referring to Figure 1.48 for the suspended ground floor, the compound U-value is

$$\frac{1}{U} = \frac{1}{U_f} + \frac{1}{U_g + U_x} W/m^2°K$$

where U_f is for the suspended floor, U_g is for the ground and U_x is for the wall and ventilation.

$U_f = 1/(R_f + R_i)$ includes surface resistances and insulation.

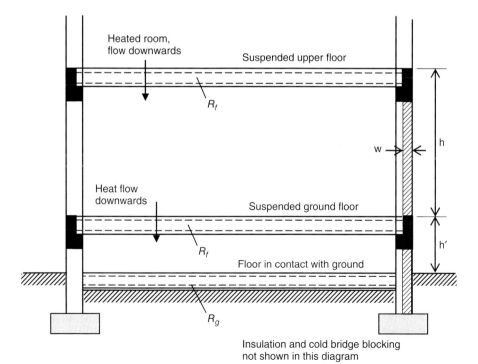

Figure 1.48   Notation for contact ground slab, and suspended ground and upper floors. (Adapted from ISO 6946 (2007) and ISO 13370 (2007).)

Equivalent thickness of ground space $d_g = w + \lambda(R_{si} + R_g + R_{se})$ where R_g is the thermal resistance of insulation on the base of the floor

R_{si} (downward) and R_{se} are the internal and external surface resistances.

and $U_g = \frac{2\lambda}{\pi B' + d_g} \ln\left(\frac{\pi B'}{d_g} + 1\right)$

If the under-floor space extends to an average depth of more than $z = 0.5$ m below ground level,

U_g is modified as:

$$U_g = U_{bf} + \frac{z P U_{bw}}{A}$$

If $d_t + 0.5z < B'$ $U_{bf} = \frac{2\lambda}{\pi B' + d_t + 0.5z} \ln\left(\frac{\pi B'}{d_t + 0.5z} + 1\right)$

If $d_t + 0.5z \geq B'$ $U = \frac{\lambda}{0.457 B' + d_t + 0.5z}$

Equivalent thickness for walls of basements below ground level $d_w = w + \lambda(R_{si} + R_w + R_{se})$

where R_w is the total resistance of the wall as specified.

$$U_{bw} = \frac{2\lambda}{\pi z}\left(1 + \frac{0.5 d_t}{d_t + z}\right) \ln\left(\frac{z}{d_w} + 1\right)$$

If $d_w > d_t$, use the above, but otherwise if $d_w < d_t$ replace d_t with d_w.

$$U_x = \frac{2 h' U_w}{B'} + \frac{1450 v \varepsilon f_w}{B'}$$

where wind shield factor $f_w = 0.02$ for city centre, 0.05 for suburban, otherwise 0.1.

h' = height of the upper surface of the floor from outside ground level.

U_w = U – value for wall $= 1/R_w$

ε = under – floor ventilation opening area / perimeter =

1.4.3.3   Upper floor (known as "internal-internal")

Referring to Figure 1.48 for the suspended upper floor, the compound U-value is:

$$\frac{1}{U} = \frac{1}{U_f} + \frac{1}{U_x} \, W/m^{2°}K$$

where $U_f = 1/R_f$

where $R_f = R_{f,slab} + \Sigma R_i + 2 \times R_{si}$

and $U_x = \frac{2 h U_w}{B'} + \frac{1450 v \varepsilon f_w}{B'}$

h = storey height

1.4.3.4   Thermal resistance R_f values for precast floor units

Referring to the example shown in Figure 1.49, and referring to ISO 6946:2007

$$1/R_f = f_{solid}/R_{solid} + f_{void}/R_{void}$$

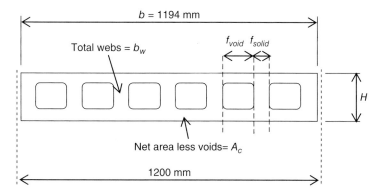

Figure 1.49   Idealised cross-section through hollow core floor unit (see Figure 2.20 for actual profile).

Table 1.9   Surface or air resistances used in U-value calculations.

	Surface or air resistances (m² °K/W)
R_{si} (internal) downwards	0.17
R_{si} (internal) upwards	0.10
R_{se} external	0.04
R_a unventilated air space > 100 mm	0.22*
R_a unventilated air space > 300 mm	0.23*

*Interpolated between

where f_{void} = ratio of number × breadth of hollow core / total breadth = n b_c/b and the remainder $f_{solid} = 1 - f_{void}$

$R_{solid} = H/\lambda_c + R_{si} + R_{se}$ (or 2 × R_{si} downward for internal-internal)
where λ_c = thermal conductivity for concrete taken as 1.7 W/m°K
Table 1.8 gives $R_{si} = 0.1$ upwards heat flow and 0.17 downwards heat flow
$R_{se} = 0.04$ m²°K/W for both directions

$$R_{void} = H_f/\lambda_c + R_{si} + R_{se} + R_a$$

where H_f = equivalent thickness of top and bottom flanges = $(A_c - b_w H)/b$
Table 1.9 gives R_a for unventilated air space > 100 mm dimension deep voids for heat flow downwards = 0.22 m²°K/W. R_a for 300 mm deep voids = 0.23 m²°K/W. Note that for concrete slabs in contact with the ground H/λ_c and H_f/λ_c are neglected.

Example 1 – For 200 deep unit at internal-internal surface:

$$f_{void} = 9 \times 85/1200 = 0.638, \text{ and } f_{solid} = 0.362$$
$$R_{solid} = (0.200/1.7) + 0.17 + 0.17 = 0.458 \text{ m}^2°K/W$$
$$H_f = (144031 - 381 \times 200)/1194 = 57 \text{ mm}$$

$$R_{void} = (0.057/1.7) + 0.17 + 0.17 + 0.22 = 0.593 \text{ m}^{2}{}^{\circ}\text{K/W}$$
$$1/R_f = 0.362/0.458 + 0.638/0.593 = 1.866 \text{ W/m}^{2}{}^{\circ}\text{K}$$
$$R_f = 1/2.482 = 0.536 \text{ m}^{2}{}^{\circ}\text{K/W} \quad \text{(including surface resistances)}$$

Example 2 – For 200 deep unit at internal-external surface:

$$R_{solid} = (0.200/1.7) + 0.17 + 0.04 = 0.328 \text{ m}^{2}{}^{\circ}\text{K/W}$$
$$R_{void} = (0.057/1.7) + 0.17 + 0.04 + 0.22 = 0.463 \text{ m}^{2}{}^{\circ}\text{K/W}$$
$$1/R_f = 0.362/0.328 + 0.638/0.463 = 2.482 \text{ W/m}^{2}{}^{\circ}\text{K}$$
$$R_f = 1/2.482 = 0.403 \text{ m}^{2}{}^{\circ}\text{K/W} \quad \text{(including surface resistances)}$$

Example 3 – For 200 deep unit at internal-soil surface:

Ignore the contribution from the concrete slab

$$R_{solid} = 0.17 + 0.04 = 0.21 \text{ m}^{2}{}^{\circ}\text{K/W}$$
$$R_{void} = 0.17 + 0.04 + 0.22 = 0.43 \text{ m}^{2}{}^{\circ}\text{K/W}$$
$$1/R_f = 0.362/0.21 + 0.638/0.43 = 3.209 \text{ W/m}^{2}{}^{\circ}\text{K}$$
$$R_f = 1/3.209 = 0.312 \text{ m}^{2}{}^{\circ}\text{K/W}$$

1.4.3.5   Worked examples for U-values for the three conditions

Calculate U-values for 200 mm deep hollow core slab (Figure 1.49, also 2.20) with a 50 mm structural topping plus 75 mm sand/cement finishes with $\lambda_c = 1.7 \text{ W/m}^{\circ}\text{K}$. Polyurethene foam insulation between the topping and finishes is 40 mm thickness with $\lambda = 0.025 \text{ W/m}^{\circ}\text{K}$. The wall thickness = 200 mm and there are flanking walls at the edges of the slabs. The thermal resistance of the walls is found elsewhere as 1.5 W/m°K. The building is located in a suburban setting ($f_w = 0.05$) with a design wind speed at 10 m height of 4 m/s. Soil type is clay. The floor area = 100 m^2 and perimeter = 40 m. Then $B' = 100/0.5 \times 40 = 5.0$ m.

Case 1 – Ground floor slab on soil
From 1.4.3.4 (Example 3) for the surface resistances $R_f = 0.312 \text{ m}^{2}{}^{\circ}\text{K/W}$
Finishes $R_f = 0.075/1.7 = 0.044 \text{ m}^{2}{}^{\circ}\text{K/W}$
Topping $R_f = 0.050/1.7 = 0.029 \text{ m}^{2}{}^{\circ}\text{K/W}$
Insulation $R_f = 0.040/0.025 = 1.600 \text{ m}^{2}{}^{\circ}\text{K/W}$
λ_s for clay = 1.5

$$d_t = w + \lambda_s(R_f + R_i) = 0.200 + 1.5(0.312 + 0.044 + 0.029 + 1.600)$$
$$= 3.178 \text{ m}$$

$$d_t < B' \text{ use } U = \frac{2 \times 1.5}{\pi 5.0 + 3.178} \ln\left(\frac{\pi 5.0}{3.178} + 1\right) = 0.28 \text{ W/m}^2 {}^\circ\text{K}$$

Case 2 – Suspended ground floor (internal-external)

Height of the upper surface of the floor from outside ground level $h' = 0.5$ m, and the under-floor height $z = 0.3$ m. Take the area of ventilation opening = 1500 mm^2 per m run. Then $\varepsilon = 1500 \times 10^{-6} = 0.0015$ m^2/m.

Table 1.7 $R_{se} = 0.17$ (Downwards) and $R_{se} = 0.04$ (Both Directions).

R_g = zero (no ground insulation)

Effective ground space $d_g = w + \lambda(R_{si} + R_g + R_{se}) = 0.200 + 1.5(0.17 + 0 + 0.04) = 0.515$ m

But under-floor space $z < 0.5$ m, then

$$\frac{1}{U} = \frac{1}{U_f} + \frac{1}{U_g + U_x}$$

where $U_g = \dfrac{2\lambda}{\pi B' + d_g} \ln\left(\dfrac{\pi B'}{d_g} + 1\right) = \dfrac{2 \times 1.5}{\pi 5.0 + 0.515} \ln\left(\dfrac{\pi 5.0}{0.515} + 1\right) =$

0.638 W/m$^2 {}^\circ$K   and $U_x = \dfrac{2 h U_w}{B'} + \dfrac{1450 v \varepsilon f_w}{B'} = \dfrac{2 \times 0.5 \times \left(\frac{1}{1.5}\right)}{5.0} +$

$\dfrac{1450 \times 4.0 \times 0.0015 \times 0.05}{5.0} = 0.220$ W/m$^2 {}^\circ$K

From 1.4.3.4 (Example 2) $R_f = 0.403$ m$^2 {}^\circ$K/W

This case has 100 mm mineral wool insulation to the soffit of the slab with $\lambda = 0.04$ W/m$^\circ$K

$$R_f = 0.403 + 0.044 + 0.029 = 0.476 \text{ m}^2 {}^\circ\text{K/W}$$

$$R_i = 0.100/0.04 = 2.500 \text{ m}^2 {}^\circ\text{K/W}$$

$$U_f = 1/(R_f + R_i) = 0.336 \text{ W/m}^2 {}^\circ\text{K}$$

$$\frac{1}{U} = \frac{1}{0.336} + \frac{1}{0.638 + 0.220} = 4.14 \text{ m}^2 {}^\circ\text{K/W (note the major contribution of } U_f \text{ in this)}$$

$$U = 0.24 \text{ W/m}^2 {}^\circ\text{K}$$

Case 3 – Suspended upper floor (internal-internal)

Storey height $h = 3.0$ m,

From 1.4.3.4 (Example 1) $R_f = 0.536$ m^2°K/W

$$R_f + R_i = 0.536 + 0.044 + 0.029 + 2.500 = 0.476 \text{ m}^2\text{°K/W}$$

$$U_f = 1/(R_f + R_i) = 0.322 \text{ W/m}^2\text{°K}$$

$$U_x = \frac{2\,h\,U_w}{B'} + \frac{1450\,v\,\varepsilon\,f_w}{B'} = \frac{2 \times 3.0 \times \left(\frac{1}{1.5}\right)}{5.0}$$

$$+\frac{1450 \times 4.0 \times 0.0015 \times 0.05}{5.0} = 0.887 \text{ W/m}^2\text{°K}$$

$$\frac{1}{U} = \frac{1}{0.322} + \frac{1}{0.887} = 4.23 \text{ m}^2\text{°K/W}$$

$$U = 0.24 \text{ W/m}^2\text{°K}$$

1.4.4   Conclusion to FES, cooling and thermal transmission

Today the key design features and benefits of high thermal mass in multi-storey buildings, city centre offices in particular, are being fully exploited in modern construction. Although the key technology is some 30 years old, it has only been fully embraced by the precast concrete industry since the mid 1990s, as some of the examples have shown. The reader should also be directed to the case studies by (de Saulles, 2006). Precast solutions have enjoyed a double bonus; being able to enhance key principles associated with FES with simplicity of design in pinned jointed braced frames, utilizing the shear cores also as ventilation shafts, and designing large floor diaphragms as free natural ventilation passages. It has also noted a turning point in the market where there are close correlations between this type of design and reducing whole life costs - a conclusion highlighted by de Saulles (2006), particularly relating to high performance (factory cast) concrete, low operational CO_2 emissions and M&E services.

Calculations for the thermal transmittance of walls and floors has advanced to include the effect of the building environment, rather than being limited to the building elements alone, as was the case in the past. This has enabled building authorities to reduce thermal transmittance U-values to around $0.2 - 0.3$ W/m^2°K with confidence that the designers are able to meet future target emission rates for CO_2 with confidence.

One of the most significant changes in this field is the availability of literature, both in publications and the background and normative information in national and international standards, and the use of case studies on the web. The role of thermal mass design in modern precast concrete buildings and façades is now being recognised in national awards.

References

Bachmann, H. and Steinle, A. (2011). *Precast Concrete Structures*, Ernst and Sohn GmbH & Co. KG, Berlin, Germany, 260 pp.

Bljuger, F. (1988). *Design of Precast Concrete Structures*, Ellis Horwood, Chichester, UK.

Boardass, B. and Leaman, A. (2012). CIBSE Case Study: Investigation of 90s Building for Energy Performance, *CIBSE Journal*, Chartered Institution of Building Services Engineers, March, London.

Braham, D., Barnard, N. and Jaunzens, D. (2001). Thermal Mass in Office Buildings: an Introduction, *Building Research Establishment Digest 454, Part 1*, Building Research Establishment, Watford, UK.

Bruggeling, A.S.G. and Huyghe, G.F. (1991). *Prefabrication With Concrete*, Balkema, Rotterdam, Netherlands, 380 pp.

CIBSE. (2005). Climate Change and the Indoor Environment: Impacts and Adaptation, TM36, Chartered Institution of Building Services Engineers, London.

CIBSE. (2006). CIBSE Guide A – Environmental Design, Chartered Institution of Building Services Engineers, London.

De Saulles, T. (2006). *Utilisation of Thermal Mass in Non-residential Buildings*, The Concrete Centre, Camberley, UK, December, 89 pp.

EC. (2002a). Eurocode EN 1991-1-1:2002, Eurocode 1: Actions on Structures – Part 1-1: General Actions – Densities, self-weight, imposed loads for buildings.

EC. (2002b). Eurocode EN 1992-1-1:2004, Eurocode 2: Design of Concrete Structures – Part 1-1: General rules and rules for buildings. Amendment 2014.

El-Arab, I. E. (2012). The Design Principle Of Precast Concrete Structures: Fundamental design, Erection details, Precast Concrete structures, LAP Lambert Academic Publishing GmbH & Co. KG, Saarbrücken, Germany, 288 pp.

El-Debs, M. K. (2000). *Concreto Pré-Moldado: Fundamentos e Aplicações*, Publicação EESC-USP, Sao Paulo, Brazil.

Elliott, K. S. (1996). *Multi-Storey Precast Concrete Framed Structures*, Blackwell Scientific Press, Oxford, UK, 624 pp.

Elliott, K. S. (2002). *Precast Concrete Structures*, Butterworth-Heinemann, Oxford, UK, 380 pp.

Elliott, K. S. and Jolly, C. K. (2013). *Multi-Storey Precast Concrete Framed Structures*, 2nd ed., John Wiley, London, 750 pp.

Elliott, K. S. (2016). *Precast Concrete Structures*, 2nd ed., CRC Press, Taylor & Francis Group, Florida, USA, 500 pp.

fib. (2002). *Bulletin 43, Precast Concrete in Mixed Construction*, State-of-art Report, Fédération Internationale du Béton, Lausanne, Switzerland, 68 pp.

FIP. (1982). FIP Recommendations, *Shear at the Interface of Precast and Insitu Concrete*, Federation International de la Precontrainte, Wexham Springs, Slough, UK, 31 pp.

FIP. (1986). FIP Recommendations, *Design of Multi-Storey Precast Concrete Structures*, Federation International de la Precontrainte, Thomas Telford, London, 27 pp.

FIP. (1994). *Recommendations, Planning and Design of Precast Concrete Structures*, Federation International de la Precontrainte SEKO, Institution of Structural Engineers, London. 138 pp.

Goodchild, C. H. (1995). *Hybrid Concrete Construction*, Reinforced Concrete Council, (former British Cement Association), Camberley, UK, 64 pp.

Gibb, A. G. F. (1999). *Off Site Fabrication: Prefabrication, Pre-Assembly and Modularisation*, Whittles Publishing, Caithness, Scotland, 262 pp.

Glover, C. W. (1964). *Structural Precast Concrete*, C. R. Books Ltd., London.

Haas, A. M. (1983). *Precast Concrete Design and Application*, Applied Science Publishers, London.

HMSO. (2006). UK Building Regulations, Part L2A, HMSO, London.

Hopkins, M. (2014). *http://www.hopkins.co.uk/s/projects/5/88*, Inland Revenue Centre, Nottingham, UK [accessed September 2014].

ISO 2631-1. (1997). Mechanical vibration and shock - Evaluation of human exposure to whole-body vibration - Part 1: General requirements, 31 pp.

ISO 6946. (2007). Building components and building elements, Thermal resistance and thermal transmittance, Calculation method.

ISO 13370. (2007). Thermal performance of buildings, Heat transfer via the ground, Calculation methods.

ISO/DIS 10456. (2007). Building materials and products. Hygrothermal properties, Tabulated design values and procedures for determining declared and design thermal values.

Lawson, M., Ogden, R. and Goodier, C. (2014). Design in Modular Construction, CRC Press, Taylor & Francis Group, Florida, USA, 258 pp.

Morris, A. E. J. (1987). *Precast Concrete in Architecture*, George Goodwin Ltd.

National Building Frame Manufacturers' Association. (n.d.). *The National Building Frame*, UDC 693.9.691.327.

New Zealand Concrete Society and National Society of Earthquake Engineering. (1991). *Guidelines for the Use of Structural Precast Concrete in Buildings*, Christchurch, New Zealand, 174 pp.

Nilson, A. H. (1987). *Design of Prestressed Concrete*, 2nd ed., John Wiley.

Sarja, A. (1992). Industrialised Building Technology as a Tool for the Future International Building Market, *Concrete Precasting Plant and Technology B+FT*, **58**(11).

Sheppard, D. A. and Phillips, W. R. (1989). *Plant-Cast Precast and Prestressed Concrete*, 3rd ed., McGraw-Hill.

Van Acker, A. (2006). Brussels North Skyline Now Dominated by Precast Tower Buildings, *CPI Concrete Plant International*, **51**, 174–182.

Van Acker, A. Master Classes "The design of precast concrete building structures".

Elliott, K. S. Workshops "The Design and Construction of Precast Concrete Structures".

Vambersky, J. A., Workshops "Design for Precast Construction" with the handbook "Designing and Understanding Precast Concrete Structures in Buildings".

Chapter 2

Industrial Building Systems (IBS) Project Implementation

Kim S. Elliott

Precast Consultant, Derbyshire, UK

The rise and fall and recent resurgence in Industrialised Building Systems (IBS) in the past 15 to 20 years has seen some of the world's major advances in precast construction in Northern Europe, USA and China. This is because of the automated design, detailing and manufacture of components and the rapid erection of precast buildings up to about 120 m height at a construction rate of up to 800 m^2 per week, and double this for low-rise buildings. The main features of procurement, performance specification and the contractual route together with the selection process for IBS suppliers are described. The IBS supply chain comprises the complete package of concept, design, detailing, manufacture and erection. The automation for the manufacture of key components such as slabs and walls is described, together with the advantages and disadvantages of using *open* and *closed* systems by Registered IBS Providers.

2.1 Introduction

The conventional design and construction of a concrete structure - ask 95% of graduate structural engineers under the age of 30 - is to erect formwork on falsework on site and fill with reinforced cast *insitu* concrete. Although the design, detailing and construction procedures for cast *insitu* concrete are well understood and entrenched in the basic syllabus of conventional construction, that syllabus also includes structural steelwork, an off-site prefabricated or industrialised building system (IBS). It is therefore surprising to find that, whilst steelwork is openly accepted as IBS, the off-site manufacture of concrete components is still classed as "unconventional", threatening small profit margins of contractors operating in countries where the balance between expensive plant and equipment and cheap labour still favours the latter. This argument does not hold for structural steelwork as there is no alternative

Modernisation, Mechanisation and Industrialisation of Concrete Structures, First Edition.
Edited by Kim S. Elliott and Zuhairi Abd. Hamid.

to off-site manufacture, but it is becoming increasingly balanced for precast concrete construction, particularly in regions of the world where cast *insitu* has dominated the building sector.

A useful marker for the implementation of IBS in concrete structures is to study the relationship between the cost of labour and materials, that is, the effective cost of producing 1 tonne of finished reinforced concrete. One such indicator is:

C-factor = cost of 1 day skilled labour / cost of 1 m^3 of concrete delivered to site.

In this context *skilled* refers to persons trained in techniques, not necessarily with qualifications, and involved in decision making. The cost of 1 m^3 is taken as 1/6 of a 6 m^3 delivery of medium strength (C32/40) ready-mix concrete over a haulage time of about 30 minutes. The data have been obtained by private communication with eminent contractors or precasters in the years 1998–2012. The costs have not been updated for inflation, and it is assumed that this has not affected the ratio of C. Table 2.1 presents the C-factor from 15 countries or regions, in which Korea tops the chart with C = 1.9, Scandinavia not surprisingly is second with C = 1.4, Western Europe and North America = 0.7 to 1.0, South America and Australia around 0.5, and Eastern Europe and south east Asia 0.2 to 0.3. There is clearly a correlation between C-factors and typical wages, as the cost of ready-mix concrete does not vary that much.

An informative correlation is the C-factor versus the quantity of cement consumed in the production of precast concrete structural components (columns, beams, walls, stairs, floor and facades) to the total consumed in buildings (excluding infrastructure

Table 2.1   International data relating the cost of labour and materials to the consumption of cement in the precast concrete industry

Country or region	Labour per day	1 m^3 of concrete	C-factor	Cement consumed by precast in buildings*
Scandinavia	€130	€90-105	1.4	70 %
UK	£70	£60-75	1.0	35 %
Eastern Europe	PLN 130	PLN 400	0.3	30 %
Baltic (former USSR)	LAT 30	LAT 75	0.4	20 %
Southern Europe	€50	€65-80	0.7	30 %
North America	US $75	US $80-100	0.8	20 %
South America	US $40	US $70-80	0.55	35 %
Caribbean	US $50	US $110-120	0.45	25 %
Middle East	US $70	US $75-90	0.85	15 %
Korea	KRW 120k	KRW 60-70k	2.0	10 %
Hong Kong	HK$ 100	HK$ 480	0.2	10 %
South East Asia	RM40-55	RM170-190	0.25	10 %
Australia	AU$65	AU$130	0.5	20 %
South Africa	RSA 550	RSA 1400	0.4	15 %

*Median taken for all types of precast production, including structural sections, floors and facades.
1 m^3 of concrete is costed as 1/16 of a batch of 6 m^3 with about 30 minutes haulage time.

and bridges). This shows the potential for IBS, in which a higher C-factor inevitably, but not always, leads to greater use of precast structural components in IBS. Apart from the two extremities of Korea and Scandinavia, the consumption of cement in precast production is roughly aligned to the C-factor.

Another informative ratio, of equal status to the C-factor, is the plant (e.g., crane) factor defined as P = daily hire of a typical crane / cost of daily skilled labour. In this example the *crane* is taken as a 50t mobile/telescopic hired for 3 months. The P factor for Scandinavia = 6–10, UK 8–12, Europe 10–15, South America 15–20, and south east Asia 30–35. These values are almost the inverse of the C-factor such that P x C = approx 10. It should be the aim of IBS strategies to try to achieve C-factors close to 0.5, and P × C between 5 and 10, particularly in developing countries where large cranes can be prohibitively expensive for large scale precast erection.

National strategies have to be formulated to overcome the perceptions that the cost of producing and erecting concrete using expensive plant and machinery, that is, the P × C quotient, and then erecting the same with large capacity (expensive) cranes and complicated connections is higher than conventional cast *insitu* work. This strategy must take account of Government policies, for example, subsidies for certain manufacturing processes, such as structural steelwork, and well publicised sustainable benefits for green materials such as structural timber and glue-laminated timber components. The strategy should focus on the need to achieve economies of scale in both government and private projects, and to exploit the use of high performance materials and techniques, such as prestressed high strength concrete, ultra-thin (approx. 15 mm) sections reinforced with steel or polymer fibres, architectural (coloured, patterned) concrete, and the combination between structural and thermal/sound efficiency. These examples when combined should reach the inevitable conclusion of whether IBS is suitable for concrete construction.

The main aim of this chapter is therefore to assess IBS as the most favoured mode of design and construction of concrete buildings (IBS is the obvious choice for concrete bridges), and to show the implementation of the key stages of conceptual/scheme design, computerised detailing and design, semi- or fully automated factory production and on-site erection of the prefabricated components.

2.1.1   Definition of IBS

The literature refers to IBS in terms of prefabrication, mass production of building components, and off-site production. Here *off-site* can also mean *on-site* but remote from the construction works, such as on-site stack-casting of wall panels, beams, columns and r.c. slabs. The most informative definition, given by Zuhairi and Kamarul (2011) is "IBS is a construction technique in which components are manufactured in a controlled environment (on or off site), transported, positioned and assembled into a structure with minimal additional site work." Key phrases missing from this are "repetition", "standardisation" and "manufacturing efficiency", as the success of IBS requires all of these.

IBS is classified into four main systems:

- precast concrete frames and facades
- steelwork frames

- timber frames
- steel formwork and prefabricated reinforcement, carpet mesh, and so on.

IBS and the implementation of the four systems must evolve to incorporate global standards and practices. The *fib* (and formerly FIP) has set the global standards and practices for IBS concrete frames and bridges, certainly in Europe and its Dependencies, with the publication of the *fib* Model Code (*fib*, 2013), and the previous version published in 2000, are recognised as the forerunner to the structural concrete Eurocodes.

2.1.2   Advantages of IBS

The main advantages are: quality, faster completion time, clean construction sites and sustainability.

2.1.2.1   Quality

Figures 2.1(a) and 2.1(b) uniquely show the differences in accuracy and quality between cast *insitu* and precast concrete using the example of a stair flight. This highlights a major objective in implementing IBS. The precast staircases in the photograph were accurate to ±3 mm in length, including the sloping dimension of the flight, and ±2 mm in width, depth and tread. The cast *insitu* steps varied from −20 to +5 mm in width and ±10 mm in depth, had problems of compaction, honey combing and grout less, even before the formwork was removed to reveal the situation in the soffit. Surface hardness for the precast stairs would give a figure of about 45 in a rebound hammer test, compared to 35–40 on *insitu* concrete of this quality. In the struggle to overcome prejudice against their use, precast components require the competitive edge that quality brings; IBS building components need to be:

- of high quality in terms of dimensional accuracy, concrete strength and surface hardness ensuring prolonged durability
- aesthetically pleasing, even if plain concrete
- cost effective
- environmentally friendly
- not give end users problems
- acceptable to future needs, by replacement, renovation or extension
- able to maintain health and safety requirements.

2.1.2.2   Faster completion time

Faster construction times have been a major selling point for multi-storey precast concrete buildings, particularly those with a small footprint to height ratio, for example, floor plan < 400 m^2 and height greater than 20 m. Buildings with a large plan area can be constructed in several areas at the same time by flooding the site with labour and resources. Faster construction of medium to high-rise frameworks lends itself to continuity of following trades, where first fix M&E services follow-on only one or two

(a)

(b)

Figure 2.1   (a) High quality of precast, self-compacting concrete staircases and landings; (b) Quality issues with cast *insitu* staircases.

Table 2.2   Erection times for fixing 6-m and 12-m span precast and composite floors equating to 1000 m² per floor for four- and eight-storey buildings

Floor type	6 m span		Erection time (weeks)		12 m span		Erection time (weeks)	
	Depth (mm)	Self weight (kN/m²)	4 storey	8 storey	Depth (mm)	Self weight (kN/m²)	4 storey	8 storey
Ribbed slabs, e.g., double tee with topping	250 + 50	3.9	6.7	16.7	600 + 75	6.2	4.2	12.5
Hollow core floor units (no topping)	200	3.3	5.3	13.3	350	4.6	3.3	10.0
Hollow core floor slab with topping	150 + 50	4.1	7.6	15.7	300 + 75	6.2	5.0	11.9
Half-slab (50% precast *insitu*)	125 + 125	5.8	10.7	26.7	–	–	–	–
Beam and block (with concrete screed)	225 + 75	4.3	9.7	–	–	–	–	–

Topping = 50 mm thickness + 10 mm allowance for camber at 1.4 kN/m² for 6 m span and 75 mm thickness + 10 mm allowance for camber at 2.0 kN/m² for 12 m span. Slab design based on 1.5 and 5.0 kN/m² dead and live load. Internal exposure. 1 hour fire resistance.

floors behind the erection of the precast components, impossible to do without IBS methodologies. This leads to shorter completion times and building occupancy. The 36-storey building shown in Figure 1.17 was awarded to the Belgian precaster Ergon on the premise that construction rate would be faster than the equivalent fire protected steelwork frame. The speed of construction benefited from the long-span prestressed floors being compatible with the projecting details from the precast beams. Later in this chapter two case studies will show the importance of the design and construction of the floor slab, which comprises 68% to 77% of the total mass of the structure. Table 2.2 gives an indication of the time (in weeks) to erect an area of floor slab, taken in this example as 1000 m² per floor, for 4- and 8-storey frames. In Chapter 3, Table 3.1 gives a wider range of construction rates for precast concrete skeletal frames, highlighting the importance of repetition and larger frame dimensions of beam and slab spans.

The consistently faster completion of prefabricated structures is due to the combined use of standardised components, CAD/CAM design and detailing solutions and simplified erection processes, benefiting also in fewer site workers and less cluttered and cleaner construction sites.

2.1.2.3   Clean construction sites

The reduced volume and weight of raw materials, formwork, falsework and casting equipment on precast sites is a major feature of IBS. The IBS designer should aim for at least 75% off-site manufacture, a figure that can be improved on by designing out much of the cast *insitu* work, such as structural toppings on slabs, gaps between precast floor units and staircases. Seasonal variations are less critical to site progress, and totally nullified at the factory, even for outdoor production. Depending on the circumstances of the design, size, and complexity of the building and the accessibility of the construction site, IBS has the following approximate savings over cast *insitu* construction at the site (Elliott & Jolly, 2013):

- scaffolding material and labour to erect scaffold   80–90%
- shuttering and formwork   90–95%
- delivery and pouring wet concrete*   75–95%
- delivery and fixing of loose reinforcement   90–95%
- time of construction of superstructure (above foundations)   25–50%
- total construction time   10–30%
- site labour on superstructure   75–90%
- total site labour   50–75%

*Lower value where structural floor screeds are used.

2.1.3   *Sustainability of IBS*

There are several aspects of IBS that have the potential to contribute to different aspects of sustainability, reducing energy demand and reducing waste. Recent estimates (Zuhairi & Kamarul, 2011) claim that the amount of environmental impact from material transportation activities at one-third of the total environmental impact of construction. IBS components may be rationalised and maximised to reduce transportation journey lengths and time, and together with well planned "just-in-time" logistics have the potential to reduce energy demands for fixed design parameters.

The *fib* Bulletin 67 "Guidelines for green concrete structures" (*fib*, 2012b) "is a practically oriented guide to reduce the environmental impact of concrete structures throughout their full life cycle. Focus is placed on CO_2 as an environmental indicator. It provides an overview of selected available green concrete technologies, followed by suggestions on how the environmental impact can be accounted for. These accounting techniques can be used to optimize concrete structures in terms of their environmental impact. Benchmark data is provided, and several examples of how the guide, or parts of it, can be used to account for and optimize the environmental impact of concrete structures." (Extract *fib*, 2014).

The focus of attention in the controlled environment of factory production includes:

- high-strength reinforcement (e.g., rebars $f_y = 600$ N/mm^2 require 30% less material than $f_y = 460$ N/mm^2, but the energy to produce both types is almost the same).
- high strength prestressing steel (e.g., new helical strand $f_{pu} = 2100$ N/mm^2 compared to former strengths around 1580 and 1770 N/mm^2).
- high strength rapid hardening cements grade 52.5R, together with superplasticisers and air-entrainment, require less cement and lower water contents to guarantee higher strength concrete at earlier age leading to thinner or shallower components.

- Portland slag cement Type CEM II B-S (containing up to 35% GGBS (a by-product material ground granulated blastfurnace slag)), Blastfurnace cement Type III/A (up to 65% GGBS, or Portland fly-ash cement Type CEM II B-W (containing up to 35% PFA (a waste material pulverised fuel ash/fly ash))
- low viscosity self-compacting concrete requires no compaction energy and simpler moulds closer to the ground.
- recycled (crushed) precast concrete waste, for example, at ends of hollow core production lines, or faulty components, for reintroduction into new components up to 20% RCCA (recycled concrete coarse aggregate).
- reclamation of mixing and wastewater for reintroduction into new precast components, saving 20% virgin water, as shown in the Figure 2.2.
- reduced concrete cross-section, mainly in ribbed and hollow core floor units, leading to 30–50% less material consumption and lighter handling, transportation and fixing demands.
- better design and accuracy of formwork leading to the optimisation of cross-sections.
- thermally insulated precast sandwich panels offering U values in the region of 0.2 W/m² °C.

2.1.4   Drawbacks of IBS

Looking into the performance of IBS in different regions of the world, it is informative to find relative success is related to the C-factors given in Table 2.1. IBS achievements

Figure 2.2   Reclaiming mixing and waste water for reintroduction into precast concrete components.

are clearly greatest in regions of high labour cost, where the return on capital investment has the lowest amortization, with typical periods of payback as low as 5 years. This is clearly not the case in Asia cast *insitu* construction has not been overtaken by off-site fabrication of concrete, steelwork or timber. A number of IBS implementation snags include (Zuhairi & Kamarul, 2011):

- developing and sustaining market demand for IBS sufficient to generate economies of scale.
- inability to achieve acceptance of IBS in the construction community.
- developing standard plans and standard component drawings for common use.
- apprenticeships and one-the-job training in the area of IBS moulds, casts and assembly of components.
- vendor development programs.
- readiness of designers' and consultants' practices, that is, IBS concept design is not taken into consideration at the outset of projects.
- cost of using IBS exceeds conventional methods of cast *insitu* construction, particularly in the face of cheap foreign labour.

The supply chain for IBS and their components is not obvious to clients and designers. Therefore, consultants will not design buildings and/or components they cannot find in the market, while producers will not produce the components as they do not find consultants designing them. This chapter aims to address the above difficulties and to prepare the ground to implement IBS.

2.2   Routes to IBS procurement

2.2.1   *Definitions*

The following terminology and abbreviations is used to describe the functions of designers, manufacturers and contractors:

Client – building developer, local government or independent commercial organisation responsible for the initiation and sometime procurement of the project.

Architect – an individual or organisation responsible to the client for the architectural concepts and form and functional design of the buildings.

Precast company (PC) - an organisation responsible for design, detailing, manufacture and erection of precast components and buildings.

Precast Consulting Engineer (PCE) – either (i) in-house designer employed by the precast producer, and/or precast contractor, or (ii) independent organisation or individuals, or government officers offering a dedicated service to the precast producer/contractor

Precast Concrete Detailer (if different from the Precast Consulting Engineer) (PCd) – an independent organisation or individuals sub-contracted to the PC or PCE to carry out detailing and prepare project drawings.

Precast Concrete Manufacturer (PCM) – may be either a part of the PC or a separate organisation responsible for manufacture only.

General Contactor (GC) – a contractor responsible for the whole project, which encompasses the precast building(s) and/or components.

Precast Erector (PE) – may be part of the PC or a separate contractor responsible for erecting components or complete structures only.

Sub Contractor to the PE (SC-PE) – a supplier and/or erector of specific building components, such as floor slabs or façade panels

Specialist Supplier to the PE (SS-PE) – a supplier of specific products to the PE or less frequently to the SC-PE, such as fittings, temporary stability equipment.

2.2.2   Preliminaries

The essential implementation issues are the options for the selection of one of the following routes to satisfy the Preliminaries and produce Tender Documents:

i.   a sole representation PC, in which in-house design, detailing manufacture and possibly construction is carried out by a nominated company. Such an organisation is usually registered with a national federation as a Registered IBS Provider (RISP), such as the PCI in USA, FEBE in Belgium, British Precast in UK, although the term RISP varies in different countries;

ii.   a PC with design and manufacturing capabilities and a nominated PE for construction. The PC may still be classed as a RISP without its own contracting division;

iii.   an independent PCE, PCM and PCE responsible to the management contractor, sometimes known as the Contract Administrator, who may often be the Architect;

all or any of which may have detailing and construction capabilities or utilise separate detailers PCd and/or sub-contractors SC-PE.

The main features of procurement, performance specification, contractual route and for the above selection process are summarised in Chapter 2 and the Appendix of "Precast Concrete Framed Buildings – Design Guide" (Elliott & Tovey, 1992). The Preliminaries to be satisfied by any of the above options should include:

a. Site Visit. Ascertain local conditions before tendering by visiting the site or at best by reference to the client's site plans and other supplied information concerning access and hard standing.

b. Obscurities. These include secrecy of design and detailing, more prevalent in the past where patented connection systems were the norm, but still have relevance today. These should be resolved by the PC or PCE before tendering.

c. Standard Building Contract Without Quantities should be allowed for, including the clients drawings and description of work, but the project is complex enough for a bill of quantities at this stage.

d. Tender Analysis. To be submitted in detail within a stated period upon request.

e. Attendances. To be defined.

f. Construction or Erection Statement. Submit outline method and sequence of frame erection.

   g. Technical Data. As applicable within the context of the project, to be submitted with tender, including current national codes of practice, standards and product standards, with relevance to national Building Regulations.

   h. Manufacturers and suppliers. Precaster to provide list of all products and suppliers, with names and details of sub-let work and bought-in items.

   i. Quality Control. Precaster to describe procedures, implementation, staff and responsibilities.

   j. PCM references. Should be current at time of tender.

   k. PCM representatives and PE supervisors and leaders to be identified.

   l. PC or PCM can be appointed with written permission of the architect/contract administrator or selected from a list of (typically three) names.

2.2.3 Project design stages

The scheme, project and construction design stages are accomplished in five steps as follows. These do not include the preparation of the client's brief and visualisation or the architect's details:

1. preparation of the schematic building layout
2. design calculations
3. layout drawings
4. component scheduling (shop drawings)
5. construction drawings

The specification for concrete and some production methods, such as prestressing, are included within the above steps, although several will cross over, for example, characteristic concrete strength may be influenced more by production requirements than by design or durability criteria.

Step (1) involves the interpretation of the architect's drawings, together, but not independently of the client's requirements, and involves the major conceptual design issues. This stage is often known as the *scheme* - the preliminaries preceding the *project* design stages (2) to (4) and *construction* stage (5). Table 2.3 presents a summary for the decision making throughout these stages (Gibb, 1999) - from the client's brief to demolition and decommissioning for modular buildings and normal and volumetric off-site fabrication. The reader will note how the *conceptual design* phase is more populated with items than the other stages.

The architect's drawings are received in packages containing key components of form, function and geometry together with a specification involving dimensional freedom and/or limitations, materials, services (mechanical, electrical, plumbing, air-conditioning, etc.) from which the precast scheme solution, together with the use of other secondary structural materials such as *insitu* concrete, steelwork, masonry or timber, and so on, is developed. Scheme and project engineers may be one and the same group, or may be affiliated to different parts of the PC, or other PCE with the relevant expertise. Whichever is the case, it is important that the scheme engineer is (a) fully conversant with PCE's design practice, (b) aware of PCM's operations (even if the project's PCM is not yet appointed), and (c) aware of the construction

Table 2.3 Decision-making matrix for off-site fabrication (adapted from Gibb, 1999)

Project Stages	Actions and Decisions			
	Project	Off-site Fabrication (OSF)	Modular Building (MB)	Volumetric Off-Site Fabrication (VOSF)
Clients' brief	Establish clients' needs. Develop design brief. Decide procurement route. Risk assessment. If partnering, team members will already be in place.	Establish overall project OSF strategy. Consider if standard solution will meet clients' brief, e.g., modular building (MB).	Obtain input from MB suppliers on standard solutions. Evaluate track record of MB supplier	Obtain input from VOSF suppliers on standard and bespoke solutions. Appoint VOSF supplier (precise timing depends on procurement route). Establish outline site logistics, i.e. weight, sizes, craneage.
Concept design	Develop concept design. Appoint construction manager. Value engineering management. Obtain input on constructability. Develop cost plan and outline programme. Obtain outline planning consent. Develop outline health and safety (H&S) plan. Appoint Design & Build (D&B) contractor.	Consider if an adapted MB system will meet clients' brief. Consider volumetric and non-volumetric applications. Convene interface meetings to agree strategy.	Consider including on-site construction work in with MB package. Finalise all client's requirements.	
Detail design	Develop detailed design. Obtain full planning consent. Develop H&S plan. Review solutions against project aims. Appoint Main Contractor (if lump sum contract)	Review detail and ensure original OSF strategy is met. Convene interface meetings to agree details.	MB supplier does detailed design	Finalise all client's requirements. Supplier does unit detail based on consultant/designer's detailed design.

Production information	Approve production information. Finalise detailed overall on-site programme.	Production information by MB supplier. Determine programme for on-site works.	Production information by MB supplier. Determine programme for on-site works.
Manufacture	Obtain detailed installation programme. Convene interface meetings to finalise on-site coordination. Off-site quality and progress inspections.	Manufacture by supplier. Prepare site: if work not done by MB supplier obtain MB supplier's acceptance of site.	Manufacture by supplier. Prepare site ready for installation. Obtain supplier's acceptance of site.
Construction/ assembly	On-site work. Practical/final completion.	MB supplier usually installs units with own craneage.	Installation usually by supplier.
Maintenance facilities	Handover and use occupation. Defects liability period, release of retention, etc. Some contracts include operate/maintain clauses.		
Demolition and decommission	Ensure full information regarding future demolition is available in H&S file.	Modular building are particularly suitable for relocation.	Demolition may need particular attention doe to OSF units.

operations of the PE and certain SS-PE, such as prestressed hollow core or double-tee floor units that may form a major part of the building.

The schematic building layout drawings forms the basis of the conceptual stage in which the potential for the use of a *closed precast system* or an *open and bespoke* solution is often decided. A *closed* system is usually defined as one where the major structural components, principles and methods of connections, means of structural stability and the design against progressive collapse, and several other smaller items, are already defined, pre-designed and very often pre-detailed. This was very much the case in the 1960–1970s era where standardised connection methods such as the billet or cleated connectors (see Figure 3.9) formed the core of *closed* systems. *Closed* systems belonged chiefly to sole representative PC or PCM, many of which were RISP, capable of delivering a full design-manufacture-erection service. The principles behind connection design and detailing were often unique to a RISP; many were patented or were difficult to develop by calculation, having been proven chiefly by full-scale tests. Chapter 7 of Elliott & Jolly (2013) gives extensive background information and design methods.

Experienced RISP are advantageous at this stage in being able to identify the parts of the structure that do not lend themselves to prefabrication, either economically or structurally, opting to use either conventional cast *insitu* concrete as a supplement or replacement to precast concrete, or utilise other structural materials such as steel-work, timber or masonry. The advantages of so-called "mixed" construction are given in the *fib* Bulletin (*fib*, 2002), or "hybrid" construction (precast with *insitu* or other forms of concrete) by the UK Concrete Centre (Goodchild, 1995). The non-precast areas are therefore identified, such as roof rafters (steel truss, glue-laminated beams), foundations and other ground works (culverts, piling, ground beams, and retaining walls), and staircases that are some way awkward to prefabricate or disruptive to the construction programme.

A key feature that is common to both *closed* and *open* systems, and yet has had many varied solutions over the years, is the vertical stabilising system, which must be determined at the outset. Skeletal structures may be designed either as unbraced (sway) or braced, according to guidelines shown in Table 2.4. Uni-directional methods may

Table 2.4   Approximate number of storeys using various stabilising systems used to aid scheme design

Approximate number of storeys	Design principle	Stabilising method or components
≤3 to 4 [Note 1]	Unbraced	Cantilever column action
≤4 to 6 [Note 1]	Unbraced	Deep column cantilever action
≤7 to 10 [Note 2]	Unbraced	Semi-rigid beam-to-column connections
≤7	Braced	Brick infill shear walls
≤10	Braced	Precast concrete cantilever shear walls
≤12	Braced	Precast concrete infill shear walls
≤15	Braced	Precast concrete cores
12+	Braced	*Insitu* concrete cores
3 to 15	Partially braced	Braced, and unbraced in upper two to three storeys

[Note 1] Depends on storey height and allowable sizes of columns
[Note 2] Depends on storey height, building arrangement and wind pressure

be used with bracing in one orthogonal direction and not in the other, for example, semi-rigid beam-to-column connections in one plane and infill shear walls in the cross-plane. Divisions between unbraced and braced structures depends on (a) the availability of shear walls or cores, for example, if positions are available around lift shafts and stair wells they may be used in buildings of any height, (b) magnitude of sway and second-order deflections and bending moments in columns (i.e. deflections due to slenderness and imperfections), (c) ground conditions and foundation modulus to resist sway moments and rotations, and last but not least (d) the design decisions made by the PCE.

Figures 2.3 to 2.5 present a case study based on a 16-storey skeletal frame building in Belgium. The construction photograph in Figure 2.3 indicates a curved edge profile made from straight -line components in two semi-elliptical office blocks, connected

Figure 2.3   Case study skeletal frame at Ellipse building, Brussels. (Courtesy of Ergon Belgium.)

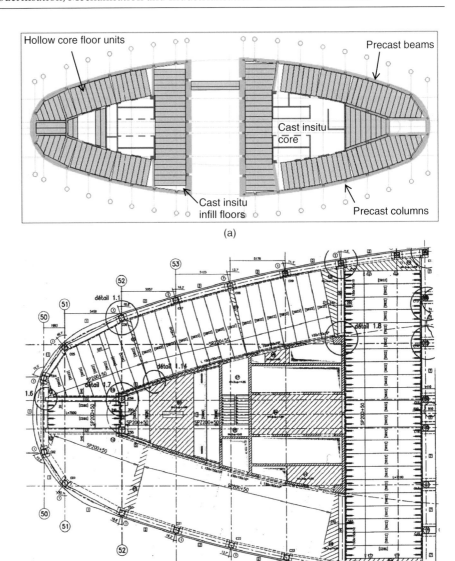

(a)

(b)

Figure 2.4   (a) Case study CAD-generated scheme layout for Covent Garden, Brussels, indicating the precast solution with cast *insitu* cores and precast columns beam and floors (adapted courtesy of Ergon, Belgium); (b) Case study structural CAD drawing for the scheme in Figure 2.5. (Courtesy of Ergon, Belgium.)

by walkways. Each block is therefore designed independently for stability and structural movement. Aware of the knowledge, shown in Figure 1.38, that the most efficient solution in terms of maximum spans versus minimum floor depth is using long-span prestressed floors spanning perpendicularly to the façade of the building (rather than spanning beams in this direction and floors parallel with the façade) the scheme is instantly developed as shown in the CAD impression in Figure 2.4(a). The first issue is stability in resisting an over-turning moment of about 50000 kNm for each half of

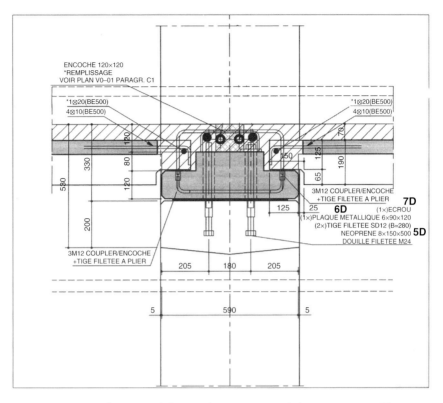

Figure 2.5 Case study structural drawing for cross-sectional elevation at internal beams at North Galaxy, Brussels (Figure 1.17). (Courtesy of Ergon, Belgium.)

the building based on (say) ultimate wind pressure $= 1\,kN/m^2$, plus a horizontal sway force of 280 kN per 16 floors based on a floor area of about $700\,m^2$ times (say) gravity dead + imposed load $= 10\,kN/m^2$ at a drift ratio (sway/height) of 1/400). The given solution was to use cast *insitu* cores of about 10 m × 12 m plan. Spanning 6.3 to 7.6 m and subject to imposed dead, partitions and live loads of (say) 1.5, 1.0 and $4.0\,kN/m^2$, composite floor slabs could be specified as 150 to 200 mm deep prestressed hollow core units (hcu) with 75 mm structural topping (total floor depth = 235 to 285 mm). 400 mm deep × 600 mm wide reinforced concrete edge beams with clear span of 4.8 m support hcu and external façade walls and curtain walling. Column size varies over the height of the building, but at ground level is about 600 × 600 mm depending on the compressive strength of the precast column, where up to grade C95/105 (cylinder/cube compressive strength) has been used, although C50/60 is more common.

Part of the structural drawings for the floor layout is shown in Figure 2.4(b), where the effect of the elliptical curvature results in small triangular areas of *insitu* concrete infill between hcu. The final layout of precast components is remarkably close to the scheme drawing, as are the cross-sectional design of beams and columns. Figure 2.5 gives an example of full structural working details for the construction of an internal beam supporting ribbed floors with halving joints onto the ledges. Together with general floor-to-floor elevations, and some component schedules showing the main rebars and links (as shown in Figure 2.6 not from this case study), these drawings give the total manufacturing and construction information.

Figure 2.6   Poor design, detailing and execution of precast components and connections.

Tender drawings should give sufficient information for the client to judge the merits of the structure, including:

- floor plans, dead and imposed loads, and choice of flooring (prestressed or rein-forced hcu, ribbed slabs (double/triple-tee), filigree voided slab, etc.), positions of large voids and service holes (say > 600 mm), and whether composite slabs with structural topping are necessary or preferred.
- continuous slab or cantilever options, designed by spanning floor units onto cantilevered beams or by cantilevered slabs with negative moment capacity.
- positions of walls (not partitions) or cores necessary for structural stability.
- cross-sectional elevations of the major areas of the structure, with local or detailed variations highlighted, showing structural zones for slabs and beams, column and wall sizes enabling service routes to be identified.
- cross-sections of key components (beams, slabs) and bearings. Connection detail would not be shown at this stage.
- non-precast areas affecting the design of the super-structure, such as rafters, stair-cases and balconies. Foundations, piling, and so on, are not shown.

Elliott and Jolly (2013) make an important observation: "At the completion of this exercise the designer should be confident that temporary stability is guaranteed no matter how the site erector may choose to build, and that the areas of insitu concrete may be allowed to cure and develop strength before further construction proceeds. To this end the scheme drawing(s) is annotated for internal use during the billing stage with information to assist the estimator to assess craneage and site costs in addition to the relatively simple task of pricing the precast concrete components. Preparing a final scheme and quotation is usually an iterative procedure: seldom are original layouts accepted after the first attempt. The points of contention are usually positions of shear walls, depths of beams and voids for service routes."

2.2.4   Design and detailing practice

Unlike the design of cast *insitu* concrete, which is based on optimising architectural and structural dimensions and form, and the design of structural steelwork, where standard serial sizes are available and connection details are often resolved by the contractor, precast design solutions are greatly dependant on specific manufacturing and construction methods of the PCM and PE with SC-PE. Precast design is ideally, but not exclusively, suited to *closed* systems, in which an iterative "knowledge loop" involving design-detail-manufacture-construction-design is preferential to any other option, that is, the RISP designer should have production and construction experience, and should understand the implications of design. Figure 2.6 is an example of where this is not so. Design decisions are influenced more by manufacturing constraints, repetition, handling, temporary frame stability and construction sequence, than by economy of materials or form.

Section 2.4 and Chapter 6 will show how design office practice is linked to manual or automated manufacturing operations, for example, bar straightening and bending, welding and cage assembly, connector positioning, casting, curing and the requirements for handling (concrete strength at 18 hours, lifting devices and anchorages, reinforcing for double curvature). Repetitious casting of major structural components lends itself to both *closed* and *open* systems, even where the latter include bespoke and specialised components, such as prestressed floors, beams, wall panels (symmetrically prestressed) and sometime stairflights and landings. Designers attempt to utilise, as far as possible, a range of predetermined structural components whilst maximising architectural and functional requirements. Designers may take advantage of high-strength concrete (40 and 100 N/mm² compressive cube strength at 1 and 28 days, respectively), low water/cement ratio (0.35 to 0.4) for improved durability and reduced cover to rebars, and dimensional accuracy and stability, all of which may lead to lower partial safety factors for materials, for example, $\gamma_m = 1.4$ (from 1.5) in Eurocode EN-1992, Part 1, Annex A.2 and A.3 (EC, 2002a).

Discussions need to be held with RISP construction personnel, or with the GC and PE (and PE with SC-PE) to take advantage of the relationships between design and handling, optimising the use and capacity of the crane, or specifying the type of crane (tower, crawler, telescopic) according to the design of the components. The construction sequence greatly influences the total project time, although the differences between *closed* and *open* systems are blurred by the dominant effect of building height, namely, vertical transportation time of the crane hook and the difficulties in propping large components (>4 tonne) at several storeys height, and floor layout. Table 3.1 in Chapter 3 shows typical rates of construction for precast skeletal frames of 4, 8 and 12 storeys based on *closed* systems in which the construction techniques and connection types were familiar to well established fixing gangs. There are no data for *open* systems as these would depend greatly on the types of connections and the nature of bespoke components, but would be expected to decrease by up to 10% for low rise and 5% for high-rise frames. The storey heights of 4, 8 and 12 are chosen as providing demarcation between low-rise (4–5 storeys can be accessed directly from the ground), medium-rise (up to 8 storeys it is possible to negotiate construction from the ground and the completion and stability of lower storeys is essential to progress) and high-rise (around 12 storeys where temporary frame stability becomes a critical issue). The data do not include wall frames (wall panel(s) with floor slabs), although construction rates are similar up to 12 storeys. The rates assume wide-slab floor units (e.g., 1.2 to 2.4 m wide) without a structural topping, the latter will delay completion of the framed structure by about 10 to 12 working days as the topping is placed only two floor levels below that of the precast. For a 4-storey structure of 4000 m^2 floor area the delay may represent a 40% increase, as shown in Table 2.2, whereas for 8 storeys it is around 15% and may not even lie on the project's critical path.

2.2.5   Structural design calculations and project drawings

Design calculations should commence after agreement that the scheme drawings and details concur with the client's intentions and/or architect's latest drawings, but this is not always feasible due to time constraints and the appointment of other disciplines such as M&E consultants or contractors. The use of software and spreadsheets has made the transition between scheme and project design much easier over the past 20 years, with less reliance on hand calculations. The same can be said for structural drawings, which can commence early because of the increased use of CAD facilities, and is not the problem it was as recently as the late 1980s. Recent advances in CAD and scheduling of two-dimensional units, such as wall panels or floor slabs, have resulted in the automation of unit schedules directly from design software, shown for example in Figure 2.7.

Typical calculation sheets for the design of prestressed hollow core slabs and reinforced beams are shown in Figures 2.8 and 2.9. These are based on Eurocodes EC0 [EN 1990 (EC, 2002a)], EC1 [EN 1991 (EC, 2002b)] and EC2 [EN 1992 (EC 2004a,b)] together with the National Annexes for each user country in the EU. Similar versions to other structural concrete codes such as the American ACI-318:2011 or Australian AS3600:2009 can be envisaged. One topic impeding progress is the position of large voids in floor slabs, say greater than 400 mm in 1200 mm wide discrete components such as hcu where steel trimmer angles are required to support the load either side of the void and the adjacent units receive additional load. Figure 2.10 shows a typical situation for a wide void framed out by steel trimmer angles. Smaller holes less than 400 mm can be compensated by increasing the number of tendons in the remaining width. Holes less than about 150 mm size can be pre-formed at the factory or cut on site with the permission of the SS-PE. Holes in double-tee units are positioned in the shallow flanges between the strengthening webs. These do not have a major influence on design as additional reinforcement can be added around the edges of the holes. However, horizontal holes in the webs must be planned in advance, as the pretensioning tendons have to be diverted or the number increased to account for the hole.

The design of most precast components is carried out by in-house or commercial software packages resulting in predetermined design and detailing information. In-house packages are often an extension of previous generations of tabulated/graphical design data and standardised drawings. The main distinguishing feature of precast design software is that design output may be in coded form to be compatible with the chosen method for scheduling components. In addition to components and connections, structural calculations also include:

- design against progressive collapse and measures taken to avoid accidental loading, see also the *fib* Bulletin 63 (*fib*, 2012a). This is provided either by the steel

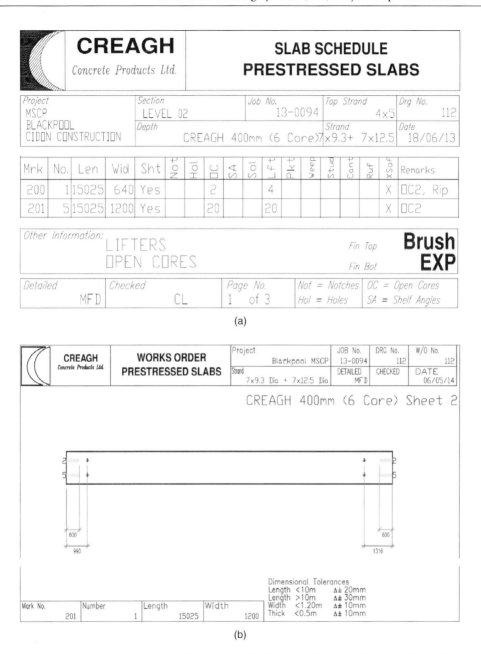

| CREAGH | SLAB SCHEDULE |
| Concrete Products Ltd. | PRESTRESSED SLABS |

Project	Section		Job No.	Top Strand	Drg No.
MSCP	LEVEL 02		13-0094	4x5	112
BLACKPOOL	Depth			Strand	Date
CIDON CONSTRUCTION	CREAGH 400mm (6 Core)			7x9.3+ 7x12.5	18/06/13

Mrk	No.	Len	Wid	Sht	Not	Hol	OC	SA	Sol	Lft	Pkt	Weep	Stud	Cant	Ruf	XSof	Remarks
200	1	15025	640	Yes			2			4						X	OC2, Rip
201	5	15025	1200	Yes			20			20						X	OC2

Other Information:			
LIFTERS		Fin Top	**Brush**
OPEN CORES		Fin Bot	**EXP**

Detailed	Checked	Page No.	Not = Notches	OC = Open Cores
MFD	CL	1 of 3	Hol = Holes	SA = Shelf Angles

(a)

CREAGH Concrete Products Ltd.	WORKS ORDER PRESTRESSED SLABS	Project		JOB No.	DRG No.	W/O No.
			Blackpool MSCP	13-0094	112	112
		Strand		DETAILED	CHECKED	DATE
			7x9.3 Dia + 7x12.5 Dia	MFD		06/05/14

CREAGH 400mm (6 Core) Sheet 2

	600				600	
	990				1316	

Dimensional Tolerances
Length <10m   Δ± 20mm
Length >10m   Δ± 30mm

Mark No.	Number	Length	Width	Width <1.20m Δ± 10mm
201	1	15025	1200	Thick <0.5m Δ± 10mm

(b)

Figure 2.7   (a) Extract from schedule of 400 mm deep hollow core floor units (courtesy of Creagh Concrete, N. Ireland). Showing 5 no. unit ref. 201 × 15025 mm length × 1200 mm width, with 5 × 4 no. 20 no. open cores (OC) and 20 no. lifters (Lft). Pretensioning uses 7 no. × 9.3 mm plus 7 no. × 12.5 mm diameter helical strands in the bottom and 4 no. 5 mm wires (called strand in the schedule) in the top. The positions and cover to the strands and wires would be given in standard drawings elsewhere; (b) Manufacturing drawing for ref 201 hollow core floor unit scheduled in Figure 2.9. (courtesy of Creagh Concrete, N. Ireland). The unit has 6 cores of which the second and fifth cores cut open in the factory for a distance of 600 mm. The reason the lifting points are not symmetrical is to enable to unit to tilt during erection allowing one end to touch down just before the other end.

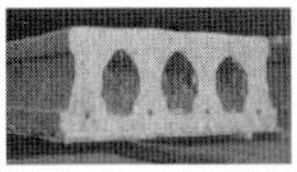

	Job no. 1234
Prestressed concrete hollow-core floor units	Made by: Elliott
designed to Eurocode EN 1992	Checked by: Zuhairi
	Date : 1 January 2015

Precast unit input data

Depth of hollow core unit h	200	mm
Nominal breadth b	1200	mm
Breadth at top b_t	1154	mm
Breadth at bottom b_b	1197	mm
Total breadth of webs b_w	474	mm
Depth of top flange h_{fb}	30	mm
Number of cores	11	
Breadth of cores b_c	60	mm
Gross area of concrete A_c	151640	mm^2
Height to concrete centroid y_b	97.9	mm
First moment of area S $(x10^6)$	4.566	mm^3
Second moment of area I $(x10^6)$	668.3	mm^4

Concrete data

Design compressive strength f_{ck}	45	N/mm^2
Mean compressive strength f_{cm}	53	N/mm^2
Transfer strength $f_{ck}(t)$	30	N/mm^2
Mean flexural tensile strength f_{ctm}	3.80	N/mm^2
5% characteristic $f_{ct,0.05}$	2.66	N/mm^2
Young's modulus E_{cm}	36283	N/mm^2
Transfer tensile strength $f_{ctm}(t)$	2.72	N/mm^2
Young's modulus $E_{cm}(t)$	32837	N/mm^2
Type of aggregate	Gravel	
Type of cement	CEM I 52.5R	
Precast concrete density	24.5	kN/m^3

Structural Data

Effective span L	7600	mm
Bearing length L_b	100	mm

Steel tendons

Yield strength for strand f_{pk}	1770	N/mm^2
Young's modulus of tendons E_{ps}	195000	N/mm^2
Tendon relaxation class	2	
1000 hour relaxation	2.5	%

Environmental and Fire Data

Exposure classification (indoor)	XC1	
Fire resistance R	60	mins.
Relative humidity in service	50	%
Design working life	50	years
Age at installation	28	days

Prestressing data

Initial prestress / ultimate ratio η	0.700	
Age at detensioning	20	hours
Relative humidity at transfer	70	%

Prestressing data

	Row 1	Row 2	Row 3
Number of tendons per row	10	0	0
Cover to each row	30	0	0
Diameter at each row	9.3	0	0
Total area of bottom tendons	520	mm^2	
Initial prestress force F_{pi}	644.28	kN	
Axis height to steel centroid y_s	34.65	mm	
Eccentricity z_{cp}	63.23	mm	

Compound section properties

Modular ratio m-1 = (E_{ps} / E_{cm}) -1	4.37	
Transformed area of tendons	2275	mm^2
Compound $y_{b,co}$	96.9	mm
Compound I_{co} $(x10^6)$	677.3	mm^4
Compound modulus $Z_{b,co}$ $(x10^6)$	6.986	mm^3
Compound modulus $Z_{t,co}$ $(x10^6)$	6.572	mm^3

Calculation of prestress losses

Initial prestress $f_{po} = \eta \, p_{yk}$	1239	N/mm^2
Initial relaxation loss	4.95	N/mm^2
Stress at level of tendons at release	8.07	N/mm^2
Ditto after deduction of self weight	5.57	N/mm^2
Elastic shortening loss	33.06	N/mm^2
Prestress force at transfer F_{pmo}	624.52	kN
Stress at level of tendons at install	6.01	N/mm^2
Creep coefficient at install	0.87	
Creep loss at install	26.42	N/mm^2
Prestress force at install F_{pmi}	610.78	kN
Stress at level of tendons incl. dead	3.83	N/mm^2
Further creep coefficient to life	1.64	
Further creep loss to life	31.18	N/mm^2
Shrinkage loss to life	95.39	N/mm^2
Long term relaxation loss	31.15	N/mm^2
Prestress force at install F_{po}	528.76	kN
Total losses	17.9	%

Prestress

	Top	Bottom	
At transfer at supports	-1.89	+9.78	N/mm^2
Limits	-2.72	+18.00	N/mm^2
In service at mid-span	-1.62	+8.38	N/mm^2
Limits	-3.80	+20.25	N/mm2

Serviceability moment of resistance M_{sR}

Top fibre limit = 21.87 x 6.572	143.75*	kNm
Bottom fibre limit = 12.18 x 6.986	85.09	kNm
*to be used with quasi-permanent load		

Ultimate moment of resistance M_{Rd}

Effective depth d	165.4	mm
Depth to neutral axis x	32.1	mm
Lever arm z	152.5	mm
Final strain in tendons εp	0.019478	
Final stress in tendons fp	1511	N/mm^2
Moment of resistance M_{Rd}	119.86	kNm

Figure 2.8 Example of design procedures to Eurocode for prestressed hollow core floor unit for serviceability and ultimate moment of resistance.

	Reinforced concrete inverted tee beam designed to Eurocodes EN 1990 and EN 1992	Job no. 1234
		Made by: Elliott
		Checked by: Zuhairi
		Date : 1 January 2015

Precast unit input data

			Concrete and steel rebar data		
Total depth of beam h	500	mm	Design compressive strength f_{ck}	40	N/mm^2
Upstand depth h_u	200	mm	Mean compressive strength f_{cm}	48	N/mm^2
Breadth at top b_t	350	mm	Mean flexural tensile strength f_{ctm}	3.80	N/mm^2
Breadth at bottom b_b	600	mm	5% characteristic $f_{ct,0.05}$	2.66	N/mm^2
Gross area of concrete A_c	250000	mm^2	Young's modulus E_{cm}	35220	N/mm^2
			Type of aggregate	Gravel	

Structural Data

			Type of cement	CEM I 52.5R	
Effective span L	6500	mm	Precast concrete density	25.0	kN/m^3
Bearing width l_w	300	mm	Yield strength f_{yk}	500	N/mm^2
Bearing length l_b	150	mm	Young's modulus E_s	200000	N/mm^2

Environmental and Fire Data

			Floor occupancy	Offices	
Exposure classification (indoor)	XC1		Ultimate combination ψ_0	0.7	
Fire resistance R	60	mins.	Frequent ψ_1	0.5	
Cover to links	30	mm	Quasi permanent ψ_2	0.3	

Service loads

			Ultimate shear design		
Self weight of beam	6.25	kN/m	Clear span L_c	6350	mm
Floor slab	18.00	kN/m	Shear span $L_c - 2d$	5451	mm
Floor finishes, ceiling and services	12.60	kN/m	Ultimate $V_{Ed} = 76.01 \times 5.451/2$	207.1	kN
Floor partitions	6.00	kN/m			
Floor imposed live load Q_k	15.00	kN/m	**Design of shear links (single leg in upstand)**		
Total dead load G_k	42.85	kN/m	v_1	0.504	
			$z = 0.9d$	406	mm
Ultimate loads. EN 1990 equations			$(b z v_1 f_{ck}/1.5) / V_{Ed}$	8.98	
6.10(a) = 1.35 x 42.85 + 0.7 x 1.5 x 15.00 = 73.60			$1.0 < \cot \theta < 2.5$	2.50	
6.10(b) = 0.925 x 1.35 x 42.85 + 1.5 x 15.00 = 76.01			A_{sw} / s per leg	241	mm^2/m
Ultimate $M_{Ed} = 76.01 \times 6.500^2/8$	401.4	kN	Minimum links A_{sw} / s	204	mm^2/m
			Use H8 at 205 mm spacing	244	mm^2/m
Flexural design			Reduce to minimum H8 at 240 c/c	208	mm^2/m
Cover to main bars	38	mm	$V_{Rd,s}$ minimum links	178.7	kN
Try 25 mm main bars. ys =	50.5	mm	Distance from end to minimum links	900	mm
Effective depth d	449	mm			
Breadth of upstand b_t	350	mm	**Span / depth check**		
$K = M_{Ed}/f_{ck} b_t d^2$	0.142		$\rho = A_s/b_t d$	0.0156	
$z = d \min\{0.95; 0.5+\sqrt{0.25-K/1.133}\}$	384	mm	$\rho = \sqrt{f_{ck}}/1000$	0.0063	
$x = (d – z)/0.4$ check $< h_u = 200$	165	mm	L/d ratio basic	14.85	
Area rebars A_s	2407	mm^2	Modified by As provided / design	1.02	
Use 5 no. H25 bars	2455	mm^2	L/d ratio	15.14	
Spacing of bars = (600 – 76 – 125)/4	100	mm	d required	429	mm
			d provided	449	mm
Curtailment $\geq 50\%$	1228	mm^2			
Curtail to 3 no. H12 + 2 no. H25	1321	mm^2	**Crack width and bar spacing**		
d at curtailment	451	mm	Tensile area around bars $A_{c,eff}$	54750	mm^2
			$\rho_{eff} = A_s/ A_{c,eff}$	0.0448	
Anchorage length of main bars			Crack spacing s	224	mm
Design stress in bars	426	N/mm^2	$M_s = (G_k + \psi_2 Q_k)L^2/8$	250.1	kNm
f_{bd} for main bars	3.68	N/mm^2	Service stress in bars σ_s	272	N/mm^2
L_{bd} required (in diameters)	28.9		Mean strain in bars ε_s	0.001160	
$\alpha_1 = 1, C_d = 38, \alpha_2 =$	0.92		Crack width $c_w \leq 0.4$ mm	0.26	mm
L_{bd} design (in diameters)	26.7		Allowable clear spacing in main bars	210	mm
$L_{bd} = \max\{10, 26.7\} \times 25$	667	mm			

Figure 2.9   Example of design procedures to Eurocode for reinforced concrete beam.

Figure 2.10　Steel trimmer angles around large voids in hollow core slabs.

mesh and site placed bars in the structural topping, or by additional tie bars in the joints between floor slabs and perimeter and internal beams, in addition to vertical suspension ties in the columns and/or walls.

- positions and provisions for expansion or construction joints to cater for thermal movement and residual shrinkage after installation.
- horizontal floor diaphragm action, satisfied either by the same structural topping or horizontal tie bars defined in bullet point 1 in combination with the interface action between discrete precast floor units.
- structural stability against horizontal wind pressure and/or notional sway/drift loads (note the Eurocode is additive, where as other codes is replacement).
- temporary stability of the framework and floor diaphragm.
- foundation loads, moments and shear forces.

Until recently structural layout drawings were all prepared manually, but almost all drawings are now made using CAD methods, unless because of the one-off nature of some components the system is not an economic use of drafting skills. However, this argument is rarely applicable to precast buildings where there is an intention to replicate as many details as possible. An example of part of a floor layout drawing for a car park is shown in Figure 2.11(a), together with cross-sections of the perimeter spandrel beams (Figure 2.11(b)) and internal inverted-tee beams (Figure 2.11(c)) on which hcu are laid to fall. Their purpose is firstly to show the architect, engineer and other trades the details of the structure and its connections, and secondly to transmit to site information necessary to erect the structure and flooring, as shown in the notes to Figure 2.11(b). Precast layout drawings tend to resemble structural steelwork

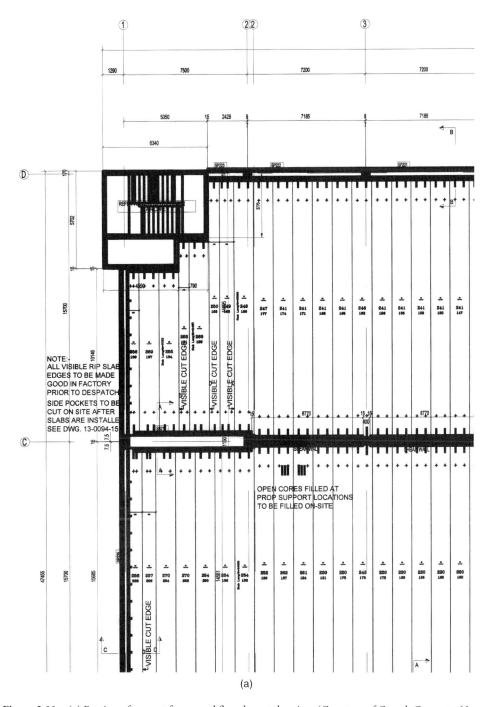

(a)

Figure 2.11 (a) Portion of precast frame and floor layout drawing. (Courtesy of Creagh Concrete, N. Ireland.); (b) CAD detailing of cross-section A-A through perimeter spandrel beam. (Courtesy of Creagh Concrete, N. Ireland.); (c) CAD detailing of cross-section B-B through internal inverted-tee beam. (Courtesy of Creagh Concrete, N. Ireland.)

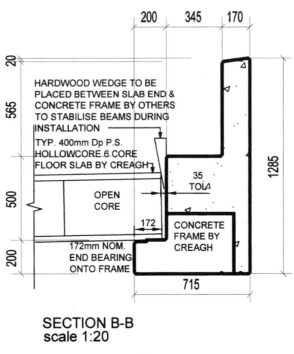

SECTION B-B
scale 1:20

(b)

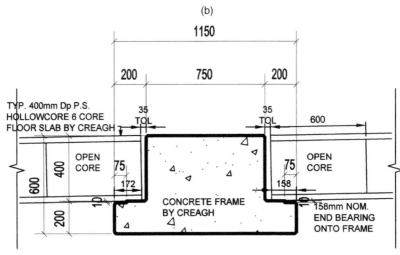

SECTION A-A
scale 1:20

(c)

Figure 2.11   *(Continued)*

layouts more than drawings for *insitu* concrete. Components are referred to individually and a reference number, which corresponds with its unit schedule number, is prominent on the drawing.

Elliott and Jolly (2013) make a further important point in the organisation and procurement of project drawings. "It is now common on larger sites for the entire building process to be managed by a Project Manager. This may be a separate management specialist, but is often a role adopted by either the architect or the precast company. The Project Manager will specify protocols (computer specification, software supplier, software revision, default settings, etc.) for a common database and CAD system that the architect, structural engineers and service engineers all agree in advance to use both in their offices and on site. He will then retain the master documents, and only his latest approved revisions are accepted as details for inclusion in the building. This facilitates alterations and additions by different contributors during construction, and leads to a single set of "as-built" drawings being available at the end of the project. Whilst this arrangement helps to avoid the blame culture that can develop when contributors' details clash, it may cause delays while approval is sought, and force non-standard and thus less efficient practices on each organisation's office."

2.2.6   Component schedules and the engineer's instructions to factory and site

This stage of the design and detailing procedures represents one of the first opportunities for the automation of drawings linking design to production via a set of "project product drawings" generated within the CAD software. The PCM will develop a huge database of parametric pictures, using object-orientated approaches to breaking down their catalogue into sets of components and ancillary details for the specific requirements of a project.

Bespoke CAD systems retrieve standard component schedules, generated directly from design input data, for example as shown in Figure 2.12(a) and (b) for the elevation and cross-section of a perimeter spandrel beam supporting façade brickwork, cavity blockwork wall and internal floor slab. The schedule includes:

- length and other linear features, such as holes, cut outs
- cross-section, either in dimensions or unit reference coding unique to the PCM
- reinforcement, individually references, or usually as a PCM reference code for both main bars and links
- beam end details such as connector type and capacity
- special requirements for surface finishes, exposed aggregates, and so on.

Component schedules showing overall dimensions and tolerances are prepared when the layout drawings are finalised or well advanced. Other information is relayed as necessary, such as special surface finishes, including exposed aggregate, retarded faces and smooth finishes, or special mixes, for example, using lightweight aggregates, selected aggregates and grading, white cement, blended cements, pigments, self compacting concrete or superplasticiser, etc. The schedule is read in conjunction with both of the project production drawings and project layout drawings. The design procedure ends with the submission of these to the factory.

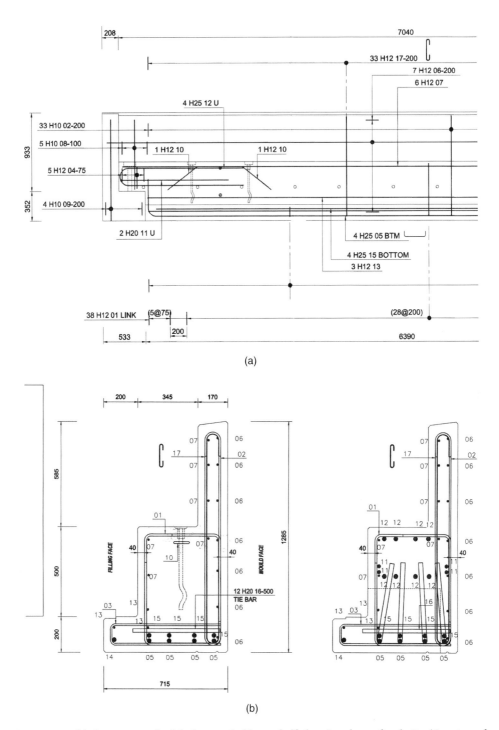

Figure 2.12   (a) Component schedule for spandrel beam: half-elevation shown for clarity. (Courtesy of Creagh Concrete, N. Ireland.); (b) Component schedule for spandrel beam: cross-sections. (Courtesy of Creagh Concrete, N. Ireland.)

2.3   Precast concrete IBS solution to seven-storey skeletal frame

A client aims to construct a seven-storey retail and office building using a total precast solution of beams, columns, slabs, staircases, stability walls and external façade panels. The architect has called for this new building to harmonise with a building similar to this shown in Figure 2.13, which is in the same vicinity.

The aim is to develop a solution in precast concrete for the building shown in plan in Figure 2.14(a), and in cross-section in Figure 2.14(b), using a skeletal framed structure of columns, beams, slabs and walls with appropriate bracing, over-clad with insulated precast concrete façade panels. The design of staircases may be used to locate the shear walls. Internal walls are only permitted in the staircase and lift shaft areas

Figure 2.13   Model prototype building used in the case study.

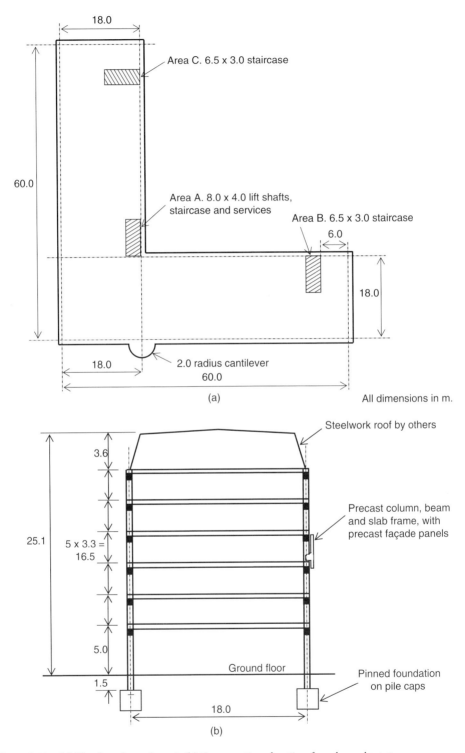

Figure 2.14   (a) Plan for scheme layout; (b) Cross-section elevation for scheme layout.

marked A, B and C. Area A is for a 4-m-wide double lift shaft and service riser, and 4-m-wide main staircase. Areas B and C are for staircases and services. There is a 2.0 m radius semi-circular cantilever balcony at the south end at the first to sixth floor. Storey heights are shown in Figure 2.14(b), in which the foundation should be a pinned joint on the pile caps. The maximum structural depth allowed is 700 mm, and the maximum column size is 400 mm.

A typical specification for loads, load combinations, materials and tolerances is:

Imposed characteristic loads:		kN/m²
Floor finishes		1.50
Services and ceiling		0.60
Partitions		1.00
Live		4.00
Roof dead load		1.25
Roof live load		0.75
Characteristic wind pressure		0.6 kN/m²
Ultimate live load combination factor* from Eurocode EC1 (ψ_o)		0.7
Ultimate wind load combination factor from Eurocode EC1		0.5

*Ultimate load $= \max\{1.35 \times \text{dead} + \psi_o\ 1.5 \times \text{live}; 1.25 \times \text{dead} + 1.5 \times \text{live}\}$

Bearing length for slabs supported on beams	100 mm
Bearing length for beams onto column corbels	150 mm
Clear gap between end of beam and face of column	20 mm
Construction tolerance for beam to column position	±15 mm
Maximum length of component	20.0 m

MATERIALS		N/mm²
Concrete	Prestressed concrete beams and slabs	$f_{ck} = 50$ (cylinder strength)
	Columns and walls	$f_{ck} = 40$
	Reinforced concrete beams, slab, cores.	$f_{ck} = 32$
	Insitu infill	$f_{ck} = 25$
Steel	Rebar and links	$f_{yk} = 500$ (grade B)
	Prestressing tendons & strand for floor diaphragm	$f_{pk} = 1770$
Exposure classification (internal, protected)		XC1
Fire resistance		1 hour
Cover to reinforcing bars and tendons, including links		30 mm

Figures 2.15(a) and 2.15(b) show a possible solution. The horizontal stability is based on the dual aspects of (i) horizontal hollow core floor diaphragm comprising 250 mm deep slab tied by internal and perimeter tendons, and (ii) precast concrete infill shear walls positioned around the lift shaft and staircases. The structure is clearly well balanced to resist horizontal loads without torsion. The designer has opted for a central line of 700 mm deep × 600 mm wide prestressed beams. These 8-m-long beams will be cambered upwards (approx. 20 to 25 mm) causing different bearing levels for the floor unit. It is preferable, but not always possible due to the location of stair and/or lift cores to line up the columns in the direction of the floor span, for example the two columns marked "columns align with floor units" in Figure 2.15(a). This prevents twisting of the floor units as beam bearings are at the same level at both ends of the floor units. This is not the case where a column is located opposite the mid-span of the beam, as marked "columns not aligned". The columns are 400 mm wide of varying depth, splices (column-to-column connection) staggered at the third floor (external) and second and fourth (internal) floor levels.

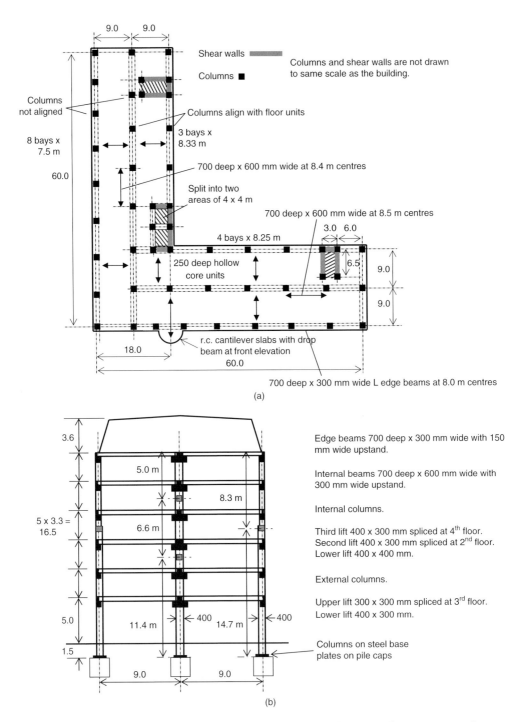

Figure 2.15   (a) Proposed precast concrete frame and floor layout; (b) Proposed cross-section and positions of columns.

2.4   Manufacture of precast concrete components and ancillaries

2.4.1   Requirements and potential for automation

The degree of automation in precast concrete factories for walls, facade panels, beams, spandrels, columns, floor slabs and stair components varies widely. At one end of the scale is fully automated table production of twin-walls or solid walls based on circulation plants with localised tilting and turning stations, idealised in Figures 2.16(a) and (b). This example production facility produces 6 panels daily and requires 20 stations with 6 curing chambers, and will be described briefly in Sections 2.4.7 and 2.4.8. This is rivalled by the fully, or semi-automated production of prestressed and reinforced floor slabs, described in Sections 2.4.2 to 2.4.5. Conversely the production of some types of beams and columns is almost entirely manual, except for concrete optimisation and batching, and the bending and manufacture of rebar cages. The use of self-compacting concrete (SCC) has reduced the manual effort, as shown in Figure 2.17, and can be delivered top loaded or pumped in automated relays into the bottom of the moulds. Moulds vary from bespoke to modular, where in the case of timber (Figure 2.18) the efficiency reduces rapidly if the piece-mark ratio, that is, repetition of components, is below about 10, as shown in Figure 2.19. Modular steel moulds may have predetermined modular widths varying from 150 mm upwards in 50 mm increments.

2.4.2   Floor slabs by slip-forming and extrusion techniques

The production of floor slabs covers the full spectrum of being almost totally automated to fully manual, depending on the type of units. Slip-formed and extruded prestressed hollow core units (hcu) and solid planks, using a semi-dry mix with low water content of around $140\,l/m^3$ and water-cement ratio (w/c) typically 0.36 to 0.4, are 80% to 95% automated depending on investment and preferences of manufacturers to carry out certain tasks manually. Only machine cast units will be discussed in this section, although it is equally possible for hcu's to be wet cast into moulds, recently casting SCC around rectangular or circular polystyrene void formers.

There are two main proprietary types:

- Type A, with oval or noncircular voids, shown in Figure 2.20, are produced by the slip-forming technique shown in Figure 2.21. Compaction is by vibration, often in two or three layers, and the machine is pulled by an external cable and winch.
- Type B, with mainly circular or near-circular voids, shown in Figure 2.22, are produced by the extrusion technique shown in Figure 2.23. Compaction is by pressing and vibrating the concrete behind the point of delivery, and the machine pushes itself forward from the rear thrust. The positions of the tendons, for both vertical cover and side cover to cores, is controlled by a set of guides, sometimes called "soldiers", as shown in Figure 2.24.

The typical width is 1200 mm, as shown in the sequence in Figures 2.25 to 2.27, except for 2400 mm and 600 mm widths representing about 10% of production. 3600 mm/11 feet widths are made in exceptional circumstances by site casting, using suction pads for lifting and handling. Depths range from 100 to 500 mm in

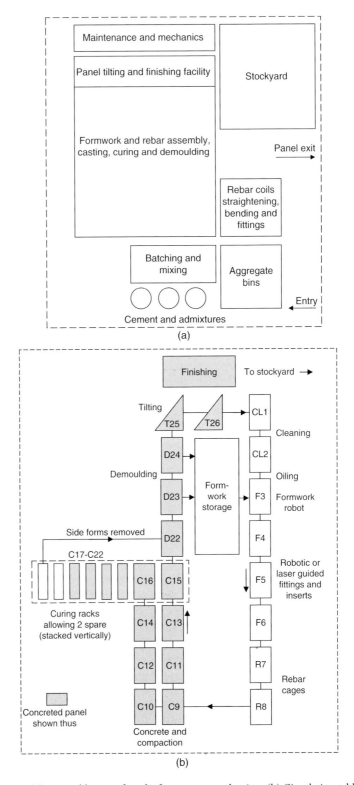

Figure 2.16   (a) General layout of works for precast production; (b) Circulation table-top formwork pallet plant within Figure 2.43.

Figure 2.17   Manual production of precast beams using SCC.

Figure 2.18   Timber mould for bespoke units and cladding panels.

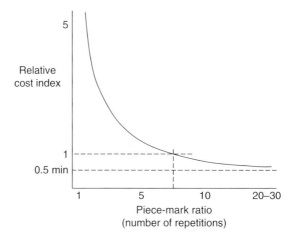

Figure 2.19   Mould efficiency versus piece-mark ratio.

Figure 2.20   Prestressed hollow core slabs in stockyard.

Figure 2.21   Slip-forming of hollow core slabs using *Prensoland* machine.

Figure 2.22   Prestressed concrete hollow core unit produced by extrusion.

Figure 2.23   Echo Precast Engineering X-liner machine for the extrusion of hollow core slabs (Courtesy of Echo Precast Engineering NV, Belgium)

Figure 2.24   Production of hollow core floor unit showing pooling of water in front of the machine, and guides to control the cover to the strands.

Figure 2.25   Automated laying of strands in hollow core slab production.

Figure 2.26   Automation of slab-cutting lines, service holes, and so on.

Figure 2.27   Lifting hollow core units from the bed and into the stockyard.

buildings (deeper units of around 700 mm are used in cut-and-cover tunnels, and new 1000 mm depth have been achieved), although 90% of the market for residential and commercial use is 150 to 250 mm, and 400 mm for car parking. The variation in depth of machine cast hcu is typically ±5 mm irrespective of depth. Elliott's data (Elliott & Jolly, 2013) shown in Figure 2.28 were collected from measurements on around 500 hcu in the UK.

Cross-section, concrete strength and surface finish are standard to each system of manufacture. The degree of prestress and depth of unit are the two main design parameters. Small variations include increased fire resistance by raising the level of the centroid of the tendons, provisions for vertical service holes, opening of cores for special fixings, cut-outs at columns, and so on. Openings and cut-outs are easily formed whilst the concrete is "green", that is, less than 12 hours old, but afterwards the operation is more expensive.

Concrete compressive cylinder/cube strength is 40/50 to 50/60 N/mm^2 at 28 days, although measured strengths are about 10–15% greater. Coarse gravel or crushed granite or limestone graded 10 (or 14) mm to 4 mm is used together with medium-grade sand or crushed fine aggregate. Lightweight aggregates are not used because of the excess wear on the machinery, and the density of aggregate required to achieve the necessary strength is quite high. Recycled concrete, crushed down from waste production to 10 mm aggregate is used at not more than 9% of the total aggregate content (see Section 3.4).

Manual work may include the final setting out or positioning and tensioning the array of tendons, such as helical strands or indented/plain wires, paid out from large spools. Figures 2.20 shows a strand-runner, carrying as many as 14 no. 9.3 or 12.5 mm

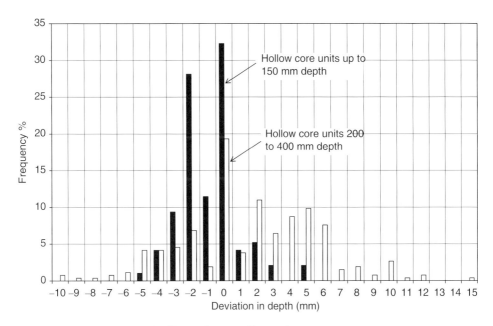

Figure 2.28 Variations in depth of hcu (Elliott & Jolly, 2013).

diameter strands or about 30 no. smaller 5 or 7 mm diameter wires, although the
number required for most slabs is typically one-half to two-thirds of this. Further
automation involves plotting end profiles and service holes and notches (Figure 2.26),
removal service holes, cutting of the units to length by diamond saw, and lifting from
the bed (Figure 2.27) prior to immediate storage in the yard. The entire operation takes
between 16 and 24 hours, depending on the manufacturer's preferences and climatic
conditions. The operations in Europe are mostly under a roof, but may be totally out-
doors in warmer climates with protection from the afternoon rains. The inside of the
machine contains void formers which, by the action of sliding forward and backwards,
and the hammer compaction of the low slump concrete, creates a voided section as
shown in Figure 2.20.

Slip-forming feeds concrete around steel formers. A profiled form is moved during
placement and the concrete is vibrated around these forms without too much pressure.
Figure 2.21 shows the operation using the *Prensoland* machine, reported to produce
200 mm deep hcu at a rate of 3 m per minute. The shape of the former is designed
to give the optimum flexural and shear performance as well as providing adequate
cover to the reinforcing tendons. Detensioning and lifting takes place 16–20 hours
after completion of casting, depending on the steam/heat curing regimes and ambient
conditions.

In the extrusion process the concrete is pressed out by rotating augers during
compaction into the required cross-section. Figure 2.23 shows the operation using
the *Echo* machine. Casting rates are about 0.2 m^3/min., or 1.4 m/min. for 200 mm
depth or 1.0 m/min. for a 400 mm depth. Providing that the rate of concrete delivery
is equal to the extrusion rate it takes around 1.5 to 2.5 hours to cast 150 m length,
or 100 m^2/hour per operator. Detensioning and lifting takes place within 12–16
hours ditto.

2.4.3   Comparisons of slip-forming and extrusion techniques, and r.c. slabs

In comparing the two techniques, slip-forming machines tend to be noisy (>85 db) but have lower maintenance costs. One problem is that the slip-forming machine may move forward in jerks, leaving a rippled surface to the sides (and sometimes the tops) of the units, with the result that the width of the unit is 2 to 3 mm too large, causing problems in tight fitting spaces. Ironically, this profile may actually improve the horizontal diaphragm capacity of the floor slab. New investors in hollow core machinery seem to prefer the slip-forming process, although the extrusion process gives a better finish and is more dimensionally accurate. Problems occur with the wear of the augers, due to contact with coarse nature of dry mix concrete, particularly crushed gravel, resulting in changes of cross-section. Logistical problems such as time consuming bed changes from one height to another, and the extent of maintenance and cost of replacement parts has lead to the development of the new hydraulic extruders and electronically controlled extruders that have lowered operational costs through parts replacement and minimised down-time between end changes.

Hollow core units are now the most widely used type of precast flooring; annual production in Europe is in the order of 30 million sq.m. representing 40 to 60% of the precast flooring market (Elliott & Jolly, 2013). This success is largely due to efficient design and automated production methods, choice of unit depth and capacity, surface finish and of course structural efficiency, as shown in the comparative load versus span graphs in Figure 2.29, where the positive advantages of prestressed concrete of say grade C45/55 over conventional r.c. of grade C25/30 are obvious, even when the r.c. slab is designed continuously with some moment distribution (i.e. $M = 0.086\,wL^2$) and the area of rebars is nearly equal to the area of pretensioning tendons.

2.4.4   Hydraulic extruder

This machine differs from extrusion by augers by using a piston driving a compactor plate that pushes dry mix concrete through a die in the compaction chamber. There are no moving parts in contact with concrete, which reduces frictional wear. The speed of extrusion and compacting force can be continuously monitored and with feedback adjusted according to the workability of the concrete. The mandrels, which form the shape of the core, vibrate at low frequencies sufficient to create a stable but smooth and accurate finish to the inside of the core – something that gives dimensional stability to the unit, avoiding misshapen or collapsed cores. An external vibrator is added to a finishing plate at the rear of the machine. Whilst the speed of casting is about the same as auger extruders, the vari-speed piston and variable compacting effort means that the rate of extrusion can be slowed according to the depth of the slab and is concordant with the rate of concrete supply, that is, the machine does not need to stop avoiding potential planes of weakness when casting restarts. Bed changes are reported to take less than 1 hour. A single power unit may be used for depths ranging from 150 to 500 mm, and due to the variable compaction effort can be used to form small beams (for beam and block flooring) as small as 150 mm deep × 80 mm wide.

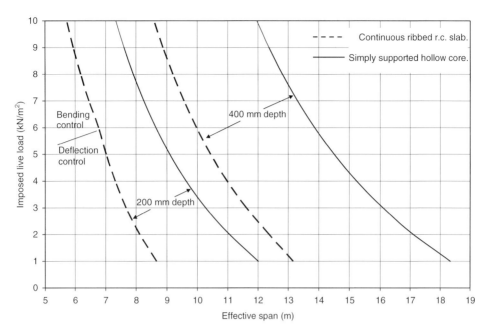

Figure 2.29 Load versus span graph for 200 and 400 mm depth psc hollow core units and reinforced concrete ribbed slabs of equal depth and reinforcement quantity designed to Eurocode EC2. Floor finishes = 1.5 kN/m². The r.c. is mostly ultimate bending controlled, except where shown for imposed load < 6 kN/m². The hcu service stress controlled with internal exposure.

Data: hcu: f_{ck} = 45 N/mm², p_{yk} = 1770 N/mm², helical strand A_{ps} = 1015 mm² (200 depth) and 1140 mm² (400 depth).

Data: r.c. ribbed slab: Top flange = 100 mm thickness, ribs width = 400 mm, f_{ck} = 25 N/mm², f_{yk} = 500 N/mm², A_s = 4 no. H20 bars = 1216 mm².

Void content for both hcu and r.c. ribbed slab = 33% (200 depth) and 48% (400 depth).

2.4.5   Reinforced hollow core slabs

Automated procedures are used for the production of semi dry-cast, that is, using w/c ≈ 0.4 to 0.45, reinforced hollow core slabs, of cross-section as shown in Figure 2.30. These units are usually 600 mm wide because of handling restrictions after curing. The units are made inverted and turned onto steel plates prior to 24-hour curing. Steel mandrels are automatically positioned to form the resulting circular cores, as shown in Figure 2.31. A rebar cage is spot welded to secure the main bars for position and spacing and flipped over by machine into the top of the inverted mould. The concrete makes two passes within about 3 minutes; the second pass is shown in Figure 2.32. Concrete compressive cylinder/cube strength is only 32/40 N/mm² at 28 days (because the slab is under-reinforced flexurally and is critical in tension) although measured strengths are about 15% greater. The filled mould is immediately turned over such that the concrete unit is upright (Figure 2.33) in readiness for drying and curing (Figure 2.34).

Figure 2.30   Reinforced hollow core slabs in stockyard.

Figure 2.31   Mandrels to form the five cores are slid into place.

Figure 2.32   Two passes of the concrete every three minutes.

Figure 2.33   Hollow core slab released from mould onto curing plate.

Figure 2.34   Reinforced hollow core slab in heated drying and curing bays.

Reinforced hollow core slabs are not as structurally efficient as their prestressed counterparts, for the reasons of a limiting span due to a flexurally cracked section, even if the partially cracked section is allowed in some codes of practice. For example, carrying an imposed dead load = 1.5 kN/m^2 and live load = 2.5 kN/m^2, a 150 mm deep × 600 mm wide r.c. unit containing 6 no. B12 mm bars (area 6 × 113 = 679 mm^2) will span 5.16 m, being long-term deflection limited at span/250 to Eurocode EC2. A 150 mm deep prestressed unit requires only 5 no. 9.3 mm diameter strands (area 5 × 52 = 260 mm^2) to achieve the same span.

2.4.6 Automated embedment machines for mesh and fabrics in double-tee slabs

Factory-cast double-tee units are primary components in buildings and car parks, having also been used as transverse slabs in some types of bridges. The top flange, typically 50 to 75 mm in thickness and reinforced with one sheet of steel mesh, is used to structurally distribute uniform, line and point loads to the structural webs that are typically 1.2 m apart (for a 2.4-m wide unit), and the mesh reinforcement is designed accordingly. The top flange is particularly vulnerable to chloride attack and corrosion problems. One solution is to replace the steel mesh with a carbon fibre grid (cfg), which is non-reactive to the depassivation of concrete. The cfg can be varied according to application to achieve different structural properties such as:

- the type and grade of carbon fibre filament
- number of filaments per rod, known as "tow strands"

- spacing of grids is from 25 to 200 mm, depending on flange width and strength requirements
- cfg thickness ranges from 1.5 to 3.5 mm depending on grid type

Recent advancement in the automated manufacture of ribbed precast concrete floor units, such as double-tee or arched floor units, has experimented using carbon fibre reinforcement, in the form of a carpet (or blanket) mesh, to reinforce shallow top flanges, often a little as 30 mm in some arched units. The cfg comprises a mesh of tow strands, between 3 and 6 mm diameter at 100 to 300 mm spacing. The tensile strength is around 3800 N/mm^2, although this is not so important in the context of its primary use, which is to resist shrinkage and other in-plane strains to prevent excessive cracking. Spacing and strength requirements are dependent upon flange depth and loading requirements of the double-tee unit. The elastic modulus of carbon fibre is slightly more than steel, 225 kN/mm^2, making strain distributions comparable with steel, although the rupture strain is only 0.016, about 1/4 that of steel, but even this is much greater than a conservative value for shrinkage strain (for precast concrete) of 0.0004. The cfg is used to replace traditional sheets of steel mesh of 4.8×2.4 m size that has a minimum depth of (e.g., A98 mesh) 2 no. 5 mm bars = 10 mm (up to 20 mm for the largest size). Even when lapped and nested the mesh has a depth of 15 mm, leading to flange thicknesses of at least 50 to 75 mm if the top cover is 25 to 40 mm for mild to aggressive corrosion protection, respectively. Conversely, cfg can be centrally bedded in a flange thickness of 25 to 30 mm, saving around 10% of the weight of the double-tee unit. The cfg tow strands are also closer to the top surface, enabling more effective crack control. The main drawback is that the cfg moves around whilst the concrete flange is cast, even when using SCC without vibration.

Introductions of cfg simultaneously lead to new technology of how to roll out the blanket style grid and position it accurately in the top flange. Traditionally, steel mesh is carried to the casting bed in lengths of 4.8 m, and lapped and nested or lapped with loose rebars. The mesh sits on plastic cover spacers, an easier task for a flat soffit compared to a curved arch, to give the required top and bottom cover. Fixing cfg by the same method is tedious and time consuming, and positional stability cannot be guaranteed; this all lead to the method of sinking the grid immediately after casting. Double-tee slabs are cast by conventional filling or by the use of a casting machine, shown for example in Figure 2.35, capable of producing 150 m length $\times$ 2.4 m width in 3–4 hours, yielding about 600 m^2/day per shift. The cfg is carried by the machine in tight coils, not possible with steel mesh, and pushed into the top flange beneath rollers as the concrete is poured.

An innovative machine, shown in Figure 2.36, is used to feed the grid beneath a series of neoprene wheels, 150 mm apart and approximately 300 mm in diameter, followed by a levelling screen to ensure correct thickness, and a smoothing blade or grooving rake to top roughness as required. Using predetermined levels the grid is pushed down to give 12–15 mm top cover, optimally positioned to resist early plastic shrinkage. Carbon fibre tow strands on grids at 50×50 mm placed at 25 mm below the surface has been shown to reduce plastic shrinkage (Ramsburg, 2008).

The technique is best suited to rectangular grids with the longitudinal tow strands at 300 mm spacing and the transverse tow strands, those being physically pushed down, at 75–100 mm spacing. Concrete flows over the top of the grid and does not prevent the transportation or blocking of coarse rounded or angular aggregate up to 20 mm

Figure 2.35   Concrete "train" used for casting double-tee floor units.

Figure 2.36   Laying carbon fibre into the top flange of double-tee units. (Courtesy of AltusGroup, Inc.)

size. The grid does not move when placed in wet concrete with proper slump having a w/c ratio around 0.35–0.4. A high viscosity mix should therefore be designed to avoid segregation by using pulverised fuel ash (PFA) or ground granulated blast furnace slag (GGBS) in proportions of around 30% by weight of cement. Viscosity modifiers may also be used, but at this point it is better to consider self-compacting concrete (SCC) which is ideally suited to this operation. The main drawback with SCC is that the finished surface may require additional floating after the initial set. Detailing information relating to the length and with of the top flange is automatically sent to the laying machine. Cfg weighs only 12 kg per 24 sq.m, four times lighter than A142 mesh (6 mm bars at 200 c/c). Not having to lay steel mesh allows earlier casting times, typically 11/2 hours in the early morning operations, or increased productivity if the timing of laying mesh is critical. Manufacturers are aiming for two casts per day from the same bed. Core samples show that the cfg stays at the preset level after hardening; manufacturer's claim (Altus Group, 2008). Other developments include automated cutting of cfg, either predetermined or by shearing at the end of each flange casting.

Full-scale static uniform load tests (Altus Group, 2008) have been carried out on 15 feet (4.57 m) wide double-tees with cantilevered flanges 82 mm thickness, sustaining 95.4 psf (4.6 kN/m^2) imposed load. The recovery was 97% of the maximum deflection. Fire tests have met the requirement for 1-hour endurance.

The relative cost of carbon fibre grids to steel mesh obviously varies with the volume of consumption, the availability of materials and the transportation and handling costs in different countries. Typically steel mesh (6 mm bars at 200 mm c/c) costs U.S. $15–20 per sq.m, whilst cfg costs U.S. $5–10 per sq.m depending on style and carbon content. The reduction in top flange thickness from 75 mm to say 30 mm saves 108 kg/m^2 of concrete, or 10.8 tonne per floor 100 m^2 floor area, making substantial savings in transportation and craneage.

2.4.7   Optimised automation

The optimum for automation depends on many factors, including:

- the balance between labour costs and machinery
- availability of skilled labour to operate computer coded equipment or perform skilled tasks manually
- availability of space in four dimensions of volume and time
- the means of installing automated plant, sometimes at different floor levels, with good access for the supply of materials
- quality control of components and of the production methods
- planned sales forecasts for the mass production of key components in large volume buildings, for example, sports stadia and residential areas
- the changing ethos for off-site fabrication
- the balance between time-cost-quality ratios; a set of functions that can be developed for any/all of the previous points.

Existing factories with highly manual operations can be adapted towards automation and its changing parameters. Carousel or circulatory systems can be readily installed to enhance or complement existing operations, without relying on total automation. Different working steps, which may differ in notionally identical manually operated plants, can be harmonised through automation. An example of this is the production of integrated precast concrete components.

In this sense the term "integrated" means that an element may perform other functions than the traditional ones of architectural or structural. The upsurge in off-site fabrication has increased the degree by which building components provide the infrastructure for services and installations, or with in built thermal and acoustic properties. Floor slabs, such as "pipe-floor" as part of precast hollow core or solid slabs (see Chapter 3, Figure 3.29) include horizontal ducts and connecting channels for service pipes. The ducts are scheduled in CAD files, converted to shop drawings and transferred to the plotter's interface for exhuming. Coordination with vertical services is achieved in cooperation with the following building and services trades. Dimensional tolerances depend of the QA system, but are typically ±10 mm allowing for tight site connections.

2.4.8   Table top wall panels

This section gives a brief introduction to these production methods – refer to Chapter 6 for greater technical detail.

The manufacture of precast wall panels using circulatory table production is widespread and highly developed worldwide. Table forms, often known as pallets, have one or two fixed edges constructed in sturdy materials up to about 200 mm in height, enabling the moveable forms to butt against the sides. Individual tables are 6 to 12 m in length. The total length of the tables varies from 60 to 140 m, depending on floor space and production output/demands, although around 100 m appears to be an optimum length. Widths (becoming the height of wall or façade panels) vary according to architectural requirements or limits on transportation, but are typically 3.5 to 4.0 m. There are usually 15 to 20 workstations, from rebar cutting, bending and cage assembly to finishing and pointing. Pallets travel gently on tracks with multiple roller blocks, whilst transverse conveyors move the pallets to other workstations and curing racks as shown in Figure 2.16(b).

Wall panels may have large door or window openings, and smaller electrical and plumbing fixings such as switches and valves are known as "back boxes", for example in Figure 2.37. Historically, positional errors of items that cannot be hidden in the finished components lead to designer's abandoning these items in favour of increased finishes. However, positioning back boxes using automated CAD/CAM systems has allowed these operations to once more fall on the critical production path, providing that it does not reduce cycle time, as well as reducing manual effort. Manufacturers now use electro-magnets to secure formwork and back boxes to steel table forms involving robotic arms, often with multi-head grippers for moving up to three to four items in one visit. Following the availability of cost effective stepper motors techniques were developed in the 1990s using semi-automated robots without a direct link to shop drawings, that is, programmed manually. The final step was to link the coordinates to CAD drawings, paving the way for reductions in manufacturing times of about 90 minutes per 6-hour production shift, or typically 25% increases in productivity.

Figure 2.37   Electrical back boxes for switches are fixed accurately with magnets. (Courtesy Unitechnik.)

A formwork robot may have three to four head grippers driven by feed belts, as shown in Figure 2.38, that can position formwork and fittings to an accuracy of ±3 mm. The steel table is not damaged or perforated by pins or screws, making cleaning and removal of laitance easier and quicker. Morning preparation of the casting beds can be cut by one hour. If more than one belt or arm is in close proximity, programming is automated to avoid clashes and increase the speed of the operation as CAM operations prepare the final fitting sequences.

Laser projection systems are also used to locate formwork and fittings on large table forms that can also have the facility to tilt in entirety. The typical clear height for a single laser to project onto several pallets at once is around 8 m, a height that is often unavailable in converted factories. Therefore multiple lasers, three or four in number, may be required with the possibility of over lapping. As shown in Figure 2.39 the laser(s) enables rapid assembly of formwork and fittings, claimed to save three hours per day.

Casting operations are carried out either directly from a hopper with one directional movement perpendicular to the direction of the table form, or by a hopper programmed to deposit concrete directly between the forms, moving around door and window openings, and so on. If vibrated concrete is specified, the table is vibrated to an amplitude and frequency to suit the dimensions and mass of the component. The noise level is usually less than 75 dB. This may be a two-part casting, depending on panel thickness and the required surface finish. Wall panels more than 200 mm thickness can be compacted using high-frequency external vibrators. A robotic screeding beam (or bridge) may be used to produce smoother finishes if required. SCC is used without vibration, so the hopper may move in linear one-dimensional motion allowing the

Figure 2.38   Formwork robot with quadruple external gripper. (Courtesy of Unitechnik.)

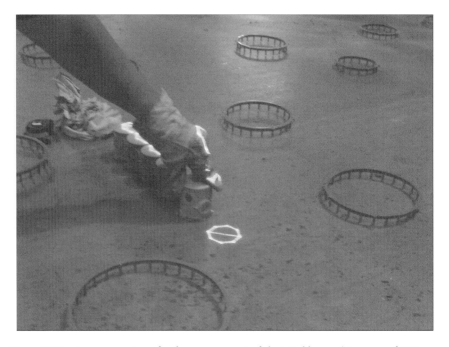

Figure 2.39   Laser projections for the arrangement of electrical boxes. (Courtesy of CPI International, 2008.)

Figure 2.40   Rack system for curing of concrete panels. (Courtesy of Vollert Anlagenbau.)

SCC to flow around 2 m sideways to either side. CAD/CAM interfaces know the precise dimensions to be filled, can calculate the optimum route to fill the mould, and can monitor the movement and provide feedback information about the positions of the forms and the completion of the casting operation. The rate of production is around 15 to 25 m² per hour for a typical single leaf wall panel of thickness of 150–200 mm, representing two to three pallets per hour, or about one-third of this for insulated twin wall.

The concrete panels are stored for curing, more recently in vertically stacked heating racks, as shown in Figure 2.40, often as many as 100 serving some 20 pallets. Larger banks of racks are usually static, but smaller ones may be mobile, with a low velocity according to the filling and emptying procedures. Segmented gate/door systems prevent too many racks being opened at once, preserving heat and humidity, some of which comes from the concrete during hydration. Some plants, that manufacture both *standard* and *special* units, often distinguished according to an outer face finish, and may split the circulation movement into two streams. Conventional wall components pass directly through the drying chambers into a pre-curing area for about two hours and do not enter storage and retrieval area. Specials are spurred off to another area (or up/down one floor level) for further processing beneath "smoothing" beams before also moving to the storage and retrieval area.

For exposed aggregate finishes, retarding agents can be applied to the table form so that the retarded concrete can be washed or scabbled off to reveal the coarse aggregate. Applying retarding agents manually can be a difficult task, especially if it is not required on the vertical edges of formwork or around door and window forms. The applied quantity of the retarding agent must be of uniform thickness. This process can be automated by a robotic spray and nozzle, specially adapted to the viscosity of the retarding agent, covering a width of between 250 to 400 mm allowing edges and corners to be precisely covered. Savings are made by reducing waste. The spraying nozzle is of course controlled by CAD/CAM operations in the same manner as before.

After 18–24 hours curing further tracks and rollers convey the cured concrete components to demoulding stations, where magnets are deactivated and released. This operation might be transversely offset from the main track system to prevent one panel from sterilising the whole system. The concrete panel is either tilted upright in a tilting frame or hoisted using two, three or four lifting sockets/hooks according to the geometry of the panel, usually with a spreader beam to ensure the sockets are loaded vertically. Although most tilting frames are about 6 m in length, entire tilting tables up to 30 m long are possible. The bottom of the tilting frame supports the fixed edge of the formwork to cater for forces during tilting and to protect free edges from damage at this early age. The edge formwork is mounted into a magazine, which automatically prepares it for reuse. Pallets rejoin the main track system and are then moved away for cleaning and inspection for damage. The pallet, once oiled beneath extraction, is transported and stored in a storage area for immediate or later reuse, whilst the debris is moved away on conveyors. The procedure is repeated as the pallet is robotically rejoined by the edge formwork, magnets, block-outs and reinforcement cage.

Production specifications and sequences are stored in flow charts and assimilated in stages, for example: critical path analysis for the priority of tracks/lines of the reinforcement, fixing formwork, casting; the time for pre-curing and/or in the drying chambers; time taken for smoothing; and time taken for cleaning and oiling pallets, all of which may be tuned to pre-set priorities or else dynamically adjusted by the controller.

Similar techniques may be used to position threaded sockets, inserts, channels or other recessed or flush fittings such as connecting plates, etc. They can be positioned robotically as shown in Figures 2.41, picked up on a wide conveyor sheet, with the time interval between fixings being around 10 seconds. These are not two-stage fittings, as the socket is both the formwork and the fitting. Wall panels and floor slabs can all be fitted for the means of suspended ceilings, air conditioning, waste and power services as shot firing and/or drilling is opposed, or even forbidden, in many countries. Preconfigured fittings can be designed that do not clash with one another or, more importantly, with robotically prepared and positioned rebars. Modern systems can install about 10 different types of sockets, ranging in diameter from M8 to M32, and in length according to design or limitations of size, as well as anchorage detail, for example, holes or pins at the rear.

Emerging technical advances have been made in manually positioned magnets to secure stainless steel/timber formwork on steel table forms. Weighing between 5 and 10 kg, roughly 50% of traditional magnets of equal adhesion, the magnets can deliver 10–20 kN of force, sufficient to secure not only the formwork up to about 500 mm in height, but also fasteners and clamps.

Fittings such as cast-in sockets are picked up using an electro-magnet and are secured to the formwork using a hot melt adhesive, even on oiled surfaces. It is easy to appreciate why this technology marries well with SCC. The operating system can optimise and reduce the cycle time. Unlike the situation with bespoke timber moulds, for example, Figure 2.18, the manufacturer does not need to have the details of the fittings well in advance of production, as items can be stock piled awaiting immediate scheduling.

Figure 2.41   Insert robot positions a recess body. (Courtesy of Unitechnik.)

2.4.9   Production of precast concrete wall panels using vertical circulation system

Enhancing the production capacity of existing plants that may have limited widths or other constraints in which to install a horizontal circulation system has lead to the novel use of vertical circulation table forms giving a straight production line. The number and size of the table forms, or pallets, typically 12 m long × 4 m wide, is governed by the hydraulic effort required to lift the pallets and the optimum number of pallets in the circulating system.

An example of this system is where single leaf walls are manufactured for low- to medium-cost housing (c. Preconco Ltd., Barbados). The production procedure is summarised in Figure 2.42 to 2.45. The pallets are stored in a buffer zone at the second-storey level (or could be stored underground as in the case in some European plants), but only at 1.2 m above the first level. They are cleaned and oiled automatically. Reinforcement cages are prepared from steel rebars of 8 to 16 mm diameter, straightened and sheared from the rolled coils. Steel forms for the edges, door(s) and window(s) are fixed onto the steel tables using electro-magnets guided by lines prescribed by an overhead laser system. A second laser marks the positions of reinforcement cage, inserts and fittings. Pallets are lowered to the vibration station at first floor level, where a concrete skip with a capacity to cover 4 m² area of wall panel follows a predetermined route to filling the mould. Conventional vibrated concrete containing a superplasticiser or SCC may be used.

On completion the pallets move horizontally into a tunnel of about 80 m length (Figure 2.43) and are lifted into the curing chamber that holds around 15 pallets. The conditions of ambient temperature and humidity help to accelerate the curing to about

Figure 2.42   Overhead laser guides the positioning of formwork.

Figure 2.43   Panels transfer to a tunnel for curing.

Figure 2.44   Tilting table allows vertical circulation of table forms.

Figure 2.45   Completed panels awaiting transportation to site.

4 hours. The pallets are transported to the tilting frames (Figure 2.44). A key feature is that the tilting frame allows other pallets to pass by the frame during demoulding, thereby minimising the width of the circulation system. The tilting frame rises to 75° taking up not more than about 1.5 m width. The precast wall panels are hoisted at two cast-in lifters and placed directly onto A-frames/racks mounted on delivery vehicles. A spreader beam may be used depending on height restrictions. The pallets are moved vertically upwards to the second level in preparation for continued use.

2.4.10   Control of compaction of concrete

The control of the compaction of vibrated concrete in large table forms, typically 6–12 m in length and 3–4 m in width, standing about 1 m above the floor level and weighing 2–4 tonnes when completed, involves expertise to produce harmonic compaction. Expertise is required in the workability (i.e. segregation, shear slippage) of concrete as well as in formwork, vibration and systems technology. Optimising the compaction may involve a mixture of finite element and computational fluid dynamics (FEM-CFD) to model the vibration of complex concrete units and guarantee the performance of the formwork, joints and seals. The design and operations functions must recognise all present and future demands, particularly if a switch to SCC is to be considered. A typical factory may have between 6 and 20 table forms, requiring 4 to 8 (or more if compartmented) vibrators per table. Vibration may be programmed singularly for shallow depths, up to about 300 mm, or dual action for deeper. As filling proceeds, the harmonics of the concrete are fed back at the interface and the amplitude and frequency is automatically adjusted to suit the mass of the fill. Temperature control is included in heated forms, for example, in tall vertical batteries for upright wall panels. Operators are still required to select the parameters for compaction depending on the density and mass, the thickness or area (or height for vertical forms) of the concrete component, together with the sequence and speed of filling. For example, a medium frequency of 2 Hz may be used to allow faster filling until the optimum vibration force is reached as the mass increases.

High frequency vibration is not new technology, but it been enhanced in recent years towards much quieter and efficient compaction by using constant speed electrical vibration exciters in one system. External synchronised vibrators compact concrete at a constant frequency in reproducible manner, minimising long wave beats.

2.4.11   Automation of rebar bending and wire-welded cages

Automated stirrups/links bending machines are used in factories with a wide ranging degree of automation; in some factories a bar-bending machine is the only automated process, whilst in others it forms a small but important part of the entire robotic production. Bending machines from the 1960s were intended mainly for two-dimensional bars such as mild steel stirrups, but since the mid 1980s have evolved to produce three-dimensional shapes (bi-directional bends), spring shapes and coils for structural columns or piles, and so on, in high tensile bar and wire with speed and accuracy. New machines are able to accept bars up to about 12 m in length. Wires and small diameter rebars (≤16–20 mm or in pairs ≤12 mm) fed from a rolled coil are first vertically and horizontally straightened between large diameter hardened steel rollers in two-plane roller configuration at 90° so as not to induce residual stress or twisting; this is a very important part of the process to ensure stirrups are not twisted across their corners. Alternatively conventional straightening with dies is used. Throughput speeds of around 100 m per minute are typical for single rollers, or more if set up in 3 to 4 parallel sets. Multi-rollers can take feed from single coils of the same diameter, or switch to different diameters in under a minute. The speed can be reduced according to bar size and complexity. The bars are automatically threaded into a second feed and pass through a series of bending heads to predetermined angles and dimensions, as shown in Figure 2.46. Some hyperbolically profiled coated steel rollers can straighten and roll-in the feed in one revolution, straightening 5 to 9 mm diameter steel tendons up to 1800 N/mm^2 or 4 to 20 mm diameter rebars of 400–600 N/mm^2 tensile strength. This has the advantage of making a line contact rather than a point contact with the rebar, reducing friction and damage. Electro-magnetic servo motors are the preferred choice.

Simpler versions are available for the automatic delivery, feeding and shearing (or flying shearing without stopping the machine) of rebars to length. The high tensile bars, ranging from 4 to 20 mm diameter, are also bent to simple shapes (hooks, bends, angles) but are not as complex as stirrups or three-d shapes. Automated lifting is used if the finished bars are too heavy for manual lifting. An automatic bundling machine, together with a bundling vehicle for the required spacing and type of tying, and conveyor completes the automated process.

The automated production of wire-welded cages involves cutting, bending and spot welding technology. Induction resistance spot welding is used for high tensile (or less frequently mild steel) rebars up to 12 mm in diameter.

Figure 2.46   Dual bending heads and drives give higher production rates. (Courtesy of CPI International, 2007.)

2.5   Minimum project sizes and component efficiency for IBS

Projects should be of a certain minimum size in order to spread out the fixed costs of setting up the key features of automation at the manufacturing facilities described above; this includes land charges and/or rental, infrastructure, circulation systems and moulds, the facilities for batching, casting, finishing and handling, construction mobilisation and overheads. This is particularly important in *open* systems that are not able to take full advantage of repetitive or standardised production. Minimum project sizes should be in the region of 1500 to 2500 m^2 floor area including 500 to 750 m linear components (beams, columns) and/or 1000 to 2000 m^2 of façade and wall panels. Non-standard components are better represented in terms of volume and complexity factor, expressed as surface area-to-volume (A/V, m^2/m^3) ratio. Table 2.5 gives some examples for reinforced, solid, wet cast (excludes machine dry cast pre-stressed hollow floor units) standardised and bespoke components in terms of the perceived (not absolute) efficiency of manufacture and construction – that is, relative difficulty and effort in man-hours of preparation, manufacture and erection, and the ease and economy of transport and craneage. Some recognition is also made of the coverage (floors, stairs) or lengths (beams, walls) or height (columns) per component. The data are presented in Figure 2.47, showing greater efficiency of standardised floor planks, rectangular beams, columns and walls, compared to ribbed floors, stair-flights, spandrel beams. There is a peak in the efficiency of mid-range components, for example, 8 m long × 0.25 m deep floor plank compared to 6 m or 10 m lengths.

The positive feasibility of precast *closed* systems is made thanks to a large volume of standardised components, leading to IBS, but one needs a long-term commitment from the market, free from the fluctuations of the mid 1980s and recent years since 2007. By contrast *open* systems are not so sensitive to changes in the market, and a lack of the supply chain for total precast concrete solutions. Therefore, *open* systems have continued to increase in recent years, possibly at the expense of *closed* systems, particularly in Europe but less in North and South America.

2.6   Design implications in construction matters

The precast superstructure, whether of standardised or bespoke components and connections, is designed and detailed to enable the safe handling, transportation and fixing, without damage to the components and framework. Here is a clear advantage for in-house PCM designers, working closely with the PE. The PCE may often hand over the responsibility for the design of lifting devices and factory tilting and site handling to the PCM, which can occasionally lead to conflicts in detailing. Detailed instructions and special provisions, and so on, are given on working drawings regarding the safe handling, lifting and storage of components, and any specific features regarding temporary and/or permanent stability.

A similar argument is made regarding the sequence in which the structure is erected and the temporary measures taken to ensure the correct transfer of load to the ground, particularly where SC-PE for floor slabs are not included in discussions relating to stability ties or the temporary horizontal stability of the floor diaphragm. Information relating to the stage of construction at which an entire, or part thereof (whether vertically or horizontally partitioned), framework is considered fully stable and may be

Table 2.5 Relationship between the efficiency of manufacture, construction and utilisation of precast components compared to geometric parameter of surface area-to-volume

Component length × breadth × depth (m) and other features (m)	A/V ratio (m²/m³)	Relative efficiency %
Floor plank : 6 × 1.2 × 0.15	15.3	95
Floor plank : 8 × 1.2 × 0.25*	9.9	100
Floor plank : 10 × 1.2 × 0.35	7.6	90
Ribbed floor : 6 × 2.4 × 0.3 (0.5 × 0.2 ribs)	16.2	65
Ribbed floor : 8 × 2.4 × 0.5 (0.5 × 0.4 ribs)	13.4	70
Ribbed floor : 10 × 2.4 × 0.7 (0.5 × 0.6 ribs)	11.7	60
Rectangular beam : 6 × 0.3 × 0.3	13.7	80
Rectangular beam : 8 × 0.4 × 0.4	10.3	85
Rectangular beam : 10 × 0.5 × 0.5	8.2	75
Spandrel : 6 × 0.3 × 0.4 (0.15 × 0.15 upstand)	14.7	70
Spandrel : 8 × 0.3 × 0.6 (0.15 × 0.35 upstand)	14.4	72
Spandrel : 10 × 0.3 × 0.8 (0.15 × 0.55 upstand)	14.2	65
Spandrel : 10 × 0.4 × 1 (0.25 × 0.75 upstand)	9.9	60
Spandrel : 10 × 0.4 × 1.2 (0.25 × 0.95 upstand)	9.7	55
Wall : 3 × 0.2 × 3	11.3	80
Wall : 4 × 0.25 × 3	9.2	75
Wall : 5 × 0.3 × 3	7.7	65
Stairflight : 3 × 1.3 × 0.2 waist (0.18 step × 0.25 going)	9.2	45
Stairflight : 3.5 × 1.3 × 0.23 waist (0.18 step × 0.25 going)	8.3	50
Stairflight : 4 × 1.3 × 0.27 waist (0.18 step × 0.25 going)	7.3	55
Column : 4 × 0.3 × 0.3	13.8	90
Column : 7 × 0.3 × 0.3	13.6	95
Column : 10 × 0.3 × 0.3	13.5	100
Column : 12 × 0.4 × 0.4	10.2	95
Column : 8 × 0.4 × 0.4	10.3	90

*datum component awarded 100 points

allowed to stand free of any external restraint is stated. The contribution to strength and stiffness from *insitu* concreted connections is assessed with respect to the time taken for the concrete to mature. Although adequate strength may be achieved in compression, the bond resistance between rebar/dowel and the concrete may take longer to fully develop.

Craneage and construction plant is discussed between the PCE, PCM, GC and PE, including SP-PE particularly if floor slabs form a major part of the critical lifting plan, which they often do. Plant should be compatible with the geometry of the structure, weight and size of components, the nature of ground and site access. There is little to distinguish between *closed* and *open* systems here. The opening gambit in some

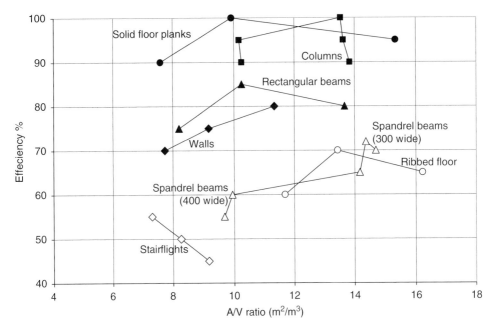

Figure 2.47   Graphical presentation of data in Table 2.2.

countries, for example, south Europe and south east Asia, is a small tower crane of around 5-tonne capacity, although this may be restrictive to longer span slabs and precast cores or walls. The next option, if available, is a 10- tonne tower crane, but the PCE needs to ensure that this machine is fully utilised and is not just working a full capacity for a fraction of the time. Telescopic cranes up to about 30-tonne capacity are a popular option on congested or difficult sites. Crawler cranes of 40- to 100-tonne capacity, with an added luffing jib for specific tasks, are popular in some parts of Europe and the USA for precast structures up to about 10 storeys.

2.7   Conclusions

This chapter has focussed on the supply chain for the integrated design-manufacture-construction of precast multi-storey framed buildings. These three parameters should not be considered sequentially but should be seen as a continuous loop where major design decisions are taken in lieu of construction or manufacturing efficiencies and safety. The in-house structural engineer in a closed precast system, or the consultant (together with the erector's representative), sit at the confluence of these operations, as shown in the illustration. The diagrammatic representation of the three operations, and the amount of overlap between them, is deliberate, and drawn roughly to scale, that is, there is more communication through the detailer than between the factory and the site.

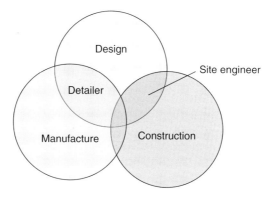

"IBS" has been introduced. It is a name that used to be synonymous with the "shoe-box" precast structures of the 1960s, but it is clear from many of the examples shown in the first three chapters of this book, that IBS now relates to the streamlining of design, manufacture and construction, and the architectural and structural features have (almost) no limitation. This chapter has shown how the main advantages of quality, faster completion time, clean construction sites and sustainability can be achieved.

The major stages of the project have been identified, so now the main conclusions for *best practice* (adapted from Goodchild & Glass, 2004) can now be teased out as:

- think precast design and construction at the earliest stage.
- seek specialist advice for the initial concept, that is, involve the precaster at this stage to evaluate site requirements, ground conditions and access, programming, grid patterns, structural zones, tolerances, surface finishes/quality, limitations of size and weight, craneage, factory and site safety.
- define responsibilities between the architect, consultant, precaster, contractor and specialist sub-contractors, for example, design, manufacture and installation of psc floor slabs.
- write new specifications appropriate to the new project, include the sustainable selection of materials and processes, for example, balance the benefits of using SCC and recycled concrete and slurry with work and energy demands.
- agree on liabilities, damages, risk, retentions, and so on.
- prepare general arrangement and detailed drawings, cross-sections and unit details to determine whether bespoke or existing table forms/moulds are to be used (refer to Chapter 6).
- finalise submissions and approvals as soon as possible.
- organise moulds, reinforcement, fixings, raw materials, casting, finishing, handling, transportation, site delivery sequences, temporary props/falsework, connection aids and *insitu* concrete work and finishes. (These are discussed in detail in Chapters 7 to 9).
- ensure feedback to complete the design-manufacture-construction-design loop.

References

Altus Group. (2008). Carbon fibre grid reinforcing revolutionizes precast double tees, *Concrete Plant International*, Cologne, Germany, No. 4, 164–169.

EC. (2002a). EN 1990 +A1:2005. Eurocode 0: Basis of structural design.

EC. (2002b). EN 1991-1-1. Eurocode 1: Actions on Structures – Part 1-1: General Actions – Densities, self-weight, imposed loads for buildings.

EC. (2004a). Eurocode EN 1992-1-1:2004. Eurocode 2: Design of Concrete Structures – Part 1-1: General rules and rules for buildings, Amendment 2014.

EC. (2004b). Eurocode EN 1992-1-2:2004. Eurocode 2: Design of Concrete Structures – Part 1-2: Structural fire design.

Elliott, K. S and Tovey A. (1992). *Precast Concrete Frame Buildings - A Design Guide*, British Cement Association, Wexham Springs, Slough, UK, 88 pp.

Elliott, K. S. & Jolly, C. K. (2013). *Multi-storey Precast Concrete Framed Structures*, 2nd ed., John Wiley, UK, 760 pp.

fib. (2002). Bulletin 19. *Precast Concrete in Mixed Construction*, State-of-art Report, Fédération Internationale du Béton, Lausanne, Switzerland, 68 pp.

fib. (2012a). Bulletin 63. *Design of Precast Concrete Structures Against Accidental Actions*, Guide to good practice, Fédération Internationale du Béton, Lausanne, Switzerland, 72 pp.

fib. (2012b). Bulletin 67. *Guidelines for Green Concrete Structures*, Guide to good practice, Fédération Internationale du Béton, Lausanne, Switzerland, 60 pp.

fib. (2013). *fib Model Code for Concrete Structures 2010*. Ernst & Sohn, Berlin, Germany, 434 pp.

fib. (2014). http://www.fib-international.org/guidelines-for-green-concrete-structures. Accessed May 2014.

Gibb, A. G. F. (1999). *Off Site Fabrication: Prefabrication, Pre-Assembly and Modularisation*, Whittles Publishing, Caithness, Scotland, 262 pp.

Goodchild, C. H. (1995). *Hybrid Concrete Construction*, Reinforced Concrete Council, (former British Cement Association), Camberley, UK, 64 pp.

Goodchild, C. H. & Glass, J. (2004). *Best Practice Guidance for Hybrid Concrete Construction*, The Concrete Centre, Camberley, UK, 64 pp.

Ramsburg, P. (2008). New, patent-pending C-GRID embedment machine leads to major productivity enhancement in double tees, *Concrete Plant International*, Cologne, Germany, No. 6, 188–191.

Zuhairi, A. H. & Kamarul, A. M. K. (2011). Implementing The Industrialised Building System, *Housing Construction*, Chapter 14, 343–355.

Chapter 3

Best Practice and Lessons Learned in IBS Design, Detailing and Construction

Kim S. Elliott

Precast Consultant, Derbyshire, UK

A shift in a large part of construction work from the site to the controlled environment of modern precast concrete factories has taken place over the past 10 years. A study by the *fib* Commission on Prefabrication in 2002 found that three-quarters of new buildings contained some elements in precast concrete, ranging from a total solution to a few simple components. The proportion of prefabrication in the superstructure, including precast concrete and steel elements, prefabricated formwork and rebar cages, has increased from about 25% in 1990 to between 40 and 60% today, depending on the nature of the work. This chapter focuses on key aspects involved in the successful use of precast concrete in buildings: high performance, self compacting and recycled concrete, standardisation, structural-architectural solutions, surface finishes, prestressed concrete for long-span beams and slabs, and the integration of building services. Each section shows best practices and makes a review of the lessons learned. Major decisions are made under informed circumstances, where the design is not constrained by the preferred practices of contractors.

3.1 Increasing off-site fabrication

Precast concrete is widely regarded as an economic, durable, structurally sound and architecturally versatile form of construction for multi-storey structures. Recent developments in high strength and rapid hardening self compacting concrete up to grade C95 have enabled IBS to be used in office buildings of nearly 40 storeys to compete with structural steel work for speed of construction, and with *insitu* concrete for span/depth limitations (Figures 1.17 and 1.18). Integrated architectural-structural precast concrete components are being used on an increasing number of prestigious

Modernisation, Mechanisation and Industrialisation of Concrete Structures, First Edition.
Edited by Kim S. Elliott and Zuhairi Abd. Hamid.
© 2017 John Wiley & Sons Ltd. Published 2017 by John Wiley & Sons Ltd.

Figure 3.1   Student accommodation at University of West of England, UK, 2009. (Courtesy of Buchan, UK.)

commercial buildings, retail centres, stadiums and college buildings. Figure 3.1 shows an excellent example of IBS where precast structures for 800 rooms were completed in one year. Designers are becoming more aware of the high-quality finishes possible in prefabricated components, but changes are now being made to the way that the traditional precast structures are conceived and designed, where the optimum use of all the components must be maximised. The requirement for off-site fabrication has increased (from about 25% in 1990 to 40 to 60% today, depending on the nature of the work) as the rapid growth in management contracting, with its desire for reduced on-site occupancy and high-quality workmanship, is favouring controlled prefabrication methods.

Prefabrication of concrete structures has a large potential for IBS in the future. The precasting industry in Europe strives to maintain a competitive edge against the increasing demands of modern society in terms of economy, efficiency, technical performance, safety, labour and environmental aspects. On-site construction is fast, competing with structural steelwork, at the rate of up to 1500 sq.m. per week for flooring, as shown in Table 3.1, and around 600 sq.m. per week for structures similar to the one shown in Figure 3.2. Note the influence in Table 3.1 of standardisation and repetition, for example, a loss of 25% in the repetition of large areas of floor units and frame components results in a reduction in fixing rates of 40%. There is also a dependence on spans and the floor areas served by fewer components, for example, doubling the floor bay area from 50 m^2 to 100 m^2 increases the speed of construction by 25 to 50%, depending on building height above four storeys.

As the Dutch coined the phrase *total football* when considering the roles played by team members, then (Elliott, 2002; Elliott & Jolly, 2013) had established the basis for *near total precast IBS* … "The correct philosophy behind the design of precast concrete multi-storey structures is to consider the frame as a total entity, not an arbitrary set of elements each connected in a way that ensures interaction between no more than the two elements being joined". Bruggeling and Huyghe (1991) state:

Table 3.1   Typical site fixing rates m² per week (European Data)

	Beam × floor span	2–4 storey	5–8 storey	9–12 storey
100% standard components on a rectangular grid.	100 m² 75 m² 50 m²	1000–1500 850–1200 750–1000	900–1300 750–1000 600–800	650–900 450–700 350–550
75% as above, 25% non-standard on a non-rectangular grid.	100 m² 50 m²	900–1250 600–900	750–1000 500–700	No data
50% as above, 50% non-standard on a non-rectangular grid.	100 m² 50 m²	650–800 400–600	400–600 300–500	No data

Excludes cladding and cast *insitu* floor toppings.

Figure 3.2   Precast concrete skeletal frame and hollow core flooring used in a 500,000 m² retail park near Rome.

"Prefabrication does not mean to 'cut' an already designed concrete structure into manageable pieces ......". Thus it is clear that all the aspects of component design and structural stability are dealt with simultaneously in the designer's mind. The main aspects include: structural form and function, frame stability and robustness, component selection, and connection design."

The first task is to establish an economical plan layout for the optimization of the minimum number of the least cost components versus overall building requirements. The minimum construction depth is achieved when prestressed concrete floors, such as the voided type of hollow core units (hcu) and double-tee units (DT), achieve their maximum span with regard to the serviceability limit state of stress and the

Figure 3.3 Prestressed concrete hollow core slabs are 400 mm deep, span 16 m and weigh just 5.2 kN/m² (compared to 10 kN/m² for a solid slab of the same depth), spanning onto 7.2 m long beams. Floors by Bison Concrete Products and beam and column frame by Buchan Concrete, 2007.

ratio of the floor span-to-beam span is around 1.5 to 2.0, as shown in Figure 3.3. The resulting span/depth ratio for the floor will be around 35:1, and the ratio of floor bay area/structural depth from 150:1 to 200:1. An most important feature is that the depth of the drop-beam below the soffit of the floor not greater than about 200 mm, leaving room for air conditioning plant and allowing office storey heights of around 3.5 m, and 3.0 m in car parks.

Figure 3.4 shows the typical imposed load versus span data for prestressed composite hcu and DT. The data include 1.5 kN/m² for finishes. Note that the hcu out-performs the DT, mainly because the serviceability stress in the DT is limited by the lower section modulus at the bottom fibre than in a same depth hcu, although deflecting and/or debonding tendons can claw back some of these losses. Additional moment capacity (at the service limit) can be achieved by utilising the compound section properties based on the transformed area of the tendons; an enhancement of about 5% for the higher tendon patterns, and by considering the self weight of the unit when calculating elastic and creep losses. Figure 3.5 shows the similar information comparing prestressed units with composite half-slabs, and different depths of slabs from 200 to 800 mm. The data are normalised in terms of a span-to-depth ratio, some of which exceed 50, the recognised the lifting limit for hcu, but not in DT. The key information here is the efficiency of the shallower prestressed units, for example, 200 mm deep hcu versus 200 mm deep half-slab, and also comparisons between 200 and 400 mm depths (also evident from Figure 1.41). Further similar comparative information on the economic use of precast, prestressed, post-tensioned and cast *insitu* concrete is available from the Concrete Centre in the UK (Goodchild, Webster, & Elliott, 2009).

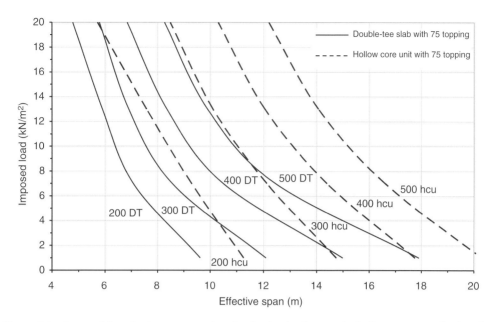

Figure 3.4   Imposed floor load versus span data for the depth of composite hollow core units (hcu) and composite double-tee slabs (DT). Total depth = precast plus 85 mm, allowing for 10 mm extra thickening due to the camber of the prestressed units.

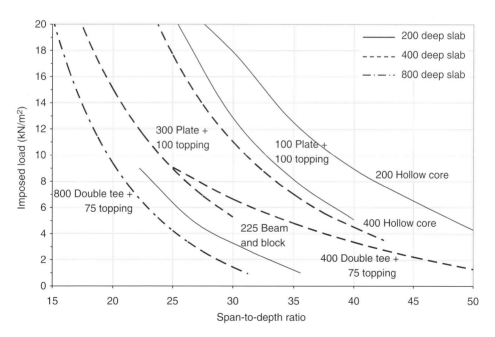

Figure 3.5   Imposed floor load versus span-depth ratio data for a range of prestressed concrete floors, and beam-and-block floors, showing the efficiency of shallower prestressed concrete alone compared to composite slabs of the same total depth or deeper units.

Changes to the way in which the construction industry should operate in a "zero waste and zero defect" environment were given in the Egan Report (Egan, 1998). The Report called for the following sustained improvement targets: 10% in capital construction costs, 10% in construction time, 20% of defects, and most importantly 20% increase in predictability, that is, less surprises on site. "The industry must design projects for ease of construction making maximum use of standard components and processes". Although the reports did not use the term "prefabrication", to many people this is what "predictability" and "standard components" means. The precast concrete industry is ideally placed to accommodate these higher demands by using experienced design teams and skilled labour in a quality controlled environment to produce high specification components.

Figure 3.6 illustrates IBS in the repetitive use of nominally identical polished granite spandrel beams, walls and columns (approximately 1000 components with about six variants) to form an integrated structural-architectural building in the convoluted shape of a shell. Even lost-form walls to the rear of the building were prefabricated to reduce site operations. High-quality architectural finishes are widely adopted for exposed precast concrete structural components, as illustrated in the integrated structure in Figure 3.7.

The precast concrete system adopted for the university building shown in Figure 3.8 is known as "hard-wall" due to the structurally and architecturally finished, and thermally insulated wall units, which are integrated with the beam and column components. This precast frame was designed as an extension to the mid-twentieth century. Portland stone building, which although slightly larger in area, took nearly two years to complete, whereas the precast structure was erected in 4 months. A main feature of the system is that the columns, which are manufactured in a single height of about

Figure 3.6   Granite aggregates polished into spandrel beams and columns to reflect the sunshine at No. 1, Spring Street, Melbourne, Australia.

Figure 3.7   Asticus Building, London, 2010. Precast beam cruciforms form the exterior structure and architectural finish. (Courtesy of John Wiley.)

Figure 3.8   Architectural precast beams and columns, with an integrated structural façade, known as "hard-wall", Portland Building, University of Nottingham, 2003.

Figure 3.9   Five-storey columns in finished concrete contain steel connectors to receive the cast-in box connector in the structural beams.

16 m and as shown in Figure 3.9, contain mechanical connectors to receive the combined beam and wall units as shown in Figure 3.10. The walls contain 80 mm insulated cavities to maintain the thermal conductivity U-value to around 0.2 W / m^2°C.

The recent shift of a large part of construction work from the site to the controlled environment of modern permanent factories that has taken place in Europe over the past 10 years has not been reflected in many other countries, or even continents. A study by the *fib* Commission on Prefabrication (*fib,* 2002) found that about 75% of new buildings contained IBS components in precast concrete, ranging from a total solution comprising over 90% precast down to a few simple floor units or structural components, often used in a hybrid or mixed structures for steel framed buildings and grandstands (Goodchild, 1995). Terrace units manufactured in high-quality self-compacting concrete into steel moulds, and long-span hcu, are used in both steel and concrete frames. Figure 3.11 shows an example in a 90,000 capacity stadium, in which 35% of the building mass was in precast concrete.

Lessons learned: the demand of modern society has forced the precast concrete industry into twenty-first-century practices for the design, manufacture and execution of precast frames, cladding and flooring. The decisions made by clients and architects are made with fruitful collaboration of the manufacturing industry. The full potential of using precast concrete must be maximized – a plain grey precast frame will not compete with a cast *insitu* frame, particularly if the footprint is large, as the contractor's viewpoint is biased against expensive cranes and favours cheap labour.

Figure 3.10   Integrated structural beams are formed together with the insulated wall panel.

This has resulted in an increase in the degree of prefabrication of the entire building process, namely, concrete, steel and timber components, formwork, prefabricated rebar cages, prefabricated services, and so on, all manufactured in a controlled environment for strength, quality and safety. Major decisions are made under informed circumstances, where the design and construction is not constrained by the preferred practices of contractors.

3.2   Standardisation

Standardisation of products and processes is widespread in prefabrication. Precast manufacturers have standardised their components by adopting a range of preferred cross-sections for each type of component. Actually it is the moulds which are standardised and therefore non-preferred shapes will incur cost penalties because of the alterations to the mould. If a non-preferred cross-section is used there should be sufficient in number, giving a piece-to-mark ratio (units per mould alteration) of 10 to 15

Figure 3.11   Precast concrete terraces and hollow core floors used in steel-framed sports stadiums.

at least (Figure 1.19). The designer can select the length, dimensions and load bearing capacity within certain limits. This information can be found in catalogues from the precast manufacturer. Wall elements have usually standard thicknesses but the height and width, doors and window openings, are free within certain limits.

The wide range of precast concrete floors used in precast skeletal structures has now reduced to five main types (Figures 3.12 to 3.14): (i) prestressed hcu, (ii) reinforced and prestressed DTs, (iii) beam (including wide beam) and block, all of which may or may not be used with a structural topping, and (iv) composite reinforced or prestressed plank floor, and (v) composite beam and plank, which must always be used with a structural topping. Floor beams are either rectangular, L-shaped or inverted-tee, as shown in Figure 3.15. Internal beams are symmetrical in cross-section, and most commonly prestressed inverted-tee, if the structural depth zone is minimised, or reinforced rectangular beams for short spans or connecting beams around stair wells and lift shafts. Columns are square, rectangular, trapezoidal and circular in cross-section (although hexagonal and octagonal columns have been used on special projects). End connections may use pockets, steel base plates of projecting rebars in grouted sleeves, as shown in Figure 3.16.

The planned use of the buildings will in most cases determine the span lengths and the direction of the floor slabs and beams, and thereby the selection of their type. In office buildings, where there are open plan areas or offices either side of a central corridor, a convenient location is to run services in either the ceiling or raised floor zones as the span of the floor elements will often be perpendicular to the main façade. This also means that the edge beams, which are not limited in depth, are fully utilised structurally. There may of course be a good reason why an edge beam is too shallow or

Figure 3.12 Standardised hollow core units in widths of 1200 mm generally (also 400, 600, 2400) and depths from 150 to 500 mm in 50 mm increments.

Figure 3.13 Standardised double-tee units in widths of 2400 mm generally and depths from 300 to 1200 mm in 50 or 100 mm increments, depending on the manufacturer.

Figure 3.14   Precast soffit part of composite plank floor of depths 60-100 mm. The spaces between the girders may contain polystyrene blocks or similar for weight saving. Lightweight aggregate concrete is often used for the topping.

Figure 3.15   Standardisation of beams is typically on a 50 mm module, but is less restrictive as timber or steel moulds can easily be adjusted.

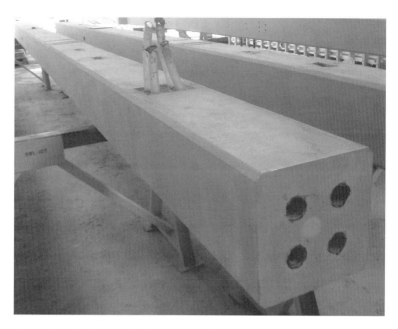

Figure 3.16   Standardisation of columns is also based on a 50 mm increment, but for the same reasons can be varied.

is not required to support the floor slab, as shown in Figure 3.17. The internal beams span in the same direction along the central spine of the building between columns, except in the case of Figure 3.18 where there is a change in direction near to a central atrium.

Lessons learned: standardisation does not necessarily restrict architectural freedom. Most of the buildings shown here were manufactured from the same set of standardised components. Consultants responsible for the design of foundations and other major parts of the structure must liaise with the precast manufacturer at the earliest possible stage in order to maximise the used of standardised components. This may require small adjustments to the grid layout in order to maintain the most popular modular grid of 600 mm and 1200 mm for floor units. Architects must respect some of these limitations and appreciate the benefits gained as shown in the example in Figure 3.19. In Figure 3.19(a) the wall is commonly positioned on the same grid as the columns, but this leads to a clear opening of 5900 mm to receive 5 no. 1200 mm wide hcu, resulting in difficult cutting of the hcu. This could easily have been avoided by moving the wall 110 mm to the left, leaving a clear opening of 6010 mm.

3.3   Self-compacting concrete for precast components

Self-compacting or self-levelling concrete (SCC) is now adopted for most wet cast work in precast factories. SCC is either mixed at source, or is delivered from nearby ready-mix plants. A simple flow table test replaces the slump test, and control cubes are used in the normal way. Colouring pigments may be added, and early strength gains means that 18 hours demoulding is possible at compressive cylinder strengths

Figure 3.17   Floor slabs spanning parallel with the façade where the edge beams do not support them. (Courtesy of T&A, Recife, Brazil.)

of about 25 N / mm^2, giving 50 to 60 N / mm^2 at 28 days. Admixture specialists have achieved the right balance between the flow and segregation and early strengths with the use of active rheology modifiers. Whereas high strength concrete focuses mainly on strength and durability, SCC also has a beneficial impact on the IBS production process. For example, in the manufacture of DTs a team of four workers (one batching and three pouring) can produce around 240 m^2 of units per day (100 m length × 2.4 m wide) with a selling price of about 20 times the cost of that labour (UK figures), a much greater ratio than in the past.

Figure 3.18 Internal beams supporting hollow core floor slabs that are spanning in different directions near to a central atrium.

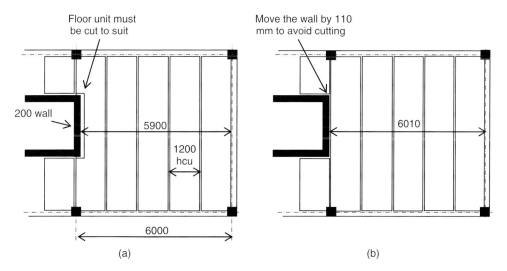

Figure 3.19 Avoiding additional work and respecting modular units. (a) the wall is positioned on the same column grid line causing difficulties in cutting floor units; (b) moving the wall and leaving gaps for tolerances.

Figure 3.20 The accuracy of precast concrete edge walls manufactured using grade C40 self-compacting concrete is reflected in the man's image, less than 2 mm undulations in a 3 m length.

All the terrace units shown in Figure 3.11 were manufactured using SCC with a labour force about one-third of the usual effort. The quality of the finish is shown in Figure 3.20, where the flatness, typically ±2 mm in 3 m length, far exceeded the requirements in the national specification. SCC needs no vibration and therefore gives a lot of advantages: moulds are placed on the ground, low noise level during casting, less mould pressure, rapid casting, easy flow around dense reinforcement and in thin or complicated cross-sections, less air pores at the surface and easy to pump. Moulds are no longer raised on isolators, and consequently much safer to step over. Care must be exercised in the use of mould oil, where patches of excess oil in a dished surface lead to the surface discolouration shown in Figure 3.21.

Figure 3.21   Excessive mould oil in a dished surface lead to some discolouration on an otherwise pale white-grey surface.

SCC is preferred for precast concrete stairs and landings, which are cast in a simple mould, laid on the factory floor. Where the landing is integral, the larger part (usually the flight) is cast first, and the mould is then tilted after 2 to 3 hours so the shorter part is horizontal ready for casting. The result is almost seamless. The surface quality is exceptional, and requires no further treatment on site, as shown in Figure 2.1(a), contrasting that of the cast *insitu* stairs in Figure 2.1(b).

Lessons learned: SCC provides very effective means of delivering high-quality, rapid- hardening concrete for precast manufacturers to exploit the full benefits of off-site IBS. The additional cost of the admixture must be balanced against the reduced cost in labour, vibration, mould fatigue, mould design, and the effects of noise. Care must be taken to avoid shrinkage cracking on the surface, and extra time may be required to finish the pieces. The architect must be aware of the benefit of using SCC, even though it may be the producer's choice to use it.

3.4   Recycled precast concrete

Waste material from precast production may be crushed and reintroduced as recycled concrete aggregate (RCA), mostly in its coarse state. This is not exclusively an IBS practice, but benefits from the knowledge of good-quality crushed material. The procedure is to first nibble and extract the reinforcement, which is sold off to waste metal merchants, and to use the crushed material as a replacement for coarse aggregate between the limits of 4 and 20 mm. A typical precast manufacturer, with a turnover of U.S. $10 to 30m will produce some 1000 to 3000 tonnes of waste material. Hollow core producers generate about 3% wastage at the ends of the production lines, and 0.5 to 2% defective units, for example as shown in Figures 3.22 and 3.23. Recent data from European hcu producer's claim to recycle 80 to 95% of the wastage, leaving between zero and 0.25% of annual waste to landfill.

There is a considerable body of research on RCA, particularly for *insitu* waste, but increasingly for precast waste by the BRE, CERIB and other European research centres. The general conclusion is that coarse RCA, when derived from virgin precast production, that is, has not left the factory site and is completely uncontaminated and hardly carbonated, may be reintroduced to fresh concrete at up to 50% replacement, providing the angularity index (this compares the weight of free fall pieces to that of perfect spheres in a vessel and is usually limited to 12 for concrete) is not greater than about 10 and the minimum size is about 4 mm. Crushing contractors may either be resident at precasting works, in which case they will reprocess other materials too, or visit the factories annually or bi-annually, crushing the waste or rejected products over a period of about 6 weeks, as shown in Figures 3.24 and 3.25. The grading is in line

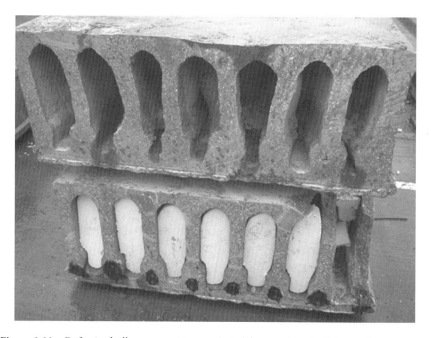

Figure 3.22   Defective hollow core units are rejected for recycling, in this case due to over dosing of an admixture ostensibly to enhance the workability of the concrete in the slip-forming machine.

Figure 3.23   Scrap hollow core units are crushed and the coarse (without fines) aggregate reintroduced in the same product.

Figure 3.24   Recycling hollow core slabs as RCA using jaw or cone crushers to give the best shape and texture to the RCA. (Courtesy of Bison Manufacturing, UK.)

Figure 3.25   Stage 2 processing into sizes, typically 14, 10 and 4 mm. (Courtesy of Bison Manufacturing, UK.)

with European limits EN 12620 (EC, 2002) for 10, 14 and 20 mm. The absorption of the RCAs around 6%, and this can be easily measured and allowed for in mix design. The maximum recommended replacement is 20% RCA, and this has been shown to have no effect on the strength or workability of concrete, achieving 60 N / mm^2 with replacements as high as 50%.

Recycled fine aggregate is more difficult to control as the grading enters the *silt* zone and particle size can be around 150 to 300 μm. It is also possible to recycle slurry water, and to filter the cement particles from the slurry using flocculating agents. Some precast concrete factories adopt a zero waste policy, in that everything is used.

Lessons learned: recycling waste production from a precast plant is time consuming and can be quite expensive, resulting in the cost of the RCA being about the same as new aggregates. However, when the landfill tax is added to the equation, and the cycle of the recycling contractor visiting precast factories on an annual rota, the benefits are more in balance. The type of crusher should be considered for different productions, that is, the cone crusher is suited to hcu waste containing gravel, but the jaw crusher to less strong concrete made from limestone. Special care is needed when RCA is introduced into machine production, such as hcu or beam and block units.

3.5   Building services

Different areas of the structure may require a higher building specification than other areas or floor levels, for example with respect to fire resistance, thermal, acoustic or vibration characteristics. Other criteria may include the requirements for unlimited freedom for vertical and horizontal service routes. It may therefore be necessary to incorporate building services into the precast units, particularly to reduce storey heights. For example, ducts, boxes or chases for mechanical and electrical fittings may be cast into wall elements, as shown in Figure 3.26, or may even be taken one

Figure 3.26   Precast concrete shear core box complete with openings and doors, and can facilitate internal fixings for services. (Courtesy of Waycon, UK.)

step further and provide the services inside the concrete to be connected on site. Figure 3.27 shows a duct for a rainwater pipe formed into a precast column in a multi-storey car park.

These facilities must be well planned ahead of production, typically 4 to 6 weeks before decisions need to be taken compared to cast *insitu*. It is necessary for the M&E contractors to liaise with the precast manufacturer early in the project, not only to ensure compliance but also to maximise the possibilities and reduce unnecessary changes on site.

Precasting offers also certain advantages with respect to building techniques. For example, thermal mass of concrete has been used satisfactorily to store thermal energy in hcu, leading to substantial savings on heating or cooling costs. Hcu's may also contain cooling passage, both along and between the hollow cores, as shown in Figures 1.44 and 3.28 in a system commonly known as TermoDeck. Used extensively in the Middle East where external and internal temperatures are in the range +40°C to 15 to 20°C, the system is almost accepted a standard solution with most manufacturers of hcu. Energy savings are claimed to be around 30%. The hollow cores may also be used to incorporate ducts and pipes in the tops of the floors, known as "Pipe floor", shown in Figure 3.29, and is popular in Northern Europe, these examples from VBI of the Netherlands.

Lessons learned: attempts to fully incorporate building services into precast elements date back to the 1950s in Eastern Europe, where a lack of accuracy and understanding of the requirements for tolerances, workmanship, and future maintenance lead to failures. The preference is often for service routes to be provided, but not complete systems, unless the manufacturer has complete control, as in the case of TermoDeck and Pipe floor.

Figure 3.27   Rainwater pipe duct in a precast column.

Figure 3.28   Air conditioning and other facilities are carried into the hollow cores of the floor slabs. (Courtesy of TermoDeck®.)

Figure 3.29   Hollow core units with conduits through the depth and width of the section.

3.6   Conclusions

1. This chapter has given an overview of the potential benefits of adopting pre-cast concrete at the choice for IBS in multi-storey buildings and stadia, and has stressed the need to make the most from industrialisation and standardisation, without losing architectural and engineering freedom.
2. An important message is to fully exploit the qualities of precast concrete, manufactured in controlled conditions. This message applies to all parties: clients, architects, M&E services, consultants, and most importantly the precast producer.
3. It has shown the key to success is to adopt an off-site construction methodology from the outset.
4. The structural efficiency of prestressed concrete, particularly for long-span hcu and shallow depth beams, has been shown.
5. Special materials, such as high-strength, self-compacting and recycled concrete, and the integration of building services, are shown to have special significance in precast construction.
6. Major decisions are made under informed circumstances, where the design is not constrained by the preferred practices of contractors familiar with cast *insitu* concrete construction.

References

Bruggeling, A.S.G. & Huyghe, G.F. (1991). *Prefabrication With Concrete*, Balkema, Rotterdam, 380 pp.

Egan, J. (1998). *Rethinking Construction*, Department of the Environment, Transport and the Regions, London.

EC. (2002). EN 12620:2002, *Aggregates for concrete*, BSI, London

Elliott, K. S. (2002). *Precast Concrete Structures*, Butterworth-Heinemann, 370 pp.

Elliott, K. S. & Jolly, C. K. (2013). *Multi-storey Precast Concrete Framed Structures*, 2nd ed., John Wiley, UK, 760 pp.

fib. (2002). Bulletin 43. *Precast Concrete in Mixed Construction*, State-of-art Report, Fédération Internationale du Béton, Lausanne, Switzerland, 68 pp,

Goodchild, C. H. (1995). *Hybrid Structures*, Reinforced Concrete Council, British Cement Association, Crowthorne, U K, 64 pp.

Goodchild, C., Webster, R. M., & Elliott, K. S. (2009). *Economic Concrete Frame Elements to Eurocode 2*, The Concrete Centre, UK, 182 pp.

Chapter 4

Research and Development Towards the Optimisation of Precast Concrete Structures

Kim S. Elliott[1] and Zuhairi Abd. Hamid[2]

[1] *Precast Consultant, Derbyshire, UK*
[2] *Construction Research Institute of Malaysia (CREAM), Kuala Lumpur, Malaysia*

This chapter reviews the links between the development and optimisation of IBS and precast concrete components and structures in particular, by taking many of the concepts described in Chapters 1 to 3 in terms of the basic and near-market research carried out over the past 25 years. The topics covered include the role of large scale precast frame testing (5-storey frame), optimisation of the automated production items such as prestressed hollow core floor slabs, composite and continuous construction horizontal floor diaphragm action, and semi-rigid behaviour of beam-column connections and their effect on frame stability. The main conclusion is that national codes are now able to exploit new R&D information that was often previously designed by "rule-of-thumb".

4.1 The research effort on precast concrete framed structures

4.1.1 Main themes of innovation, optimisation and implementation

Since the 1960s fundamental and practical research trends in the realm of concrete structures have followed a pattern of innovation-optimisation-implementation, the cycle being repeated approximately every 15 to 20 years, albeit with different aims. For example, R&D in precast concrete beam-to-column mechanical connections has attracted interest *circa* 1965 (focussing on strength), 1990s (rotational stiffness) and 2010s (frame stability). Prestressed concrete hollow-core floor units have been researched in waves, for example, shear capacity (1970s), lateral load spread (1980s), floor diaphragm action (1990s), flexible supports (2000s) and fire resistance (2000 to date). Prefabricated concrete technology has also peaked in the 1970s (alkali aggregate/silica reaction), 1990s (self-compacting SCC) and 2000s (blended cements and high-strength HSC). In most cases optimisation overlapped with innovation, for example, rheology modifiers for early strength gain using SCC, and implementation is overlapping with optimisation, for example, 90% of wet cast in prefabrication was SCC before the full extent of behaviour had been fully tested.

In most countries the research has been industry driven because it is difficult to divorce the fundamental behaviour of precast structures from the "design-manufacture-construct" philosophy, an approach which is unique to precast concrete. This has lead to the research being "near market", parcelled and channelled into a small number of themes, and finally "object orientated" through:

i.   global behaviour splintering into:
ii.   components
iii. connections
iv.   stability

much to the detriment of fundamental research and development.
R&D can be separated into:

- research effort on the behaviour of materials, structures, components and connections
- development work leading to new technologies in circulatory pallet production, extrusion and slip-forming techniques, automation in reinforcement cages, casting and finishing techniques, and so on.

The research effort can be broadly divided into about 12 themes based on the behaviour of:

- skeletal and wall frames – global structural analysis of beam-column-slab framework, and load bearing wall and wall-slab connections in panel construction.
- stability of skeletal frames – braced using three-d cores and two-d cantilever and infill walls, or unbraced developing frame action from post-tensioned or semi-rigid partial strength connections.
- progressive collapse – accidental loading, catenary action and the use of horizontal and vertical ties.
- prestressed beams for bending, shear and deflection – focussing on narrow webs and with large holes in the webs, including composite construction with cast *insitu* or precast slabs.
- beam ends – half-joints (dapped) with rebar or shear box shear and bursting resistance.
- joints and connections – compression, tension and shear joints, beam-column and foundation connections, corbels and hidden or mechanical connections.
- prestressed concrete floor slabs – hollow core, double-tee, solid plank.

- composite floor slabs, interface shear at toppings and between steel beams and precast slabs.
- continuous construction – negative continuity in prestressed floor slabs.
- horizontal floor diaphragms with or without structural toppings.
- natural frequency and peak acceleration - floor and roof systems, stadium terraces.
- self-compacting and high-strength concrete, admixtures, blended cements, recycled concrete aggregate.

Development work has lead to:

- circulatory/carousel pallet systems – table forms for wall and façade panels and twin lattice walls.
- automation of concrete mixing, delivery, compaction and finishing.
- architectural finishes – stone tablets, brick slips/halves, grit blasted, acid-etched, exposed aggregates, polished, colouring pigments and painted surfaces.
- automation of reinforcement – assembly and positioning of cages and mesh, semi-automatically welded lattices.
- slip-forming and extrusion machines – hollow-core floor slabs and floor beams, including hydraulic extruders.
- integrated M&E services in floors – cooling and heating air conditioning in cores (*TermoDeck*), pipe floor.
- insulated floor slabs – lightweight vaulted double-tee or multi-ribbed slabs.
- lightweight floors – slender prestressed sections, channel or inverted U-sections, voided floors (*Bubble deck*).
- shallow floors – precast-steel section *Slimflor*, *Delta* beam and floor slab.
- hybrid construction – cast insitu frames or columns with precast (psc or rc) beams and slabs, possibly with post-tensioned beams and slabs.
- mixed construction – precast elements and frames with structural steelwork, timber, masonry, glazing and polymer composites (pultruded sections).

This chapter aims to report on some of the most progressive of these with the objective of establishing the background leading to modernization and automation, mainly in prefabricated concrete materials, components, connections and structures – literally where the greatest advancements have been made in the past 50 years.

4.1.2   *Structural frame action and the role of connections*

There has been very little research into the behaviour of complete structures, even though the profession generally agrees that such research would enhance the status of precast frames by demonstrating their unquestionable structural integrity. The increasing awareness towards the structural integrity of structural connections has been where most of the recent research effort has been devoted. The details used to achieve robustness in a structure have a significant effect on the structural mechanisms by which horizontal forces are distributed, and vice versa. Rarely has it been possible to test entire structures, the exception being the half-scale 3-storey precast

Figure 4.1   Horizontal load testing of precast concrete diaphragm floors in a half-scale skeletal structure at University of California, San Diego's Englekirk Engineering Centre in February 2008. (Courtesy of Prof José I. Restrepo, University of California, San Diego, USA.)

structure tested at the University of California, San Diego, in 2008 (Schoettier *et al.*, 2009) as shown in Figure 4.1.

A main feature of this $17.1 \times 4.9 \times 7.0$ m structure was to develop a diaphragm seismic design methodology for precast buildings, and these tests included 16 input ground motions, monitored by 640 sensors dynamically recording the various damage limit states. The floors comprised single-span 255 mm deep double-tee units and 102 mm deep hollow-core units, both with a 38 mm topping. Subjected to peak ground accelerations of between 0.30g and 0.61g, the maximum roof drift ratio (sway deflection/storey height δ/h) recorded at the centre of the floor diaphragms was 0.23% and 1.88%, respectively; a sway deflection of h/434 is close to the recognized ratio of h/500 used in national codes.

A number of projects have been successful in demonstrating precast integrity by placing the emphasis on the continuity of connections rather than on the elements themselves. The most notable of these has been the PRESSS (Precast Seismic Structural Systems) Project (Priestley, 1996; Stanton, 1998; Nakaki *et al.*, 1999); a collaboration in seismic research between academics and professional engineers in Japan and the USA. The objectives were to develop new materials and structural systems for multi-storey precast systems in seismic zones, leading to calculation models suitable for code drafters to use. A 60% scale $9.15 \times 9.15 \times 11.43$ m five storey precast beam and column sway frame, Figure 4.2, stabilised with shear walls in one direction only, was tested at the University of California, San Diego in 1999. In the direction of the two bay sway frames, one line of beam-column connections was prestressed, whilst the other line was not. Moment resistance for the lower three stories was provided by tension/compression yielding gap connections (TCY, see Figure 4.3).

Figure 4.2   Five-storey precast beam and column sway frame under seismic load test. (Courtesy of John Stanton, University of Washington, Seattle, USA.)

The frame was designed using a so-called "direct displacement design" procedure, in which inelastic target displacements (as determined by storey height drift factors) and effective (secant) frame stiffness are the design objectives. This approach is opposite to the load-based approach normally used for seismic design where load capacity and ductility are the design objectives and deflections, rotations, strains etc.

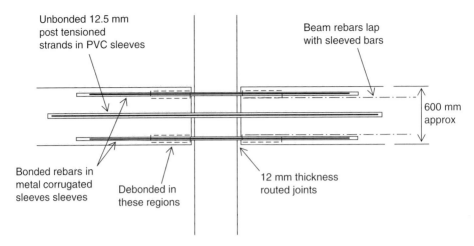

Unbonded 12.5 mm
post tensioned
strands in PVC sleeves

Beam rebars lap
with sleeved bars

600 mm
approx

Bonded rebars in
metal corrugated
sleeves sleeves

Debonded in
these regions

12 mm thickness
routed joints

Figure 4.3   Beam-column connections used in the PRESSS project.

are measured. Priestly *et al.* (1999) report that pseudo-dynamic horizontal loads were applied at floor levels giving δ/h ratio 25, some 20 times greater than that (h/500) accepted in a statical frame for elastic response at design loads and about twice the target "design displacement" for strength-based design seismic forces. Important discoveries are:

- good confirmation of the direct displacement approach used to determine the strength and stiffness of the frame.
- sway deflections of twice the seismic strength design approach value of 2% of the height were achieved without any loss of gravity load carrying capacity.
- under the dynamic loading greater forces than expected were sustained in the floor diaphragms, which translated into storey shear forces much greater than the anticipated code values.

The message for designers of precast frames, whether seismically loaded or not, is that it is possible, with attention to detail, to develop ductile precast concrete unbraced frames several storeys high. The ACI Standard 318-11 (ACI 2011), having provisions similar to those of the 2000 NEHRP Recommended Provisions (National Earthquake Hazards Reduction Program) (Hawkins & Ghosh, 2000) will allow the use of precast frames in high seismic zones of the types tested in the PRESSS building. There are several skeletal precast frame structures in high seismic zones in the USA that utilize the post-tensioned frame type of the lower three stories of the prestressed frame in the PRESSS building. The tallest structure is a 44-storey-high building located in San Francisco.

In Scandinavia, a number of shallow beam framing systems, such as the "Delta" beam shown in Figure 4.4 use plated steel beams or low-level steel corbels to support long-span beams and floors (Juvas & Pousi, 2000). A main feature is the span-beam depth ratio of around 20:1 to 25:1 achieved for office loading (5 kN/m²). Suikka (Suikka, 2000) reviews several proprietary prestressed and reinforced concrete floor systems, including the so-called "slim floor" system in which the floor slab is recessed into the depth a rolled steel section or plated beam, shown in Figure 4.5.

Figure 4.4   Shallow steel plated "Delta" beam is designed compositely at the service limit state with prestressed hollow core floor slabs. (Courtesy of Peikko.)

Figure 4.5   Shallow-floor construction where the slab is recessed into the same depth as the webs of steel beams.

4.1.3   Advancement and optimisation of precast elements

Precast concrete frames generally consist of "sturdy" elements, such as columns and walls, where technical advancement has been slow in recent years and "slender" elements, such as long-span shallow beams and voided floor slabs, where considerable research has been devoted to their optimisation, manufacture and design. Researchers have attempted to keep pace with the increased sophistication with which elements such as prestressed hollow-core units (hcu) for floors are being designed and manufactured. For 30 years the units changed very little, although some advancements were made in improving the cross-section profile. The 1980s then saw an upsurge in R&D aimed at gaining a greater understanding of behaviour, particularly in the third dimension (lateral and torsional), followed by the development of deeper units up to 700 mm depth (in the year 1995) and 1000 mm depth (2013), Figure 4.6. This activity has continued to the present day thanks to the efforts of manufacturers and IPHA, the international precast hollow-core association where research work has been carried out in the following areas:

• transmission lengths and lateral bursting stresses in the transmission zone of prestressed units
• reduced shear capacity of hcu's bearing onto flexible beams
• lateral load distribution for point and line loads in hcu slab fields
• continuity of bending moments in hcu's at interior supports
• horizontal floor diaphragm action without structural toppings
• behaviour in fire.

Figure 4.6   1000 mm deep hollow core floor slab by Nordimpiani, Italy (Sao Paulo Concrete Exhibition, 2014).

4.1.4   Shear reduction of hcu on flexible supports

Traditionally, shear tests had always been carried out on rigid supports. In this manner the shear stress in the web is two-dimensional, that is, τ_{zy}, as shown in Figure 4.7(a). However, if the units are seated on flexible supports, Figure 4.7(b), the curvature of the beam causes lateral curvature of the floor unit, resulting in transverse shear stress τ_{zx}. VTT in Finland (Pajari, 1998) carried out full-scale tests, comprising double spans of 7.2 m × 265 mm deep hcu's, five abreast, supported on a range of precast inverted tee and steel-plate beams in such a manner that both the

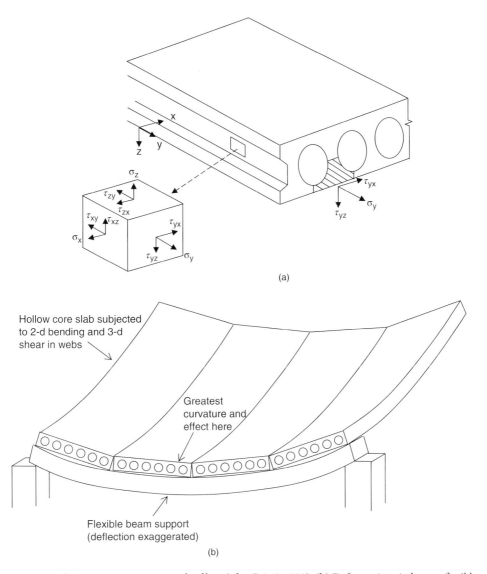

Figure 4.7   (a) Stress components in web of hcu (after Pajari, 1998); (b) Deformations in hcu on flexible bearing beams (after Pajari, 1998); (c) Testing hollow core slabs on flexible beams at VTT, Helsinki, Finland; (d) Composite beam model showing effective breadth of beam and hcu (after Pajari, 1998).

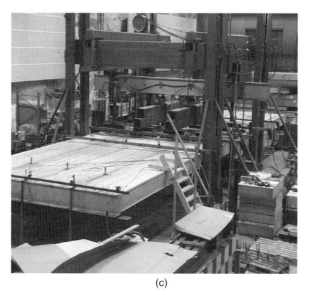

(c)

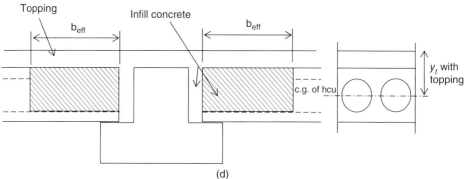

(d)

Figure 4.7   (*Continued*)

slabs and beams were in conjugate bending. Figure 4.7(c). The startling result was that the shear web capacity on flexible supports was 23% to 60% lower than on a rigid support. Referring to Figure 4.7(d), the effective breadth b_{eff} for the total compressive flange width (i.e. on both sides of the beam) is found experimentally because of the complex interactions between the hcu, infill concrete, interface shear reinforcement and beam. For a typical precast concrete inverted tee beams $b_{eff} \approx 400$ mm, and for edge beams $b_{eff} \approx 200$ mm.

The additional stress due to out-of-plane bending $\tau_{zx} = \dfrac{3\,v\,b}{4\,b_w\,h_{ct}}$

where v is the transverse shear flow $= \dfrac{y_t\,EA_f\,V_{beam}}{EI_{beam}}$

b = breath of unit, usually 1160 mm, b_w = total breath of web per hcu, h_{ct} = depth of unit minus depth of web along its narrowest part, y_t = centroidal distance to the top of the slab from the neutral axis, EA_f = axial stiffness of top flange of slab (parallel to span of beam) = $E_{cm}\,b_{eff}\,h_f$, h_f = thickness of top flange of hcu, V_{beam} = maximum ultimate <u>imposed</u> shear force in the beam, and EI_{beam} = flexural stiffness of the supporting beam or the composite beam.

The usual design equation for shear capacity (EN 1992-1-1, eq. 6.4) is thereby modified to

$$V_{Rd,c} = \frac{I_c b_w}{S_c} \sqrt{f_{ctd} + \alpha_1 \sigma_{cp} f_{ctd} - \tau_{zx}^2}$$

4.1.5   Continuity of bending moments at interior supports

In the past 20 years, practical experience has been gathered in making the simply supported ends of hcu's flexurally continuous at interior supports, both with and without structural toppings. This is common is seismic countries where the action is used more for structural integrity than for increased moment capacity. However, the negative restraint moment capacity must be considered at both the serviceability and ultimate limiting states. Figure 4.8 shows a typical detail where site-placed reinforcement of maximum diameter 6 + h / 25 (h = depth of hcu in mm) is cast insitu (grade C40 concrete minimum) into top opened cores, overlapping (not physically touching) with factory-placed upper prestressing tendons. The length of the bars should be 2 × bond length, or to the point of contraflexure under imposed live load only. At least three cores in a 1.2 m wide unit (4 if the span is greater than 6 m) should be filled. Design procedures are now published (*fib*, 1999). The negative moment of resistance is relevant only to the imposed loads, and is provided by an appropriate area of reinforcement calculated in the normal manner. Continuity leads to an increase in the span to depth ratio, typically from 30 to 35, with a small reduction (about 5%) in the *required* area of prestressing tendons. Due to the negative moment shear capacity is based on the non-prestressed section.

Figure 4.8   Continuity of negative moment is achieved by placing rebars in the top of opened cores in hcu for a distance of about 1.5 m either side of the supporting beam (Sporting Lisbon soccer stadium, Portugal).

4.1.6   Horizontal diaphragm action in hollow core floors without structural toppings

There has been extensive experimental work describing the shear transfer mechanism in cracked reinforced concrete, and the formulation of some basic relationships between the dominant material and geometrical effects which are based mainly on experiments using small specimens (Millard & Johnson, 1984). However, there is a lack of experimental data on the monotonic and/or cyclic behaviour of large-scale units, and on the correlation between the small-scale testing and the full-size diaphragm.

Tests carried out by Wahid Omar and Bensalem (Davies *et al.*, 1990; Wahid Omar, 1990; Elliott *et al.*, 1992; Elliott *et al.*, 1993) on full-scale 200-mm-deep hollow-core units, shown in Figure 4.9, found that the attainable horizontal shear stress in the floor slab exceeded the permissible design stress for unreinforced uncastellated joints. The laboratory tests were carried out in a "realistic environment" regarding materials specification, geometry and on-site practice. The results have shown interface shear stresses are in excess of the working load by a factor based on Eurocode EC2, Part 1 (EC, 2002) of at least 4.1, despite the presence of initial cracks in the interface up to 0.55 mm wide. The working load is calculated by multiplying the working stress by the net contact area in the longitudinal joint. The working stress is defined as the ultimate shear stress according of $\tau_{Rd} = 0.15$ N / mm^2 divided by γ_f of 1.5 for wind loading = 0.1 N / mm^2. Clamping forces normal to the precast units resulted in coefficients for the ratio between shear wedging and shear friction T/V of at least 5. Designers also assume that the floor plate undergoes negligible shear deflection, that is, the longitudinal shear stiffness is very large. In reality, it is in the order of $K_s = 500$ kN/mm to 1000 kN/mm for 4 m long × 200 mm deep slabs (Elliott *et al.*, 1992). Based on this experimental evidence, the maximum horizontal shear deflection between these slabs

Figure 4.9   Precast prestressed concrete hollow core floor diaphragm subjected to cyclic horizontal loading (University of Nottingham, UK).

at a design working shear stress of up to $0.15/1.5 = 0.1$ N$/$mm^2 would be approximately 0.16 mm. If this deflection was summed over a building length L (m) a total shear deflection at the mid-point would be $0.05L$ (mm). Comparing this to (Schoettier *et al.*, 2009) tests, where L $= 17.1$ m, span $= 4.9$ m and h $= 7.0/3 = 2.33$ m per floor, then $\delta = 0.05 \times 17.1 \times (4.9/4.0) = 1.05$ mm and $\delta =$ h$/2200 <$ h$/434$, showing that the floor plate is relatively very stiff.

There was considerable debate in the profession some 25 years ago whether or not a horizontal floor diaphragm could be provided solely using discrete precast floor units, such as hcu's, without the need for a structural cast insitu topping. The evidence today is that, providing the units are adequately tied to prevent their moving apart, the units alone are capable of providing a diaphragm in all situations except perhaps seismic situations. To achieve integral action a horizontal shear mechanism resulting from shear wedging, shear friction and dowel action must be generated in the longitudinal joints by the correct placement of site bars and cast insitu concrete infill (grade C25 minimum) as shown in Figure 4.10. The shrinkage cracks shown in this photograph up to 1 mm have little influence on interface shear stress or stiffness.

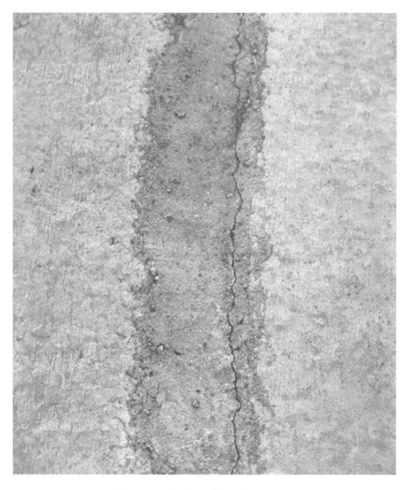

Figure 4.10   Cast *insitu* infill mortar (with small rounded aggregate) between hollow core units, showing shrinkage cracks up to 1 mm wide.

A calculation model is given (Cholewicki & Elliott, 1998) for the strength and shear stiffness of the longitudinal joint based on previous experimental (Davies *et al.*, 1990) and theoretical work (Cholewicki, 1991). Where bending and shear combine the shear stress in the longitudinal joint is $(V - \sqrt{\mu F_c})/A < \tau_{Rd}$ where V = ultimate horizontal shear force between hcu's, A = net contact area between hcu's, taken as B times $(D - 30)$ mm, B and D = length and depth of hcu, μ = friction factor = 0.7, F_c = compressive force due to diaphragm bending = $M_h/0.8$ B, and M_h = horizontal diaphragm–bending moment. From the numerous testing programmes carried out on precast floor diaphragms the results confirm, without exception, adequate interface shear performance and diaphragm action using the normal construction practice.

4.2   Precast frame connections

4.2.1   Background to the recent improvements in frame behaviour

Frame action in bending is rarely considered by designers of precast frames in the belief that connections made by site welding, grouting or bolting are insufficiently ductile, or have low rotational stiffness at low loads. The effective length factor β of a column in a pinned jointed precast skeletal frame is taken as 2.3 over the full height of the column (Cranston, 1972). However, Eurocode EC2-1-1 (EC, 2002) allows relative stiffness at the ends of the columns, expressed as the beam/connection stiffness $k = (E_{cm} \, I_{col}/l_{col})/(M_{column}/\theta_{joint})$, to be used in β. Unfortunately, this information is not applicable to semi-rigidity where M/θ is required at beam ends. In response to this, between 1985 and 2012 there were over 100 full-scale (or reasonable size model scale) tests carried out on precast beam-column connections, sub-frames, continuous beams, and so on. Figure 4.11 shows an external beam-column hidden corbel connection with a negative moment of resistance of around 85 kNm.

The leaders in this field have been (Stanton *et al.*, 1987; Priestley *et al.*, 1999; Lindberg & Keronen, 1992a; Lindbergh *et al.*, 1992b; Elliott *et al.*, 2003a, 2003b, 2005; Ferriera *et al.*, 2002, 2003; deChefdebien & Dardare, 1994). Semi-rigid behaviour is described by a moment-relative beam-column rotation $(M - \theta)$ diagram idealised in Figure 4.12. The total connection includes parts of the beam(s) and column(s) beyond the connection. The characteristics of the connection must be tailored to suit the requirements of the adjoining members. The *beam-line* method, illustrated in Figure 4.12, is therefore used to quantify this relationship at point E and determine the semi-rigid rotational stiffness S_E and bending strength M_E.

4.2.2   Moment-rotation of beam to column connections

The preferred method of making beam-to-column connections in skeletal frames is to attach the end of the beam to the column face via a concrete corbel, steel billet or stiffened steel angle. When precast floor slabs and tie bars are included the connection possesses flexural strength and stiffness to resist imposed gravity and wind loads. Figure 4.13 shows $M - \theta$ plots for single side moments (edge column situation) and balanced moments (interior column) yielding S_E. and M_E. The final rotation

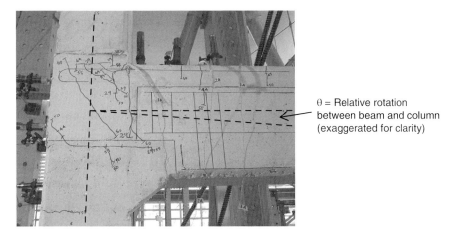

Figure 4.11 Bending and shear cracks due to negative bending moments in precast concrete beam-column connection (at CREAM laboratory, Kuala Lumpur, Malaysia).

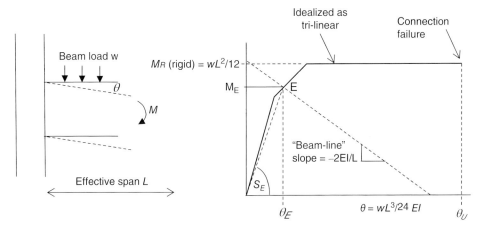

Figure 4.12 Definition of moment versus rotation parameters for connections.

of the connection should be greater than rotational capacity of the beam θ_u. Results varied enormously, for example, M_E from 18 to 238 kNm and S_E from 0.9 to 44.0 kNm/m.radian. However, dividing M_U for the connection by that of the beam M_R, and letting $K_s = S_E L / 4EI$ regression analysis lead to:

$M_U / M_R = 0.87 \sqrt{K_s}$ for single-sided connections (e.g., Figure 4.11)
$M_U / M_R = 0.62 \sqrt{K_s}$ for internal double-sided connections.

To qualify the rotational stiffness S of the connection, which may be determined by testing or by calculation, a fixity factor γ (Monforton & Wu, 1963) is adopted. This is a non-dimensional parameter varying from $\gamma = 0$ to 1 for pinned to full rigid connections, respectively, and is given by:

$$\gamma = \left[1 + \frac{3EI}{S_E L}\right]^{-1}$$

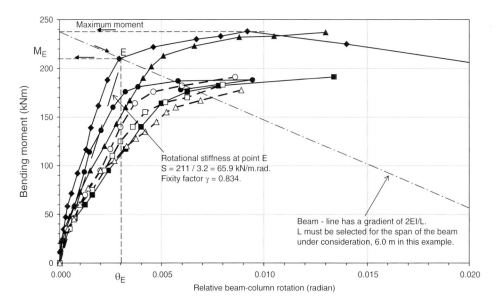

Figure 4.13 Moment versus rotation data for double-sided beam to continuous column connections using steel billet and welded plate (▲◆ symbols) connectors (Elliott & Jolly, 2013).

Figure 4.14 gives a final design chart for the above parameters, where γ can be input data into frame analysis programs. There are two important demarcation points:

$\gamma = 0.4$ distinguishes low-strength from medium-strength because when $\gamma > 0.4$ the semi-rigid behaviour provides more than 50% of full rigidity.

$\gamma = 0.67$ distinguishes medium-strength and high-strength because $M_E > M_{span}$.

giving five distinct zones, as indicated in Figure 4.14.

The rotational stiffness S_E as defined in Figure 4.12 as:

$$S_E = M_{RC}/\phi_c \text{ (note } \phi \text{ is used to define rotation in theoretical model)}$$

where $M_{RC} = z\, f_{yk}\, A_s$ is the connection resistance moment and ϕ_c is the total end relative rotation due to M_{CR}, where z may be taken as 0.9d at the connection, where d is the effective depth from the bottom of the beam to the steel bars in the top. The contribution of the dowels, cleats and welds within the body of the connection gives additional resistance of between 5 to 10 kNm depending on the details.

The total end relative rotation ϕ_c arises from two primary deformations. These are:

i. $\phi = f_{yk}\, l_e/E_s\, d$
     where l_e = embedment length defined in Figure 4.15(a).

ii. $\phi = \left[\dfrac{1}{r}\right]_{cr} l_p = \dfrac{M_{RC}\, l_p}{E_c\, I_{beam}}$
     where l_p = anchorage development length defined in Figure 4.15(b).
     $(1/r)_{cr}$ is the curvature of the beam (or beam plus slab) based on the flexurally cracked section.

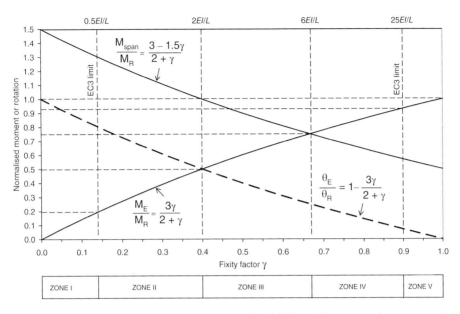

Figure 4.14   Classification system for pinned, semi-rigid and fully rigid beam-to-column connections, after the work of Ferriera (Elliott & Jolly, 2013).

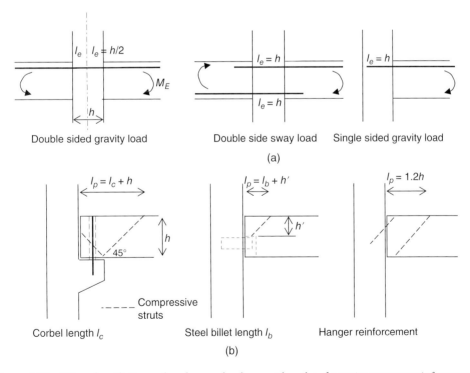

Figure 4.15   Values for effective and anchorage development lengths of negative moment reinforcement. (a) Effective length l_e (b) development length l_p.

The required moment capacity M_{ER} for the connector at ULS can be obtained from the intersection of S_E with the beam-line (see Figure 4.12) based on the flexurally cracked section of the beam I_{cr} as:

$$\frac{M_{ER}}{M_{Rd}} = \left[1 + \left(\frac{2E_{cm}I_{cr}}{L} \right) \left(\frac{\varphi_c}{M_{RC}} \right) \right]^{-1}$$

Letting $M_{RC} = 0.9 \, A_s \, f_{yk} \, d$ and $\phi_c = (f_{yk} \, l_e / E_s \, d) + (M_{RC} \, l_p / E_{cm} \, I_{cr})$ this equation can be rewritten as:

$$\frac{M_{ER}}{M_{Rd}} = \left[\left(\frac{L + 2\ell_p}{L} \right) + \left(\frac{2.22 E_c I_{cr}}{A_s E_s d^2} \right) \left(\frac{\ell_e}{L} \right) \right]^{-1}$$

In practice, consider a steel billet connection shown in Figure 4.16(a) comprising a 300×300 mm precast column and double sided 300×300 mm beams with a 150-mm-deep floor slab containing 2 no. H16 mm rebars (402 mm^2) at $d = 400$ mm. The bearing level is 150 mm from the bottom of the beam and the bearing (projecting) length is 100 mm. Concrete strength $f_{ck} = 40$ N/mm^2 and $E_{cm} = 35$ kN/mm^2, with a long-term creep factor $= 1.8$, then $E_{cm,eff} = 12.5$ kN/mm^2. Let the beam length (say) $L = 6.0$ m. This new research enables the connection to be designed for M_{RC} and hence M_{ER}, ϕ_c, S_E and fixity factor γ and hence determine the classification according to Figure 4.14.

$$M_{Rd} = 500 \times 402 \times 0.9 \times 400 \times 10^{-6} = 72.36 \text{ kNm}$$

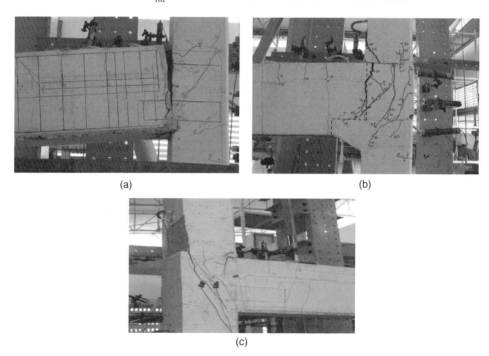

(a) (b)

(c)

Figure 4.16 (a) Steel billet RHS cast in the column showing bar fracture at ultimate capacity at top of beam (BIC tested at CREAM laboratory in Malaysia); (b) Concrete corbel (BHC tested at CREAM laboratory in Malaysia); (c) Steel insert RHS is cast in the beam (SIB tested at CREAM laboratory in Malaysia).

Table 4.1   Characteristic values for connections based on $M - \theta$ data

Ref.	Moment at point E, M_E (kNm)	θ_E (m.rad)	Secant stiffness S_E (kNm/m.rad)	Stiffness ratio K_S	Fixity factor γ
BHC	60.0	6.6	9.09	0.79	0.51
BIC	48.1	8.7	5.53	0.82	0.52
SIB	59.0	7.0	8.43	0.78	0.51

Modular ratio $m = 200/12.5 = 16.0$, $mA_s = 16.0 \times 402 = 6434$ mm². Section analysis gives height to cracked centroid $x_c = 111$ mm and $I_{cr} = 674 \times 10^6$ mm⁴

Figure 4.15(a) gives (double sided sway mode) $l_e = h_{col} = 300$ mm

Figure 4.15(b) gives $l_p = 100 + (500 - 150) = 450$ mm

$$M_{ER}/M_{Rd} = [(L + 2l_p/L) + \{(2.22\ E_{cm}\ I_{cr}/A_s\ E_s\ d^2)(l_e/L)\}]^{-1}$$
$$= [1.15 + (1.203 \times 0.05)]^{-1} = 0.818$$

Then $M_{ER} = 0.818 \times 72.36 = \underline{59.2\ \text{kNm}}$ (compare this value with the experimental result in Table 4.1 for *single*-sided BIC connector = 48.1 kNm)

$$\phi_c = (f_{yk}\ l_e/E_s\ d) + (M_{RC}\ l_p/E_{cm}\ I_{cr}) = 0.00504\ \text{rad}.$$

$S_E = 59.2/0.00504 = 11746$ kNm / rad. (compare with Table 4.1 for single-sided = 5530 kNm / rad.)

Then $K_s = S_E/(3E_{cm}I_{cr}/L) = 11746/4213 = 2.79$ and $\gamma = 1/(1 + 1/2.79) = 0.74$ is Zone IV. Therefore, the connection is suitable to be used as a semi-rigid connection satisfying the requirements of strength, stiffness and ductility. The results of this analysis are supported by the following experimental work.

4.2.3   Research and development of precast beam-to-column connections

Full-scale testing has been carried out on precast concrete beam-to-column connections in order to determine their moment resistance and connection classification. The connections are formed between 450×300 mm-deep single sided beams and a 300×300 mm edge column. The testing involved the following connection types:

i.   steel billet insert (BIC), Figure 4.16(a)
ii.   concrete corbel (BHC), Figure 4.16(b)
iii. steel insert in beam (SIB), Figure 4.16(c)

each with negative moment top reinforcement anchored into the column. Testing was carried at CREAM laboratory, Kuala Lumpur, Malaysia. The function of the billets and corbels is only for temporary support during the installation of the beam acting as a pinned connection. After completion of *insitu* concrete infill at the top of the beam using threaded inserts in the column and lapped reinforcement to the beam the connection possesses negative moments of resistance and rotational stiffness. The connections were designed to BS8110, Part 1 (1997).

The steel billet connection consists of a steel insert (RHS) of size $100 \times 60 \times 8$ mm grade 275 N/mm^2 which is filled with concrete of minimum grade 25 to stiffen the webs and prevent buckling of the section, together with a 16-mm-diameter vertical dowel to secure the beam via an angle cleat bolted to the column at the top of the beam. The top of the beam has a 860-mm-long recess to permit the hand placement of 2 no. T16 mm rebars ($f_y = 460$ N/mm^2) $\times$ 770 mm length anchored to the column using threaded splice couplers.

The design of the concrete corbel connection was based on the strut and tie model, which leads to a more accurate representation than by considering the corbel as short cantilever in bending and shear. The top connection consists of fully anchored tension reinforcement, 2 no. T20 mm bars and *insitu* topping of grade C25.

The steel insert cast in beam connection consist of fully anchored 2 no. T20 mm tension bars. The steel insert (RHS) of size $120 \times 80 \times 8$ mm, grade 275 N/mm^2 and filled with concrete as above. Confinement links are wrapped around the insert to restrain the force acting during installation prior to completion of the connection.

Vertical point loads were applied to the ends of the beams allowing the reactions to develop in the connections and columns according to their natural equilibrium and stiffness. The experiment gave the result as presented in Figure 4.17(a) to (c) moment-rotation ($M - \theta$) graph. The beam line method of analysis is used to quantify the relationship of $M - \theta$ from which the allowable moment capacity (M_E), secant stiffness (S_E) and rotation (θ_E) were obtained as given in Table 4.1 (Elliott *et al.*, 2003a).

According to classification system given in Figure 4.14 all the connections are in Zone III which is classified as semi-rigid connection with medium strength (Elliott & Jolly, 2013).

Figure 4.16 show the bending and shear cracks due to negative bending moment and shear force in the connections. For the billet connection, first cracks started to appear $M = 29$ kNm in the region of the column. The specimen exhibited flexural cracking in the beam and column regions followed by diagonal cracking in the connection itself. Based on the damage, and as shown in Figure 4.16(a) a plastic hinged has formed in the beam at face of the column, such that ultimate moment resistance of the beam was reached at $M = 98$ kNm. Splitting cracks were also observed within the connection region. In addition, the bar fractured was happened at two additional tension bars (T16) which anchored to the column using threaded splice coupler and shown in Figure 4.17. According to Jen Hua Ling (2009), steel bar is fractured when they achieved their ultimate capacity and the splice connector provides adequate interlocking mechanism to resist steel bar from slipping out.

The cracks in the corbel developed in a more gradual manner, where as shown in earlier moment-rotation plots there was dissipation energy prior to ultimate failure. Generally, the first flexural crack appeared at $M = 43$ kNm in the *insitu* concrete infill. This leads the bar being subjected to an eccentric tie force, thereby reducing the axial stiffness. For the second stage, cracks start to grow and widen in a diagonal pattern, causing splitting along the face of the beam-column intersection and diagonally at the connection until failure at ultimate moment $M = 106$ kNm.

For the SIB specimen, cracking propagated at $M = 37$ kNm near to the column. The first flexural crack appeared in the *insitu* concrete infill leading to the bar being subjected to an eccentric tie force. For the second stage, cracks developed and widened in a diagonal pattern in the connection area until failure at ultimate moment $M = 84$ kNm.

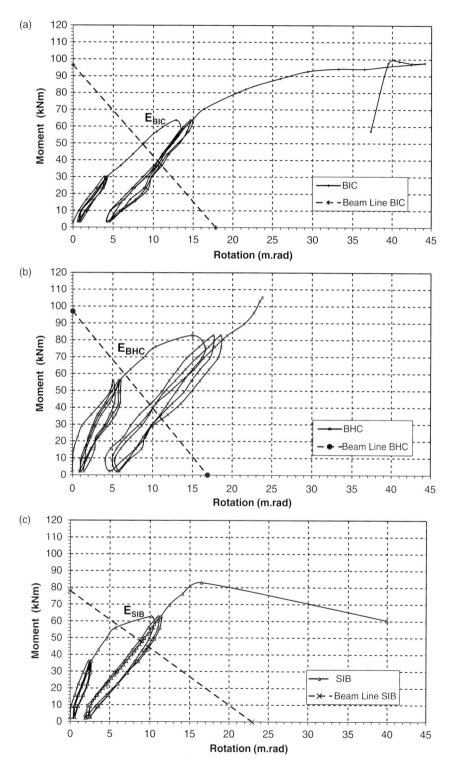

Figure 4.17   (a) Moment-rotation (M − θ) graph for steel billet in the column BIC; (b) Moment-rotation (M − θ) graph for concrete corbel BHC; (c) Moment-rotation (M − θ) graph for steel inert in the beam SIB.

In all tests, flexural cracks initiated at the column to precast beam joint interface. This happened due to relative strength and stiffness weakness of the two different materials in the joint. Shear cracks in the column, inclined at about 65° to the horizontal are due to the combined flexural tension, shear force and the horizontal resolution of the bending moment over the depth of the beam, in total sufficient to cause limited column damage.

4.2.4   Column effective length factors in semi-rigid frames

The results given in Section 4.2.2 may be used in a linear elastic, geometric second-order 2D computer program to determine β factors for columns in various types of sway frames. These values may be used to determine the second order moments (Eurocode notation) $M_2 = N_{Ed}\, e_2$, which when added to frame moments should not exceed M_E given above. A fuller explanation of this procedure is given in a frame example (Elliott et al., 2003b, Elliott & Jolly, 2013).

Figure 4.18(a) shows a ground floor sway frame. Figure 4.18(b) shows the variation in β with K_s for two common values of k. It is found that for values of $K_s < 2$ the β factors are more sensitive to changes in K_s than k. This is an important result because experiments have found K_s to be less than 2.5. The code equation for β with a fixed foundation ($k_1 = 0.01$) and fully rigid beam-column joint is:

$$\beta = \left(1 + \frac{k_1}{1+k_1}\right)\left(1 + \frac{k_2}{1+k_2}\right)$$

with $k_2 = 0.25$ modified for the semi-rigid stiffness K_s to $\beta = \left(1 + \frac{k_1}{1+k_1}\right)$
$\left(0.84 + \frac{k_2 + 1/4K_s}{1 + k_2 + 1/4K_s}\right)$

with $k_2 = 0.5$ modified for the semi-rigid stiffness K_s to $\beta = \left(1 + \frac{k_1}{1+k_1}\right)$
$\left(0.84 + \frac{k_2 + 1/3K_s}{1 + k_2 + 1/4K_s}\right)$

Example: Let $k_1 = 0.01$ and $k_2 = 0.25$, for $K_s = 1$, then $\beta = 1.18$. If the clear storey height = say 3.2 m, effective length $l_o = 1.18 \times 3200 = 3776$ mm and $b = h = 300$ mm. slenderness ratio $\lambda = l_o/i = 3776/86.6 = 43.6$ according to EN 1992-1-1, clause 5.8.3. Now $\lambda_{limit} = l_o/i$ is $20ABC/\sqrt{n}$, where we may take A = 0.7, B = 1.1, C = 0.7 for unbraced frames, and letting $n = N_{Ed}/A_c f_{cd} = 1$, then $\lambda = 10.8 < 43.6$. The precast column will design using first- and second-order moments.

4.3   Studies on structural integrity of precast frames and connections

4.3.1   Derivation of catenary tie forces

The relationships between the stiffness and ductility of individual components/connections and the behaviour of the total structure is a complex, non-linear, 3D problem with which, by the mid 1990s, some 25 years after the progressive collapse in May 1968 at Ronan Point (Griffiths et al., 1968), research institutions

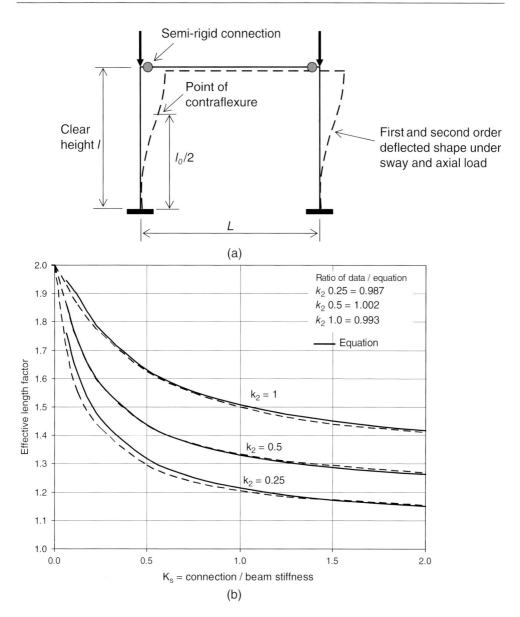

Figure 4.18   (a) Definitions for ground floor sub-sway frame; (b) Variation on column effective length factors in ground floor sub-frame with semi-rigid connections at first floor based on EC2-1-1 (EC, 2002).

were only just beginning to grapple. However, in the past 15 years a considerable body of work now exists where the partial strength and stiffness of slab-beam and beam-column connections has been enhanced by the use of continuity ties placed within the beam/floor zone. These methods have formed the basis of the *fib* document Design of Precast Concrete Structures Against Accidental Actions (*fib*, 2012), and the Unified Facilities Criteria document (UFC, 2009).

The hollow-core specimens were tested as shown in Figure 4.19, which would effectively simulate the catenary action at the remote supports (Engstrom, 1992).

Figure 4.19   Hollow core slab continuity tests by Engstrom (Courtesy Chalmers University, Sweden).

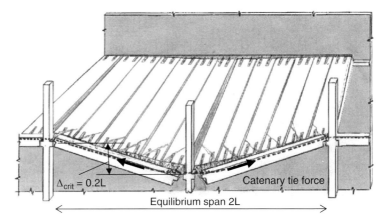

Figure 4.20   Possible scenario of the structural behaviour of a precast frame structure after sudden column loss due to accidental actions (Courtesy of Fédération Internationale du Béton [*fib*]).

The tie steel used was high tensile or mild steel bar ($f_{yk} = 400/260$ N/mm^2) or prestressing wire ($f_{pk} = 1690$ N/mm^2) and cube crushing strength f_{cu} varied from 25 to 33 N/mm^2. Initial failures occurred abruptly due to flexural tension failure in the interface between the precast unit and the *insitu* infill, after which the units were connected by the tie steel alone. The full tensile capacities of the bars were mobilised when the rotation of the slabs was at least 0.1 to 0.2 radians, with ultimate failure at 0.3 radians and the stress in the continuity bars around 530 N/mm^2.

With increasing deformation a new equilibrium state will develop as shown in Figure 4.20 (*fib*, 2012) where the deflection reaches a critical value Δ_{crit}. If the deflection exceeds Δ_{crit} the tie steel will either fracture or debond in the adjacent spans. If the sagging deflected shape of the beam is deduced from uniformly distributed loading (udl), and the tie steel is elastic-perfect-plastic, it can be shown that for a characteristic udl w acting on a beam of length 2L (= two spans L affected by the loss of an internal column) the catenary force T is given as:

$$T = 2wL \left(\frac{0.208\,L}{\Delta_{crit}} + \frac{0.25\,\Delta_{crit}}{L} \right)$$

and for a point load P acting at mid span of a beam of length $2L$:

$$T = 0.5\,P\,\sqrt{\left(\frac{L}{\Delta_{crit}}\right)^2 + 1}$$

In this equation, P is the characteristic axial force in the column from the storey above only, that is, $w(L/2 + L/2)$, because catenary action is taking place at all subsequent floors above the level under consideration. Δ_{crit} is related to the strain in the tie steel, which is a function of the type of reinforcement and detailing, and can only be obtained by testing. Test results (Engstrom, 1990) show that just prior to failure $\Delta_{crit} \approx 0.2L$. The above equations may be used to determine catenary tie forces and assist in understanding the derivation of the stability tie forces that are given in the Eurocode EN1992-1-1 (EC, 2002).

The *fib* document (*fib*, 2012) adopts an energy balance based on the elongation of the catenary tie due to deflection of the beams $a < \Delta_{crit}$ under the state of equilibrium.

The deformation energy in the tie is $W_{int} = \Delta L$ where $T = A_s\,f_{yk}$ and $\Delta L = \varepsilon_s\,L$ and the external work is $W_{ext} = p_d\,L\,a\,/\,2$ where $a = \sqrt{[(\Delta L + L)^2 - L^2]}$ and the accidental dead plus quasi-permanent load live load is $p_d = g_k + \psi_2\,q_k$ where for residential and offices $\psi_2 = 0.3$, shopping, congregation and car park $\psi_2 = 0.6$ and storage $\psi_2 = 0.8$.

Equating $W_{int} = W_{ext}$ gives $T = p_d\,L\,a/(2\Delta L)$

For example, if $L = 7.2$ m. $p_d = 50$ kN / m. $\varepsilon_{uk} = 75 \times 10^{-3}$ for rebar ductility class C500 where $f_{yk} = 500$ N / mm^2 then $\Delta L = \varepsilon_{uk}\,L = 75 \times 10^{-3} \times 7.2 = 0.540$ m and $a = \sqrt{[(0.54 + 7.2)^2 - 7.2^2]} = 2.84$ m. T $= 50.0 \times 7.200 \times 2.84/(2 \times 0.54) = 947$ kN.

$A_s = T/f_{yk} = 947 \times 10^3/500 = 1893$ mm^2. Use 4 no C25 (1963) tie bars, two either side of the beam and passing either through a slot in the column or, if the beam is sufficiently wide, on either side of the column.

Other attempts to quantify the floor membrane forces present in precast catenary systems were made following Ronan Point (Regan, 1974; Schultz *et al.*, 1978; Wilford & Yu, 1973). Full-scale horizontal catenary tests were carried out on hcu's tied across their supports using short lengths of seven-wire helical strand in the longitudinal joints between adjacent units. Catenary action was achieved at loads of up to 80% of the slab failure load, which suggested that by increasing the anchorage length and the amount of continuity reinforcement full force catenary action might be possible.

References

ACI Standard 318-11. (2011). Building code requirements for reinforced concrete, American Concrete Institute, Detroit, USA.

Cholewicki, A. & Elliott, K. S. (1998). Diaphragm Action of Hollow Cored Floors, *Betonwerk and Fertigteil-Technik*, Journal of the Federal German Association of the Concrete and Precast Building Components Industry, Part 1 – Theoretical Models, Vol. 64, No. 12, p 76–90. Part 2 – Experimental Work, Vol. **65**, No. 3, 61–68.

Cholewicki, A. (1991). Shear Transfer in Longitudinal Joints of Hollow Core Slabs, *Betonwerk and Fertigteil-Technik*, Journal of the Federal German Association of the Concrete and Precast Building Components Industry, Vol. **57**, No. 4, 58–67.

Cranston, W. (1972). Analysis and Design of Reinforced Concrete Columns, Research Report 20, Cement & Concrete Association, Wexham Springs, UK.

Davies, G., Elliott K.S., & Wahid Omar. (1990). Horizontal Diaphragm Action in Precast Hollow Cored Floors, *The Structural Engineer*, London, **68**(2), 25–33.

deChefdebien, A. & Dardare, J. (1994). Experimental Investigations on Current Connections between Precast Concrete Components, COST C1 Proceedings of the Second State of the Art Workshop, Semi-rigid Behaviour of Civil Engineering Structural Connections, Prague, 21–30.

EC. (2002). Eurocode EN 1992-1-1:2004, Eurocode 2: Design of Concrete Structures – Part 1-1: General rules and rules for buildings. Amendment 2014.

Elliott, K. S., Davies, G., & Wahid Omar. (1992). Experimental and Theoretical Investigation of Precast Hollow Cored Slabs Used as Horizontal Diaphragms, *The Structural Engineer*, London, **70**(10), 175–187.

Elliott, K. S., Davies, G., & Bensalem, K. (1993). Precast Floor Slab Diaphragms Without Structural Screeds, Concrete 2000 - Economic and Durable Construction Through Excellence, Dundee, 617–632.

Elliott, K. S., Davies, G., Ferreira, M., Mahdi, A. A., & Gorgun, H. (2003a). Can Precast Concrete Structures be Designed as Semi-rigid Frames? – Part 1, The Experimental Evidence, *The Structural Engineer*, London, **81**(16), 14–27.

Elliott, K. S., Davies, G., Ferreira, M., Mahdi, A. A., & Gorgun, H. (2003b). Can Precast Concrete Structures be Designed as Semi-rigid Frames? – Part 2, Theoretical Equations and Applications for Frame Design, *The Structural Engineer*, London, **81**(16), 28–37.

Elliott, K. S., Ferreira, M., de Aranjo, D., & El Debs, M. (2005). Analysis of Multi-storey Precast Frames considering Beam-column Connections with Semi-rigid Behaviour, Keep Concrete Attractive, FIB Symposium, Budapest, 496–501.

Engstrom, B. (1990). Connections Between Precast Components, Nordisk Betong, *Journal of the Nordic Concrete Federation*, No. **2-3**, 53–56.

Engstrom B. (1992). Ductility of Tie Connections in Precast Structures, PhD Thesis, Chalmers University of Technology, Goteborg, Sweden.

Ferriera, M. A., El Debs, M. K., & Elliott, K. S. (2002). Theoretical Model for Design of Semi-Rigid Connections in Precast Concrete Structures, Proceedings 44th Brazilian Concrete Congress, IBRACON Brazilian Concrete Institute, Belo Horizonte, Brazil, August, 100-116.

Ferriera, M. A., El Debs, M. K., & Elliott, K. S. (2003). Analysis of Multi-storey Precast Frames with Semi-Rigid Connections. Brazilian Conference on Concrete, IBRACON 2003, 45th Brazilian Concrete Congress, Brazilian Concrete Institute - IBRACON, Vitoria, Brazil.

fib. (1999). *Precast Prestressed Hollow Core Floors*, Guide to good practice, Fédération Internationale du Béton, SETO, London.

fib. (2012). Bulletin 63, *Design of Precast Concrete Structures Against Accidental Actions*, Guide to good practice, Fédération Internationale du Béton, Lausanne, Switzerland, 72 pp.

Griffiths, H., Pugsley, A. G., & Saunders, O. (1968). Report of the Inquiry into the Collapse of Flats at Ronan Point, Canning Town, Her Majesty's Stationery Office, London.

Hawkins, N.M. & Ghosh, S. K. (2000). Proposed Revisions to 1997 NEHRP Recommended Provisions for Seismic Regulations for Precast Concrete Structures: Part 2 – Seismic-Force-Resisting Systems, *PCI Journal*, **45**(5), 34–44.

Ling JH, Abd Rahman AB, Ibrahim IS, Abd Hamid Z, & Mirasa AK. (2009). Structural Performance of Splice Connector for Precast Concrete Structures. Proceedings of the 7th Asia

Pacific Structural Engineering and Construction Conference (ASPEC) and 2nd European Asian European Engineering Forum (EACEF), Langkawi, Malaysia, 4–6 August, 402–408.

Juvas, K. & Pousi, O. (2000). Tempo – A New Frame System for Concrete Elements, Internal Report of the Partek Concrete Division, 20320 Turku, Finland, 8 pp.

Lindberg, R. & Keronen, A. (1992a). Semi-rigid Behaviour of a RC Portal Frame, COST C1 Proceedings of the first state of the art workshop, *Semi-rigid Behaviour of Civil Engineering Structural Connections*, E.N.S.A.I.S., Strasborg, France, 53–63.

Lindberg, R. *et al.* (1992b). Beam-to-Column Connections in Storey Height Concrete Frame, Report 57, Tampere University of Technology, Finland, 104 pp.

Millard, S. G. & Johnson, R. P. (1984). Shear transfer across cracks in reinforced concrete due to aggregate interlock and to dowel action, *Magazine of Concrete Research*, **36**(126), 9–21.

Monforton, G. R. & Wu, T. S. (1963). Matrix Analysis of Semi-rigidly Connected Frames, *Journal of Structural Division*, **89**(6).

Nakaki, S. W., Stanton, J., & Sritharan, S. (1999). An Overview of the PRESSS Five Story Precast Test Building, *PCI Journal*, **44**(2), 26–39.

Pajari, M. & Koukkari, H. (1998). Shear Resistance of PHC Slabs Supported on Beams, *Journal of Structural Engineering*, **124**(9), Part 1: Tests, p1050-1061. Part 2: Analysis, 1062–1073.

Priestley, M. J. N. (1996). The PRESSS Program-Current Status and Proposed Plans for Phase III, *PCI Journal*, **41**(2), 22–40.

Priestly, N., Sritharan, S., Conley, J., & Pampanin, S. (1999). Preliminary results and conclusions from the PRESSS Five Storey Precast Test Building, *PCI Journal*, **44**(6), 42–67.

Regan, P. E. (1974). Catenary Tests on Composite Precast - Insitu Concrete Composite Floors, Report to the Deptartment of Environment, The Polytechnic of London, UK.

Schoettier, M. J., Belleri, A., Zhang, D., Restrepo, J. I., & Flieschman, R. B. (2009). Preliminary Results of the Shake-Table Testing for the Development of a Diaphragm Seismic Design Methodology, *PCI Journal*, **54**(1), 100–123.

Schultz, D., Burnett, E., & Fintel, M. (1978). A Design Approach to General Structural Integrity – Design and Construction of Large Panel Concrete Structures, Supplemental Report A, *Portland Cement Association*, Stokie, USA.

Stanton, J. F., Anderson, R. G., Dolan, C. W., & McCleary, D. E. (1987). Moment Resistant Connections and Simple Connections, Final Report to PCI, Specially Funded R & D Projects Nos 1 & 4, 1986, and *PCI Journal*, **32**(2), 62–74.

Stanton, J. F. (1998). The PRESSS Program in the USA and Japan – Seismic Testing of Precast Concrete Structures, *Control of the Semi-rigid Behaviour of Civil Engineering Structural Connections*, COST C1 International Conference, Liege, Belgium, 13–24.

Suikka, A. (2000). Slim Precast Floor Structures, The Second International Symposium on Prefabrication, Helsinki, Finland, 101–107.

UFC, United Facilities Criteria 4-023-03. (2009). *Design of Buildings to Resist Progressive Collapse*, US Dept. of Defence, updated 2013. 245 pp.

Wahid Omar. (1990). Diaphragm Action in Precast Concrete Floor Construction, Ph.D Thesis, University of Nottingham, UK.

Wilford, M. J. C. & Yu, C. W. (1973). Catenary Action in Damaged Structures, Paper No. 6, Department of Environment and CIRIA, The Stability of Precast Concrete Structures, Seminar, London, 1–24.

Part 2

Mechanisation and Automation of the Production of Concrete Elements

Chapter 5

Building Information Modelling (BIM) and Software for the Design and Detailing of Precast Structures

Thomas Leopoldseder and Susanne Schachinger

Precast Software Engineering, Wals-Siezenheim, Austria

Building Information Modelling (BIM) and its implementation in companies and constructions is an outstanding challenge for all participants and its importance will increase rapidly in the near future. Precast companies play an important role in the worldwide construction industry; therefore, they have to deal with BIM and its consequences to their whole value-added chain. This chapter describes the history of BIM, different implementation levels from all over the world and standard technologies that are currently used in BIM implementations. The chapter also focuses on the process to provide data for BIM. A continuous example is used to describe this process, the data and the advantages to get a better understanding of BIM in the precast industry.

5.1 Building information modelling (BIM)

5.1.1 Introduction

Building Information Modelling (BIM) is one of the seminal technologies in the construction industry. Although all suppliers and construction companies have to work together, the precast industry plays an important role in the whole BIM process. This chapter will give an overview of the history of BIM, its technologies and its consequences for the precast industry.

The BIM process will be described by one example to explain the different areas of BIM using the same construction. This construction was erected at the wine yard Scheiblhofer in Andau/Burgenland (Austria). With about 65 hectares the wine

Modernisation, Mechanisation and Industrialisation of Concrete Structures, First Edition.
Edited by Kim S. Elliott and Zuhairi Abd. Hamid.
© 2017 John Wiley & Sons Ltd. Published 2017 by John Wiley & Sons Ltd.

Figure 5.1   Event hall at Scheiblhofer wine yard. Planning of precast elements, production and delivery by manufacturer Josef Lehner e. U. AT. (Courtesy of Wine Yard Scheiblhofer, AT.)

yard belongs to the major wine producers in Burgenland producing different kinds of high quality wine (Scheiblhofer, 2016). An Event Hall was erected in 2015 consisting mainly of precast elements totalling 689 m^3, as shown in Figure 5.1.

5.1.2   History and ideas

Building Information Modelling (BIM) has its origin in the mid-1970s when the first papers about virtual building models were published (Eastman *et al.*, 1974). Van Needervan and Tolman (1992) coined the term "Building Information Modeling" but it took an additional eleven years to make this term more common in describing a new concept for the planning of constructions (Autodesk, 2003).

The National BIM Standard-United States® (NBIMS-US™, 2015) defines BIM as a *"digital representation of physical and functional characteristics of a facility. As such, it serves as a shared knowledge resource for information about a facility, forming a reliable basis for decisions during its life cycle from inception onward."*

The goal of BIM is the generation of a different kind of more or less detailed information along the whole process of design, production and erection to support the stakeholder of a construction during the whole complex process from planning to erection and for the creation of information for operation, renovation and demolition as well. Additional information that is produced during the usage of the building can be added to the BIM model. Architectural models based on ifc – files are visualized in Figures 5.2a and b.

The perfect BIM model represents the current virtual status of the building construction at all times. As a matter of fact BIM is not just a 3D geometric representation of the construction but it is the structured collection of all data which is generated along the whole life cycle of the building. 3D representation is an important part and precondition of a BIM model.

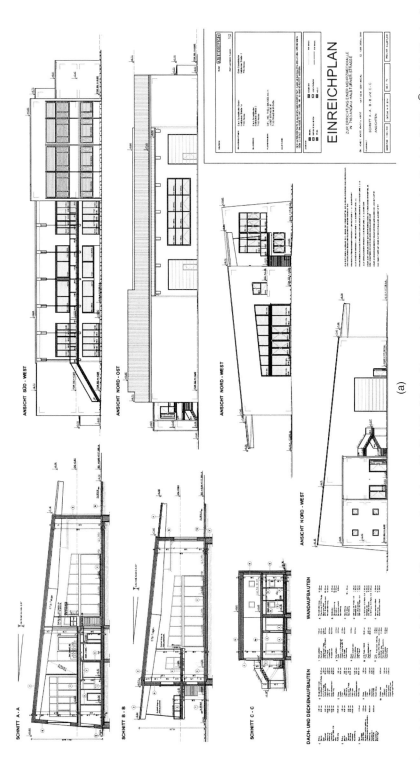

(a)

Figure 5.2 (a) Architectural drawing. (Courtesy of architect Thell, AT.); (b) Architectural model based on ifc files visualized in "Solibri® Model Checker". (Courtesy of architect Thell, AT and Solibri, FIN.)

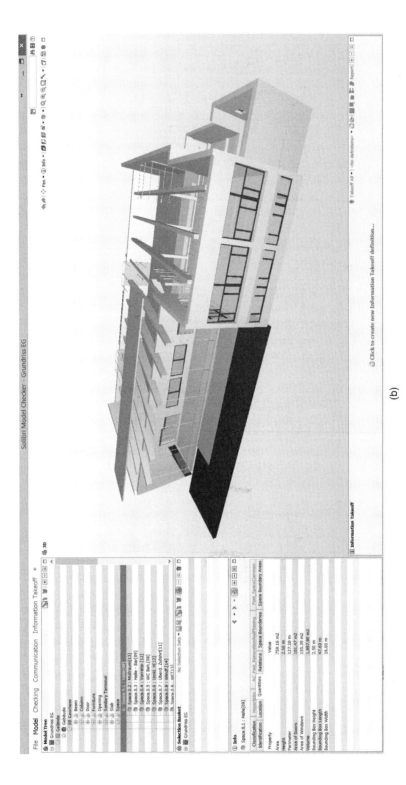

Figure 5.2   (*Continued*)

There are four different components of BIM:

1. BIM-Process: To establish a BIM solution the BIM process has to be defined which determines the way data have to be generated and put into the BIM model.
2. BIM-Organization: The organization includes all the people who are responsible for the compliance of the process and the actuality of the virtual BIM model.
3. BIM-Model: The output of the process is the model itself which provides all the data that has been generated during the entire construction life cycle. The model can be saved in a database or in file structures. The model represents the virtual 3D building which contains not only the geometry representation but also information about the relationships between construction elements, material data, time, costs, maintenance information, and so on. BIM-Model-Software enables access to relevant information that is required by the stakeholders in the process.
4. BIM-Technology: It enables software vendors to provide solutions enabling the creation of the BIM model and ensuring that the stakeholders are able to retrieve information as needed.

5.1.3   Types of BIM

Depending on the level of "Process-integration" and "Functionality-integration" four types of BIM can be differentiated (Figure 5.3):

1. "Process-integration"
   a. **Little BIM:** Just a small part of the process is covered (e.g., architectural design).
   b. **Big BIM:** The whole process, not only during design, production and erection but also during operation and demolition (Jernigang, 2008).
2. "System-integration"
   a. **Closed BIM:** A BIM solution is realized by just one software solution from one software vendor.
   b. **Open BIM:** A universal approach to the collaborative design, realization and operation of buildings based on open standards and workflows. In an Open BIM solution the suppliers, planners, architects, and so on, are able to use their preferred software for their business. By using open standards and workflows it is possible to integrate the relevant life-cycle data of the construction into the BIM model not depending on special software by using standardized definitions and interfaces (buildingSMART®, 2016).

According to those four dimensions the complexity of a BIM solution raises in one dimension from Little BIM to Big BIM and in the other dimension from Closed BIM to Open BIM. In fact only a Big BIM/Open BIM solution will cover the requirements of a seminal and practical solution.

In that context it should be mentioned that the OpenBIM program, developed by leading members of the buildingSMART® initiative aims to develop and implement worldwide standards for data exchange in the construction industry to enable worldwide Open BIM Solutions. Additionally, buildingSMART® established a system to certify software solutions according to identical standards (buildingSMART®, 2016).

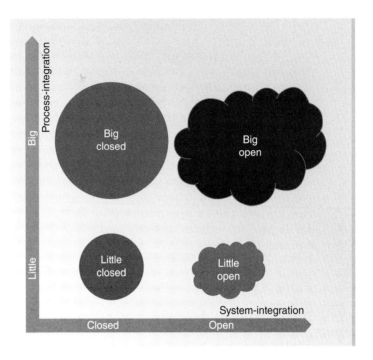

Figure 5.3   Types of BIM.

Advantages of Open/Big BIM solutions:

1. Open/Big BIM allows all participants in the process to use their preferred software system which fits their requirements best. Using defined interface standards the output can be exchanged and imported into the BIM model and into different other software solutions as well.
2. Open/Big BIM enables software vendors to implement independent software systems based on international standards for data exchange.
3. By establishing Open/Big BIM solutions it is possible to cover the whole life cycle of the construction that is not depending on the used software versions and software systems.
4. Open/Big BIM solutions increase quality by checking all relevant key factors of the construction in the digital BIM model to avoid conflicts, duplications, planning and construction errors.

By focusing on the adoption process of the construction industry it is customary to distinguish between the following levels of BIM (NBS, 2016):

- **Level 0 BIM** is characterized by 2D CAD drawings, print outs or electronic distribution of drawings. No integrated planning takes place therefore it is not really a BIM implementation.
- **Level 1 BIM** is a mixture of 2D and 3D planning and there is no managed collaboration between the different disciplines. This level is very often used in the construction industry.

- **Level 2 BIM** is characterized by different 3D CAD models separately used by each participant of the construction. A common shared model does not exist. A standard data format for data exchange between different planning teams is used for validation. In countries all around the world it is a common practice to achieve a certain percentage of level 2 BIM projects at public constructions.
- **Level 3 BIM** represents the full integration of all disciplines by using a shared model which is stored in a central database. All participants have access and can do their planning directly in this centralized BIM model.

5.1.4   BIM around the world

BIM methods and technologies are used and implemented in countries around the globe at different levels:

5.1.4.1   Asia

"Building Information Modelling (BIM) is identified as a key technology to improve productivity and level of integration across various disciplines across the entire construction value chain" (Building and Construction Authority (BCA), 2016) in Singapore. In May 2012 BCA published the Singapore BIM Guide Version 1.0 which was updated in 2013 to version 2 (Building and Construction Authority (BCA), 2013). In this guideline the relevant information about BIM projects in Singapore are defined. As a key technology it is very important for construction and precast companies to implement BIM standards according to this guideline.

Referring to the SmartMarketReport (Dodge Data & Analytics, 2015) China is one of the fastest growing construction countries of the world for BIM implementations by the use of contractors. Pursuant to that study the number of contractors doing over 30% of their projects with BIM will increase by 108% in the next 2 years. In that forecast China is under the top 5 fastest-growing countries in the implementation of BIM solutions. The China BIM Union which is a sub-organization under the China Association for Engineering Construction Standardization coordinates all activities concerning the implementation of BIM technologies and standards (China BIM Union, 2016).

South Korea has an expected growth of 126% in the two-year forecast of Dodge Data & Analytics (2015), therefore it will be a rising market for BIM technologies as well.

According to Dr. Zuhairi, Executive Director of the Construction Research Institute of Malaysia (CREAM), the Malaysian construction industry has been identified as a very important and productive sector that drives the Malaysian economy. Under the Eleventh Malaysia Plan (11th MP), transforming construction is part of seven focus areas that have been identified as key drivers to accelerate the momentum of economic growth. In response to the national agenda, the construction industry has aggressively embraced new technologies and modern construction methods in improving the efficiency of construction project implementation (Economic Planning Unit Malaysia, 2015). Information and communications technology (ICT) as the future direction will drive the industry towards better performance and remain competitive globally. Building Information Modelling (BIM) recognized as advanced ICT transforms the construction industry by increasing efficiency, productivity and quality. Currently, a number of projects in Malaysia have started using BIM. The National Cancer Institute (NCI) in Putrajaya is the first government project that adopts BIM. Although many benefits can be gained by the adoption of BIM, the pace of adoption of BIM in Malaysia is still low as the construction industry players see BIM as "disruptive technology" (Zahrizan, Ali, Haron, Marshall-Ponting, & Abd, 2013). Taking cognisance of several issues and challenges in implementing BIM, the Construction Industry Development Board (CIDB) Malaysia has taken proactive initiatives to promote and support the adoption of BIM in the construction industry. A minimum of 40% implementation rate of Level 2 BIM for public projects above RM 100 million is recommended by CITP. CIDB will also collaborate with various key stakeholders to develop national BIM guide and standards as well as develop a BIM roadmap for Malaysia. The roadmap highlighted seven critical areas, known as pillars, encompass strategies and initiatives to materialise the BIM agenda in Malaysia. These pillars comprise of (CIDB Malaysia, 2015b):

- Standard and Accreditation
- Collaboration and Incentives
- Education and Awareness
- National BIM Library
- BIM Guidelines and Legal Issues
- Special Interest Group (SIG)
- Research and Development (R&D)

CIDB currently are in the midst of developing BIM roadmap for Malaysia's construction industry.

5.1.4.2   Europe

In 2014 the European Parliament recommended a change in the European public procurement law and defined BIM as a mandatory technology for public constructions. "For public works contracts and design contests, Member States may require the use of specific electronic tools, such as of building information electronic modelling tools or similar" (European Parliament, 2014).

Starting in 2016 all members of the European Community should support BIM solutions during the realization of public constructions. In some countries (e.g., UK, Finland, Netherlands, Denmark, Norway) it has been mandatory to use BIM technologies in public constructions for several years.

The BIM strategy of the British government is remarkable as well. Already in 2011 the British government announced their BIM strategy to use digital technologies in constructions to reduce costs around 15 to 20% and emissions by up to 50% (Cabinet Office, 2011). Additionally, the British construction industry should achieve major competitive advantage in the international markets by implementing BIM standards.

Germany announced its official BIM initiative (in coordination with the "Digital infrastructure" initiative) at the end of 2015 (BMVI, 2015). The master plan of the German government which has been published in December 2015 defines that starting in

2020 all new public infrastructure projects have to be planned by using BIM solutions. In this context three steps have been defined:

- 2015–2017: Preparation phase
- 2017–2020: Extended pilot phase (Level 2 BIM)
- 2020– …: Level 2 BIM for new planned projects

As a consequence, digital planning and construction will be established as the national standard for public constructions in Germany in the next years.

Finland has a long BIM tradition. It focuses on the implementation of an open and software independent BIM concept that can be used in all phases of infrastructure projects which should be completed in 2025. The Building Smart Finland Infrastructure Business Group was founded to develop standards and support the BIM process in Finland. "Built Environment Process Re-engineering (PRE)" was an initiative by Rym Ltd, a private and non-profit company owned by different companies, universities, cities and government agencies. Between 2010 and 2014 technologies and concepts supporting Open BIM solutions (Salonen, 2015) have been developed. Six thematic work packages have been defined and worked on to proceed in developing BIM in Finland.

5.1.4.3   North America

In May 2015 the National Institute of Building Sciences BuildingSMART alliance® released the NBIMS-US™ Version 3 (NBIMS-US™, 2015). This standard covers the whole life cycle of constructions and describes the relevant BIM topics in the U.S. market. Additionally major public construction authorities (e.g., U.S. General Services Administration, 2003; New York City, 2012) request BIM models as a mandatory precondition in public projects. With the new standard it is expected that BIM activities will increase in the United States.

In Canada many organizations like the buildingSMART Canada®, Canada BIM Council® or the Institute for BIM in Canada® are supporting BIM implementations in Canada and publish guidelines and standards for the Canadian market.

5.1.5   BIM and precast structures

The process of precast construction is accompanied by many different interfaces:

- Import of architectural Computer Aided Design (CAD) data to the Precast CAD system
- Import of built-in parts to the Precast CAD system
- Calculation of quantities of precast elements and used materials and export to the Precast Enterprise Resource Planning System (ERP) for time scheduling, material management, invoicing, and so on.
- Status tracing along the whole process, starting with the calculation phase and ending in the invoicing phase, by using 3D status-visualizing tools gaining the

status from different kinds of software systems (Precast ERP system, 3D CAD System, Master Computer, Logistic System, On-Site-Construction Software, etc.)
- Simulation of the erection of a building to determine optimized production, mounting and delivery of units
- Interfaces between planning and production systems for correct production and status tracking

Since BIM covers all of these topics, it is very important for precast companies to implement BIM to fulfil the requirements in BIM-relevant markets. This can be done in three areas:

1. Only a 3D Precast CAD system is able to generate all relevant data for a BIM model. The CAD system should be able to import and export IFC data structures (for more details see Section 5.2)
2. Production technologies should be able to generate production data as additional information for the BIM model
3. BIM data should be created along the whole process starting from the pricing of a project until mounting of precast elements in one system for all precast-relevant topics
4. It should be possible to simulate the erection process to derive optimized information for production and delivery, as shown in Figures 5.4a to c.

5.2   Technologies

5.2.1   Industry foundation classes (IFC)

5.2.1.1   Definition

The National Institute of Building Science (2007) defines the Industry Foundation Classes (IFC) data model as "*definitions, rules, and protocols that uniquely define data sets which describe capital facilities throughout their life cycles.*"

IFC is a data standard developed by BuildingSMART®, which describes a format where information between different software systems are shared along the whole life cycle of a construction by different project members. By using IFC deep data interoperability is achieved and IFC data files can be easily imported into the BIM model. There are IFC data structures for geometric information of buildings and for any other necessary information of constructions as well.

This data structure allows software vendors to implement BIM software with IFC interfaces to provide data exchange without losses. The implementation can be done not depending on internal data structures.

In March 2013 BuildingSMART® published IFC4 (2013) as the latest version of IFC. This version was also released as an official ISO standard in ISO 16739 (2013).

5.2.1.2   Layers and model view definitions

IFC is a very complex format defined in different hierarchical layers (Figure 5.5). Each layer describes object classes which can be referenced to object classes in higher layers

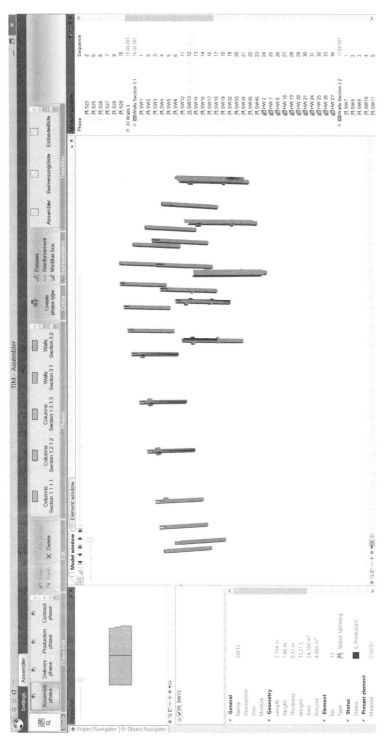

Figure 5.4 Erection sequence of precast elements, simulated by "TIM technical information manager". (Courtesy of Precast Software Engineering, AT and Josef Lehner e. U., AT.)

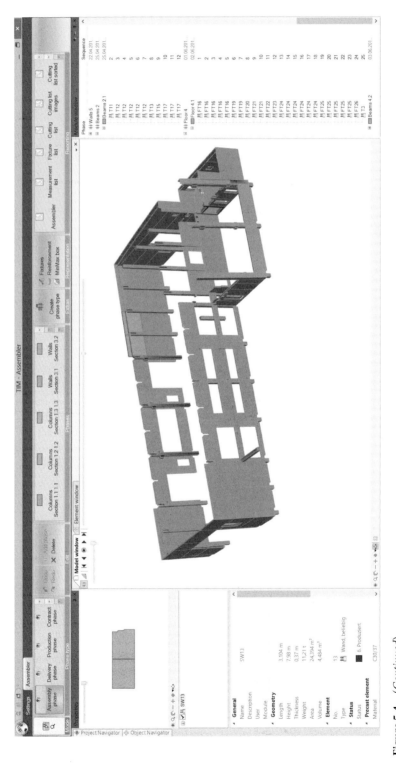

Figure 5.4 *(Continued)*

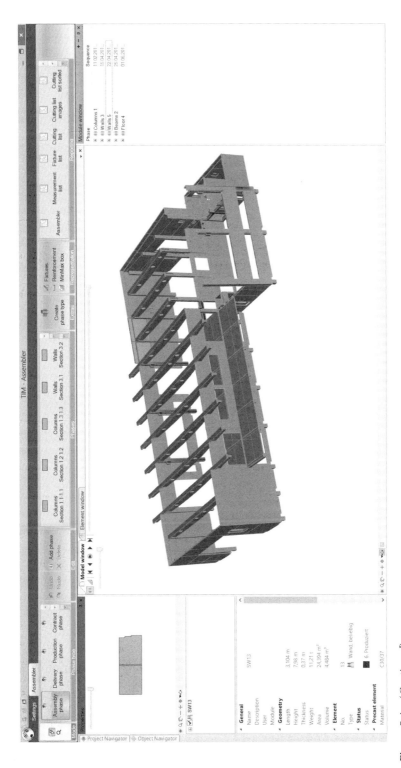

Figure 5.4 *(Continued)*

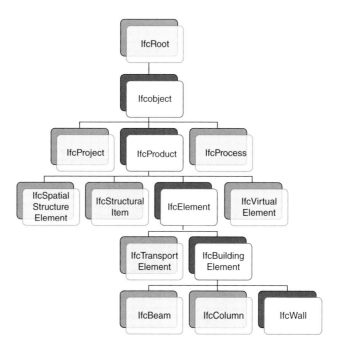

Figure 5.5   Part of the IFC data model structure.

(e.g., the object class IfcWall is part of the class IfcBuildingElements, which is part of object class IfcElements which is part of IfcProduct which is part of IfcObject). The result is a hierarchical data model where properties have to be defined just once and can then be used in various object classes on lower levels.

On the other hand, Model View Definitions (MVD) have been created as a subset of the complete IFC structure. MVDs are a consistent representation of the IFC model specially created for subject-specific exchange of data in a BIM model. For example the Model View Definition PCI-001 for the precast concrete industry describes, based on IFC, digital data which is necessary in the context of the precast workflow including information about material, shape, components and field location of each precast element in a construction (Eastman *et al.*, 2009). However, PCI-001 did not reach an official status of BuildingSMART®. Complex precast products (e.g., multi-layered walls with different layers of concrete, isolation, tiles), build-in-parts such as thermal break parts and complex reinforcement cannot be represented in the current version of IFC4 as necessary. Therefore an extension of IFC4 for the precast industry is absolutely essential. Some precast software vendors and members of BuildingSMART® are just planning to extend IFC4 according to that special requirements of the precast industry.

5.2.2   IFC data file formats and data exchange technologies

5.2.2.1   Definitions

Software systems that support BIM technologies and IFC standards have to create files in one of the following formats (buildingSMART®, 2016):

```
1   ISO-10303-21;
2   HEADER;FILE_DESCRIPTION(('ViewDefinition [CoordinationView_V2.0]'),'2;1');
3   FILE_NAME('C:\\TIM ExtraFiles\\IFC Export\\Scheiblhofer.ifc',
4   '2016-02-16T16:08:41',('Thomas Leopoldseder'),('Baumeister Stahlbetonteile LEHNER',
5   '3300 Amstetten, S\X2\00FC\X0\dlandstra\X2\00DF\X0\e 1, 07472/603-0, Fax DW22'
6   ),'The EXPRESS Data Manager Version 5.02.0110.b05 : 16 Jul 2014',
7   'Allplan 2016.1 02.02.2016 - 13:28:27','');
8   FILE_SCHEMA(('IFC2X3'));
9   ENDSEC;
10
11  DATA;
12  #1= IFCAXIS2PLACEMENT3D(#32,$,$);
13  #4= IFCOWNERHISTORY(#20,#23,$,.ADDED.,$,$,$,1455635004);
14  #5= IFCPROJECT('1D5PNksuX9HRlqIJtiqvWf',#4,'Scheiblhofer',$,$,$,$,(#11),
15  #39);
16  #11= IFCGEOMETRICREPRESENTATIONCONTEXT($,'Model',3,1.000000000000000E-5,
17  #1,$);
18  #14= IFCPERSON($,'Leopoldseder','Thomas',$,$,$,$,$);
19  #16= IFCORGANIZATION($,'Baumeister Stahlbetonteile LEHNER',$,$,$);
20  #20= IFCPERSONANDORGANIZATION(#14,#16,$);
21  #23= IFCAPPLICATION(#16,'2016.1','Allplan','Allplan');
22  #24= IFCSIUNIT(*,.LENGTHUNIT.,$,.METRE.);
23  #25= IFCSIUNIT(*,.LENGTHUNIT.,.MILLI.,.METRE.);
24  #26= IFCSIUNIT(*,.PLANEANGLEUNIT.,$,.RADIAN.);
25  #27= IFCMEASUREWITHUNIT(IFCPLANEANGLEMEASURE(0.0174532925199433),#26);
26  #28= IFCDIMENSIONALEXPONENTS(0,0,0,0,0,0,0);
27  #29= IFCCONVERSIONBASEDUNIT(#28,.PLANEANGLEUNIT.,'DEGREE',#27);
28  #30= IFCSIUNIT(*,.AREAUNIT.,$,.SQUARE_METRE.);
29  #31= IFCSIUNIT(*,.VOLUMEUNIT.,$,.CUBIC_METRE.);
30  #32= IFCCARTESIANPOINT((0.,0.,0.));
31  #34= IFCGEOMETRICREPRESENTATIONSUBCONTEXT('Body','Model',*,*,*,*,#11,$,
32  .MODEL_VIEW.,$);
33  #36= IFCGEOMETRICREPRESENTATIONSUBCONTEXT('Axis','Model',*,*,*,*,#11,$,
34  .MODEL_VIEW.,$);
35  #37= IFCGEOMETRICREPRESENTATIONSUBCONTEXT('Profile','Model',*,*,*,*,#11,
36  $,.MODEL_VIEW.,$);
37  #39= IFCUNITASSIGNMENT((#25,#29,#30,#31));
38  #41= IFCSITE('126fX40457bAwDE5rTh7PB',#4,'Liegenschaft',$,$,#46,$,$,
39  .ELEMENT.,$,$,$,$,$);
40  #46= IFCLOCALPLACEMENT($,#51);
41  #49= IFCCARTESIANPOINT((0.,0.,0.));
42  #51= IFCAXIS2PLACEMENT3D(#49,$,$);
43  #53= IFCBUILDING('2U9dbI1Ez8Yfvj39vKL09b',#4,'Geb\X2\00E4\X0\ude',$,$,
44  #54,$,$,.ELEMENT.,$,$,#64);
45  #54= IFCLOCALPLACEMENT(#46,#55);
46  #55= IFCAXIS2PLACEMENT3D(#56,#58,#60);
47  #56= IFCCARTESIANPOINT((0.,0.,0.));
48  #58= IFCDIRECTION((0.,0.,1.));
49  #60= IFCDIRECTION((1.,0.,0.));
```

Figure 5.6   Extract of an .ifc file.

- .ifc: This is the default exchange format (Figure 5.6) and is based on the STEP physical file structure according to ISO 10303 (2002)
- .ifcXML: This is the XML (Extensible Markup Language) version of the ifc-file (Figure 5.7) and normally 300 to 400% larger than an .ifc-file. It can be generated directly in a BIM application or can be converted from an .ifc-file.
- .ifcZIP: This is a compressed .ifc or .ifcXML file by using the PKzip 2.04g compression algorithm. It reduces the size of the file up to 95% in case of .ifcXML files and up to 80% in case of .ifc-files.

By creating one of those file types according to IFC or MVD standard as an output of an architectural CAD system, Precast CAD system, and so on, it is ensured that the BIM model software is able to read and visualize the information.

5.2.2.2   Data exchange technologies

In Open BIM different kind of software systems produce ifc-files for different purposes (e.g., architectural ifc-file, precast element ifc-file, sanitary installation ifc-file). These files have to be imported into the BIM model software to enable the advantages of an integrated BIM solution.

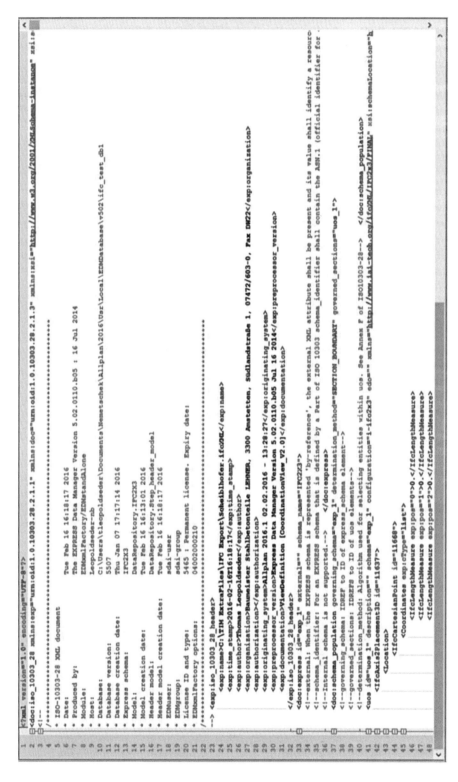

Figure 5.7   Extract of an .ifcXML file.

Files can be sent by email or transmitted by using shared file systems or web services. Web services are a very convenient and standardized way to exchange data between software systems connected by a network.

In the context of an Open BIM solution ifc-files can be transferred to stand-alone BIM model software systems or BIM cloud solutions, where all BIM participants use IT infrastructure provided by a storage solution provider. Another possibility of a cloud solution would be that a BIM solution provider offers BIM model software and data storage as a service (software as a service [SaaS]).

The transfer can be done automatically according to the status of the information (e.g., if the status of a precast drawing has been changed to "approved" the web service is able to transfer the drawing and its information automatically to the BIM model software). The advantage is an automatic way of data exchange without requiring user interactions. All participants of the BIM model software can get use of actual data and are able to check the data with their data and information.

5.2.3   BIM model software

As explained in Section 5.2.2 an Open BIM solution needs BIM model software which is able to integrate different ifc-files into one model. The BIM model software can be centrally installed on one server (e.g., in the office of the general contractor of the construction) or can be a cloud solution provided by a BIM cloud supplier. It will visualize the ifc-files in a 3D viewer with all detailed information given in the ifc-files, which is an easy way to check the model in many different ways. An example of conformity check between the architectural and precast model is shown in Figure 5.8.

The usage of BIM model software for an Open BIM solution has many advantages:

1. Clash detection: Finding of clashes between different trades, for example, Figure 5.9 (e.g., planning of cut-outs in a precast wall by the precast engineering department and the planning of water installation done by the technical department of the plumber).
2. Completeness check: Within the model missing items, wrong connections or duplicated parts can be easily found. Automatic checks can be done to identify problems.
3. Version control system: The organization of change management is one of the most challenging issues during a construction project. Renovation tasks produce a lot of new or changed information about the construction. BIM model software is able to manage the different versions and states of a construction in a digital way and makes sure that all participants have access to the latest version of data and information.
4. Data mining: A lot of data will be collected throughout the whole process. Only the BIM model software is able to provide up-to-date statistics and reports to all participants.

Section 5.3 describes, by means of a precast construction, some important steps to gain an Open BIM.

Figure 5.8 Conformity check between architectural model and precast model – visualized in "Solibri® model checker", courtesy of Solibri, FIN. (Courtesy of architect Thell and Josef Lehner e. U.)

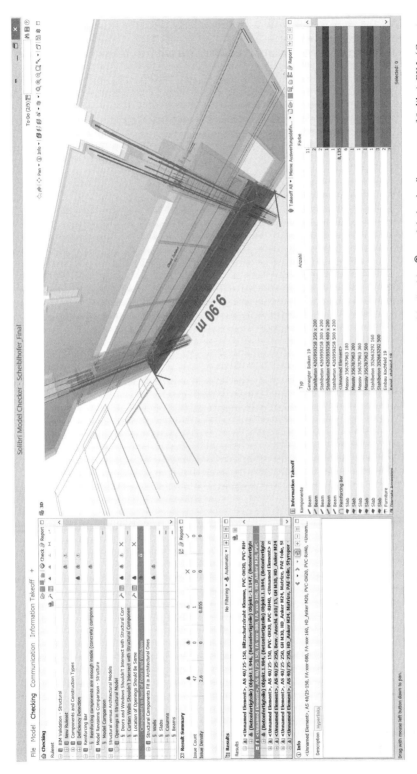

Figure 5.9 Clash detection showing that the beams are not in the correct position, visualized by "Solibri® model checker", courtesy of Solibri, FIN. (Courtesy of architect Thell and Josef Lehner e. U.)

5.3   BIM in precast construction

5.3.1   *Project pricing for precast structures based on 3D models*

Precast fabricators are usually invited by the general contractor to tender as suppliers for a pre-defined part of the entire construction project. The systematic determination of project scope, expenditures and effort is essential for calculation purposes as well as for submission of comprehensible tenders.

The application of project pricing based on 3D models encourages a systematic tendering process by:

- comprehensible and visualised tenders
- documentation of the offer's fundamentals
- bill of quantities

The initial process depends on the data and information provided by the contractor. Nowadays, the precast fabricator often receives 2D drawings only – when BIM is not applied as a method for the overall construction project. In this case the 3D model has to be created by the precast company. Contemporary CAD systems offer comprehensive options for importing 2D data. The 3D model should then be created according to the precast fabricator's product range and under consideration of the manufacturing facilities and equipment in an easy way by the sales person normally without profound technical knowledge.

The 3D precast model of the construction can be created with the following steps:

1. Import the architectural drawings (e.g., PDF, DXF, DWG, ifc) into a 3D precast CAD system.
2. Make use of predefined 3D precast objects (walls, floors, beams) to generate a 3D precast model of the full construction without detailed technical knowledge.
3. Create a 3D visualisation for the customer (e.g., 3D PDF; Figure 5.10(a)).
4. Calculate volumes and masses for each type of precast element in the 3D precast CAD and export those data to a calculation software system to generate quotations almost completely automatically (Figure 5.10(b)).

The result is the exact scope of supply and services provided by the precast fabricator, including a graphic representation of the offered components which lead to a better understanding for supplier and customer. All changes during the sales process will be documented and can be used internally and externally.

After receiving the order, the final 3D precast model, material information, quantities etc. can be exported to the BIM database by using the ifc-interface of the precast CAD system. In that case the BIM process starts immediately with the sales process and the BIM model can be used as a reference model for further technical engineering.

5.3.2   *Technical engineering*

The technical engineering is the main task accomplished in the Precast CAD, carried out with careful consideration of the production facilities, workflow and available machinery.

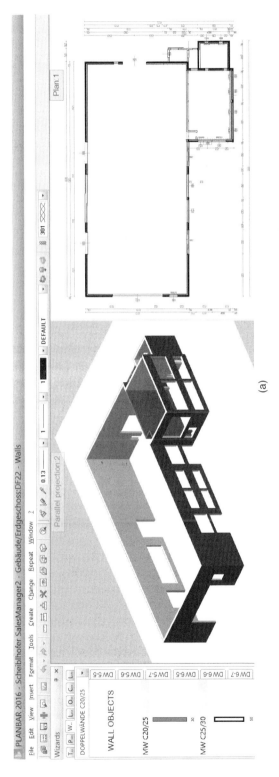

(a)

Figure 5.10 (a) Quick and easy creation of basic 3D wall model from 2D floor plan - generated by planbar® SalesManager. (Courtesy of Precast Software Engineering, AT; architectural data courtesy of architect Thell.); (b) Report of quantities to determine the billing areas, generated by planbar® SalesManager. (Courtesy of Precast Software Engineering, AT.)

Report Walls

planbar[2016]

Project:	SalesManager
User:	sschachinger
Date / Time:	02.03.2016 / 19:32
Notes:	

Type	Concrete grade	Thickness	Openings	m²
MW	C20/25	30cm		233,42
Billable area				215,71
Opening	<1 m²			
Opening	<2,5 m²		2	
Opening	>2,5 m²		4	-17,96
MW	C25/30	50cm		1684,58
Billable area				1305,54
Opening	<0,25 m²			
Opening	<1 m²		4	
Opening	<2,5 m²		3	
Opening	>2,5 m²		21	-379,04

(b)

Figure 5.10   (*Continued*)

Based on the Sales 3D model or architectural drawings the structure is broken down into producible precast elements of the desired type, such as various types of slabs, walls and structural precast elements as shown in Figure 5.11. Every single precast element is defined by its mark number and the attributes describing its features.

In the next step the connections of the precast elements among themselves and with parts of the construction that shall be cast *insitu* are determined. All required built-in parts are created; in many cases catalogues of 3D objects provided by suppliers of built-in parts can be imported to the Precast CAD and used during construction in a very easy way. Reinforcement is then designed according to structural requirements and in accordance with the production facilities (Figure 5.12).

At the end of the technical engineering process information about all elements (type of concrete, reinforcement, type and quantity of built-in parts, weight, dimensions, etc.) can be derived from the model not only for the production but also for mounting and accounting.

Prior to production it is important to check the precast model against the architectural model and other models such as the planning of air conditions, heating, and so on. The BIM model can be used to find problems and differences between all kinds of models. It is the task of the BIM manager to evaluate the models, examine

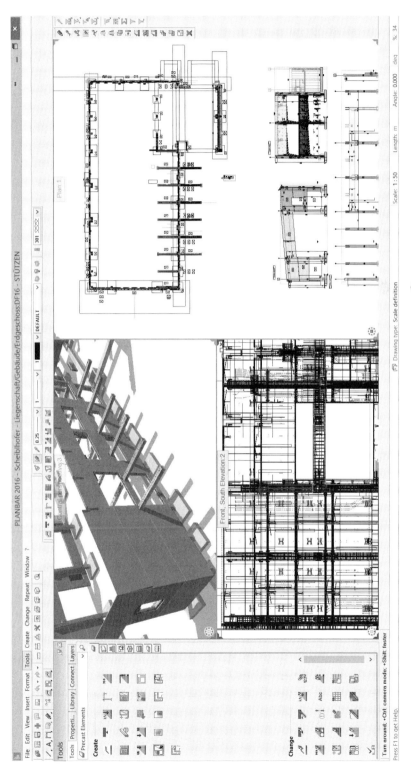

Figure 5.11 Part of the construction with detailed planned precast element – generated by planbar®. (Courtesy of Precast Software Engineering, AT and Josef Lehner e. U, AT.)

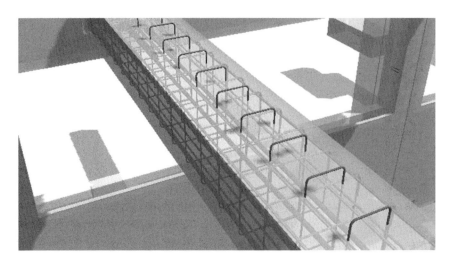

Figure 5.12   Beam with reinforcement – generated by planbar®. (Courtesy of Precast Software Engineering, AT and Josef Lehner e. U.)

for consistency and discuss changes with the different suppliers to make sure that all problems are fixed before production and construction commences. BIM model checker software is able to generate reports for different kinds of suppliers, for example, as shown in Figure 5.13.

While the model checker is able to identify problems in a model or between different models another check should be done by the technical department. By simulating the production of reinforcement it is possible to provide optimized producible reinforcement depending on the production possibilities, as shown in Figure 5.14. Changes can be done easily and will be provided to the BIM model. Based on that information production data can be created for reinforcement machines without errors.

5.3.3   Production data and status management

Based on the finalized 3D precast model production data has to be generated. According to the delivery and mounting sequence for walls and floors stacks have to be created. A stack is defined as a number of precast elements which will be produced in the same production cycle and delivered to the construction. This sequence is very important for the logistics of an automatic precast plant (for more details see Chapter 6). During stacking the dimensions of transport utilities will be checked to find an optimized combination with minimum quantity of transport devices. Stacking information is part of the production data for the master computer and will be added to the BIM model (Figure 5.15).

During the BIM process the status of the project, the stacks and the elements will change according to the progress. BIM model software is able to visualize this status in the 3D model to support engineers, project managers and construction companies in planning and organizing the construction (Figure 5.16).

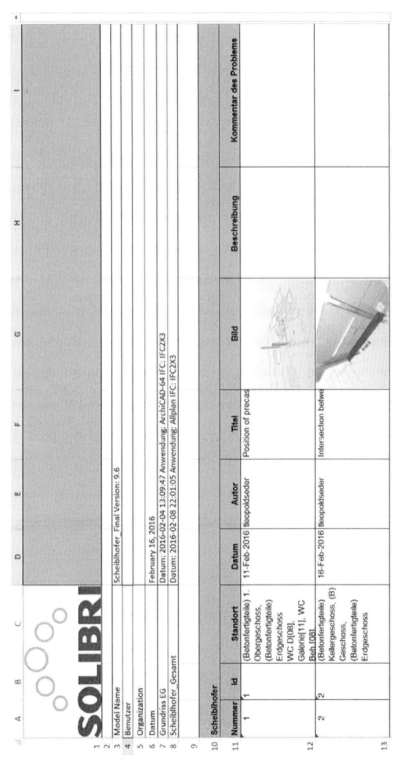

Figure 5.13 Example of a report containing some errors which have to be fixed, created by "Solibri® model checker". (Courtesy of Solibri.)

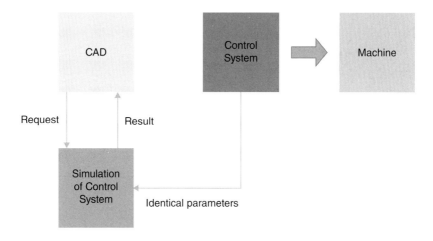

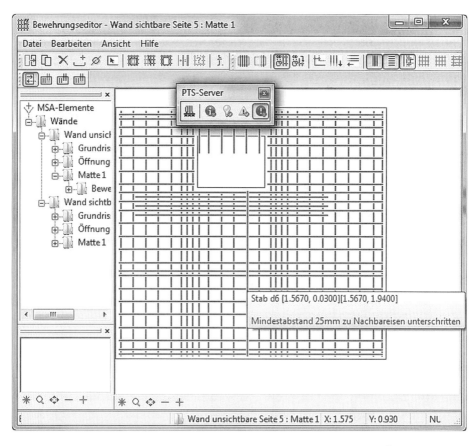

Figure 5.14   Schematic diagram of simulation and visualization of result in planbar®. (Courtesy of Precast Software Engineering, AT.)

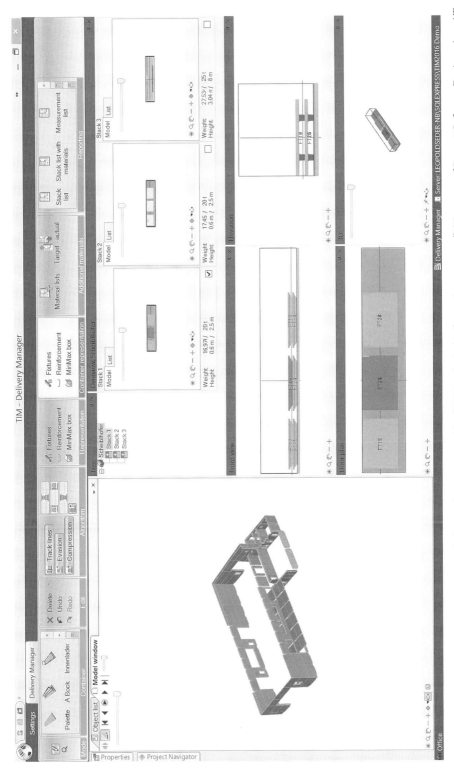

Figure 5.15 Stacked elements which are ready for production created by "TIM technical information manager". (Courtesy of Precast Software Engineering, AT and Josef Lehner e. U, AT.)

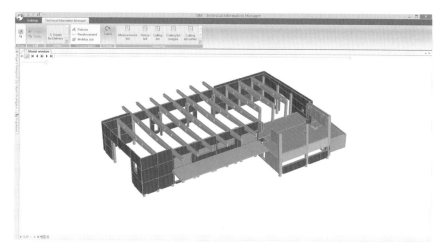

Figure 5.16   Visualisation of status of elements in the 3D model by "TIM technical information manager". (Courtesy of Precast Software Engineering, AT and Josef Lehner e. U, AT.)

Figure 5.17   Tracing and setting of status on tablet PC (mTIM mobile TIM solution). (Courtesy of Precast Software Engineering, AT.)

5.3.4   *Logistics, mounting, and quality management*

The production, stocking and delivery of precast elements has to be planned according to the mounting sequence. A BIM model allows the simulation of this mounting process to derive optimized delivery and production cycles for the precast elements as described earlier. Not only the planning team but also the mounting team should have mobile access to the BIM model by using smartphones or tablet PC (Figure 5.17). In that case the erection team is able to retrieve information about planning and production status and detailed and up-to-date drawings, times schedules, and so on. Since collisions have been checked in the model they will not occur during erection. Time of erection is reduced and problems can be minimized.

During the construction process data for the BIM model (e.g., delivery date and time, documentation of defects, data about mounting) can be collected and stored in the BIM database (Figure 5.18).

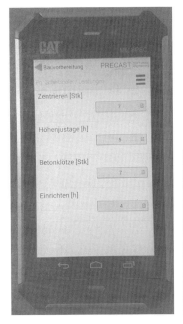

Figure 5.18   Working process on mobile devices (mTIM mobile solution). (Courtesy of Precast Software Engineering, AT.)

5.4   Summary

BIM is an integrated concept that leads to more transparency in the complex process of planning, erection and operation of constructions. Many countries define BIM as a mandatory technology for constructions in the public area expecting more efficiency, less costs and a higher adherence of scheduling. Many software vendors support BIM technologies and an increasing number of architects, construction companies, technical engineering offices and companies of the producing industry are working on implementing BIM strategies in their organizations.

But the lack of knowhow and the absence of legal principles and international standards complicate the introduction of BIM in companies and markets.

Nevertheless Building Information Modelling will sustainably change and improve the worldwide construction industry in the next few years and precast has to be part of it.

References

Autodesk. (2003). Building Information Modeling. Available at: http://www.laiserin.com/features/bim/autodesk_bim.pdf [accessed July 1, 2016].

BMVI. (2015). Building Information Modeling (BIM). Available at: http://www.bmvi.de/SharedDocs/DE/Pressemitteilungen/2015/152-dobrindt-stufenplan-bim.html [accessed July 1, 2016].

Building and Construction Authority (BCA). (2013). Singapore BIM Guide, Version 2, Building and Construction Authority, 5 Maxwell Road, #16-00 Tower Block MND Complex, Singapore 059110.

Building and Construction Authority (BCA). (2016). Singapore. Available at: http://www.bca .gov.sg/bim/bimlinks.html [accessed July 1, 2016].

buildingSMART. (2016). Available at: http://www.buildingsmart.org/standards/technical-vision [accessed July 1, 2016].

Cabinet Office. (2011). Government Construction Strategy, Cabinet Office, London, UK, Available at: https://www.gov.uk/government/uploads/system/uploads/attachment_data/file/ 61152/Government-Construction-Strategy_0.pdf [accessed July 1, 2016].

CIDB Malaysia. (2015a). Construction Industry Transformation Programme 2016–2020. doi:10.1007/s13398-014-0173-7.2.

CIDB Malaysia. (2015b). Workshop Report (Series 2) of Building Information Modelling (BIM) Roadmap for Malaysia's Construction Industry. Available at: http://www.cidb.gov.my/ cidbv4/images/pdf/announcement/BIM/roadmapreport2.pdf, [accessed September 2, 2016].

China BIM Union. (2016). Available at: http://www.chinabimunion.org/html/aboutUS/index .html, [accessed July 1, 2016].

Dodge Data & Analytics. (2015). SmartMarketReport – The Business Value of BIM in China, Dodge Data & Analytics.

Eastman *et al.* (1974). Eastman, C., Fisher, D., Lafue, G., Lividini, J., Stoker, D. and Yession, C., An Outline of the Building Description System; Institute of Physical Planning; Carnegie-Mellon University. Available at: http://eric.ed.gov/?id=ED113833 [accessed July 1, 2016].

Eastman *et al.* (2009). Eastman, C., Sacks, R., Panushev, I., Menon, M. and Aram, S., Precast Concrete BIM Standard Documents – Volume 1 – Model View Definitions, 2009. Available at: http://dcom.arch.gatech.edu/pcibim/documents/Precast_MVDs_v2.1_Volume_I.pdf [accessed July 1, 2016].

Economic Planning Unit Malaysia. (2015). Eleventh Malaysia Plan 2016-2020, Anchoring Growth on People.

European Parliament. (2014). Directive 2014/24/EU of the European Parliament and of the Council of 26. Feb. 2014 on public procurement and repealing Directive 2004/17/EC.

IFC4. (2013). Industry Foundation Classes, IFC4 Official Release, buildingSMART International Limited, 1996–2013. Available at: http://www.buildingsmart-tech.org/ifc/IFC4/final/ html [accessed July 1, 2016].

ISO 10303-21. (2002). Industrial automation systems and integration – Product data representation and exchange – Part 21: Implementation methods: Clear text encoding of the exchange structure, International Organization for Standardization, ISO Central Secretariat, CP 401, Geneva, Switzerland. Available at: http://www.iso.org/iso/home/store/ catalogue_tc/catalogue_detail.htm?csnumber=33713 [accessed July 1, 2016].

ISO 16739. (2013). Industry Foundation Classes (IFC) for data sharing in the construction and facility management industries.

Jernigang, F. E. (2008). Big BIM little BIM: the practical approach to building information modelling: integrated practices done the right way!, 2nd ed., 4 Site Press, Salisbury, MD; USA.

Josef Lehner e. U. (2012). Precast Company, Amstetten, Austria. Available at: http://www .lehner-beton.at.

NBIMS-US™. (2015). National BIM-Standard Fact Sheet published by National Institute of Building ciences and buildingSMART® alliance. Available at: https://www .nationalbimstandard.org/files/NBIMS-US_FactSheet_2015.pdf [accessed July 1, 2016].

NBS. (2016). NBS England. Available at: https://www.thenbs.com/knowledge/bim-levels-explained [accessed February 16, 2016].

National Institute of Building Science. (2007). National Institute of Building Science and buildingSMARTalliance; United States National Building Information Modelling Standard – Version 1 – Part 1: Overview, Principles, and Methodologies.

New York City. (2012). New York City Department of Design + Construction, BIM Guidelines, July 2012. Available at: http://www.nyc.gov/html/ddc/downloads/pdf/DDC_BIM_Guidelines.pdf [accessed July 1, 2016].

Salonen, A. (2015). The Finnish Way to Speed Up the Change, BIM European Summit 2015, Barcelona. Available at: http://rym.fi/wp-content/uploads/2015/03/BIM_PANTALLES_Salonen-2015-02-12.pdf [accessed July 1, 2016].

Scheiblhofer. (2016). Wine Yard Scheiblhofer, Andau, Austria. Available at: www.scheiblhofer.at [accessed July 1, 2016].

Thell. Architect Dipl.-Ing. Werner Thell, Frauenkirchen, Austria.

U.S. General Services Administration. (2003). National 3D-4D-BIM Program. Available at: http://www.gsa.gov/portal/content/105075 [accessed July 1, 2016].

Van Needervan, T., van Nederveen, G.A., and Tolman F. P. (1992). Modeling Multiple Views on Buildings, *Automation in Construction*, **1**(3), 215–224.

Zahrizan, Z., Ali, N. M., Haron, A. T., Marshall-Ponting, A. and Abd, Z. (2013). Exploring The Adoption of Building Information Modelling (BIM) in the Malaysian Construction Industry: A Qualitative Approach, pp. 384–395.

Chapter 6

Mechanisation and Automation in Concrete Production

Robert Neubauer

SAA Software Engineering GmbH, Austria

This chapter covers the methods of industrialization and automation for the concrete prefabrication industry. Starting with the development of these factories at a glance, it focuses on the CAD-CAM and building system requirements first, because they are essential for proper and successful automation. Realized automation solutions in different areas of the production process will be the major part of this section. Because parts of the so-called Industry 4.0 System, where machine control systems are linked to master computers and to each other for optimizing productivity, have been used and are necessary in these factories for many years, the IT integration of the prefabrication production process is also included. Finally the chapter highlights some limitations of automation in this production system.

6.1 Development of industrialization and automation in the concrete prefabrication industry

To understand the development of industrialization and mechanization of the concrete prefabrication industry, it could be meaningful to focus a little bit into the history of development within this sector. Some of the mentioned words and expressions will be explained later on in the following sub-chapters.

Besides having forming tables, fixed forms, moulds and concrete mixing plants that have been widely used in more or less all parts of Western and Eastern Europe, the major advancements in mechanization was developed in the 1970s in Germany and Austria by inventing the first idea of pallet circulations plants. Inventing one of the first flexible precast concrete construction systems to build housing (such as the Austrian company Mischek) it was necessary to efficiently produce concrete walls and

Modernisation, Mechanisation and Industrialisation of Concrete Structures, First Edition.
Edited by Kim S. Elliott and Zuhairi Abd. Hamid.
© 2017 John Wiley & Sons Ltd. Published 2017 by John Wiley & Sons Ltd.

Figure 6.1   An early pallet circulation system for precast concrete wall panels.

floors just in time, against a background of higher construction demand and cheaper production. First, flexible and mechanized shuttering systems and moving-table systems with curing chambers were developed for variant production of these kinds of walls. Until this moment there was nearly no real automation, except that the pallets were moved around by an electric control system connecting motors and sensors with concrete delivery, casting, demoulding and finishing work. Figure 6.1 shows a typical example of this type of work.

Even at that time were the first mechanization steps, such as a semi-automatic shuttering machine, to place the steel forms for up to two rectangular elements. This could be called the first shuttering robot, shown in Figure 6.2.

Because uniformed precast concrete elements used in eastern and central areas of Europe were growing more and more unpopular, partly because of low quality in terms of workmanship and durability, and their connections and an enhanced demand of architectural flexibility were unsolved, the development of modern construction systems by flexible production methods, industrialization and automation was starting in the centre of Europe.

Several different companies from Austria and Germany and the Technical University of Vienna joined forces to develop the first flexible automatic concrete prefabrication plant in 1987, which from the concept was outstanding and was leading minds into the future. There were several other methods being developing in parallel.

6.1.1   Stationary flexible forms, tables and formwork in a prefabrication plant

Stationary flexible forms and forming tables were the first pieces of industrializing the precast concrete prefabrication industry, as shown in Figure 6.3.

Having tilting tables and heatable steel tables, eventually equipped with a high-frequency compaction system and formwork mounted by magnets, it was for the first

Figure 6.2   First shuttering-machine within a pallet-rotation plant.

time possible to produce high-quality solid walls and floors in a predictable daily out-put of the prefabrication plant. Due to the tilting system, the demoulding of walls was possible without damage to the edges and surfaces, and the ground steel surface generated a perfect surface at least at one side of the concrete element.

Nowadays tilting tables are still in use in the industrialized prefabrication indus-try, especially as a startup plant or for production of heavy solid or sandwich walls. Combining multiples of these tables one after the other and having connecting table pieces to join two of these tables for one longer table, makes these tilting table plants very flexible. It is possible to support this production method by using cleaning and plotting machines, overhead concrete buckets, heating systems with curing monitor-ing systems and to integrate these production machines with reinforcement prepa-ration or mesh welding systems by a usual MES – Manufacturing Execution System, with work preparation functionality. These machines and master computer systems will be mentioned later in more detail to understand their functionality. Tilting table plants could also be easily supported by laser projector systems to display the con-tours, cut outs and position of precast embedments onto the steel tables, as shown in Figure 6.4.

It is necessary to mention that in regions where the double-wall system is very well known and used – a concrete wall consisting of two thin shells of concrete, connected by truss girders filled with concrete on the construction site – a special system of tilting tables is used. It is usually called the "flapping pallet" (Figure 6.5) and consists of two tables to congruently produce the first and second shells of these elements and put them together like big toaster. For more detailed information about this concrete see Chapter 5 and Section 6.2.2.1.

Figure 6.3   Old tilting table.

Of course there are some other production tables used on the world market, but many are outside the scope of this chapter which focuses on mechanization and automation.

6.1.2   Long-bed production

Because flooring systems like the hollow core floor (Figure 3.12), double-tee units (Figure 3.13) and filigree floor (Figure 3.14) were getting increasingly popular between 1965 and 1985, there was the need to enhance production of these prefabricated elements.

Since acceptable transportation measures limits the width of these semi-floor plates to 2.4 m or 3.0 m (depending on country, 95% of automatic plants are 2.5 m), long-bed

Figure 6.4   Modern tilting table plant.

Figure 6.5   Flapping pallet for double-wall production.

Figure 6.6   Floor production on long bed with cross shuttering.

production lines were built to satisfy these measures. Usually in one production hall there would be 4 to 15 steel plated lines mounted on the floor, with a length of 50 to 150 m, occasionally 200 m in the Middle East. The ground steel long beds are usually equipped with two fixed side forms of the necessary height, between 60 and 100 mm, having a demoulding angle of 3–5°. With this configuration the producer could reduce the formwork effort. In the length direction the slabs are separated from each other by steel or plastic cross shuttering, supporting bent or straight reinforcement bars, as shown in Figures 6.6 and 6.7.

To enable longer span or shallower elements, prestressing technology was used instead of static reinforcement. This was easily integrated into long-bed production by just having a suitable foundation, stressing heads/anchorages and the prestressing equipment, for example as shown in Figure 6.8. Cross shutters were slightly adapted

Figure 6.7   Prestressed floor production on long bed with plotting device.

Figure 6.8   Hollow core slab production (using Elematic extrusion method).

or replaced by polystyrene shutters. It was also possible to mix the production of prestressed and non-prestressed elements within one production hall, of course on different long bed lines. Equipping these lines with automatic cleaning, plotting and cross shuttering machines, as well as automatic machines for pouring concrete and cutting the prestressing strands, was leading this kind of precast concrete element production into the stage of industrial and automatic production cycle.

Last but not least, it is necessary to mention the production of hollow core slabs, the most widely known precast concrete element all over the world. This is produced on long beds too, but due to the construction system the width of the production lines is usually 1.2 m, occasionally 2.4 m. Because of the use of dry-cast concrete there is no need to have side forms, but prestressing equipment is an integral part of this production concept. After slip-formers were used at the beginning of this technology, extruders where invented in the 1980s to have a more controlled casting process. Automatic cutting devices as well as plotters and aspirator systems for voids and cut-outs were step by step automating this system of production. The following sections describe the automation tools for pallet and flooring production.

6.1.3   Pallet circulation plant

To have really industrialized and optimized the production of prefabricated concrete elements, it seemed feasible to have borrowed the fabrication methods of the automotive industry for this branch too. After the first manual plants with moving forming-tables in the 1980s the first precast concrete floor plant was created in Germany. In this plant the manual work stages could be done at certain fixed stations where all the palettes were passing by one after the other, that is, the work moves to the worker rather than vice versa. Tools, precast embedments (cast-in channels, sockets, boxes) and other necessary resources are located at the workstation where they are need. In between the stations where some manual work takes place, a few automatic machines assured the quality and precision of the final product.

Key features of this plant, summarised in Figure 6.9, are:

- Pallet circulation plant, with pallet size of 11.0 m × 2.5 m and fixed side forms on both sides of the pallet, including deforming angle.
- Manual and automatic transport of the pallets from station to station
- Plotting machine with manual operation, very soon later called MRP Machine – a machine to clean the pallet, plot all necessary outlines and place the cross shutters onto the certain positions.
- Workstation for manual formwork, mounting of precast embeds
- Reinforcement straitening, cutting and positioning for spacers, cross reinforcement, length reinforcement and the necessary truss girders for the so called "filigree floor"
- Workstations for connecting and finishing the elements content
- Concrete spreader
- Curing chamber with automatic stacking crane
- Workstation with demoulding crane to lift of and stack the hardened elements
- Run-off truck to transport the stacks to the stock yard

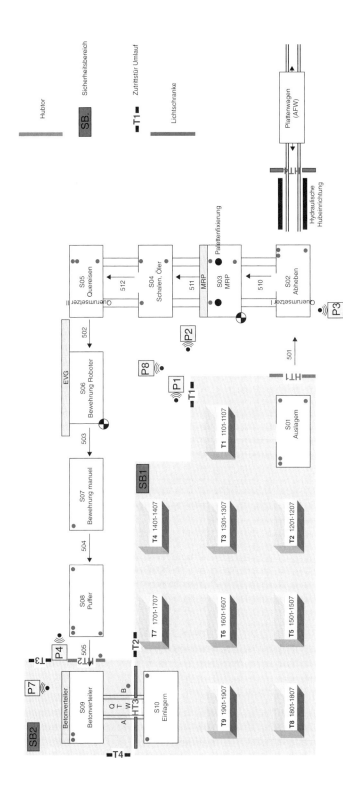

Figure 6.9   Part of the first circulation plant with semi-automatic plotting machine.

Each pallet, which starts its way into production, receives different data with reference to the elements nested onto this pallet. Work preparation and nesting as well as production planning was and is done by manufacturing execution systems called the "master computer" in this area of business. Doing so it was the first time possible to produce different element shapes in lot sizes of one, extracted directly from the architectural drawing. Very soon after this first industralization new machines were integrated to do more and more automation. In later sections these machines will be described more in detail.

The production of wall systems made it necessary to develop new staged production systems. For the double-wall system the curing chamber (Figure 6.10) was changed to a flexible and random accessible system, supported by a rack crane and sectional doors to make heating of the system more energy efficient. Element or pallet turning devices (Figures 6.11 and 6.12) eased the production of double walls to place the first shell easily and with high precession into the second production shell. The new multi-functional concrete prefabrication plant was invented (Figure 6.13).

Within the following years, the high demand of prefabricated concrete elements lead to the development of the circulations plant systems delivered by different manufacturers in central Europe to more and more complex production systems, having multiple production lines for different complexity and automatic production machines within the process, as shown in Figure 6.14.

The development of solid walls, sandwich walls, multi-layered concrete elements and special parts is still driving the industry forward to accomplish the needs of industrial production.

Figure 6.10   Curing chamber with rack crane.

Figure 6.11   MFA with pallet-turning device.

Figure 6.12   Vacuum element-turning device.

Figure 6.13   MRP machine.

Figure 6.14   Highly automated plant in 2013.

6.1.4   *CAD-CAM: the path to automation*

In 1987 the first complete CAD-CAM driven concrete prefabrication plant was installed in Germany (a project of several German companies delivering machinery, an Austrian company for CAD system, CAM system and control systems together

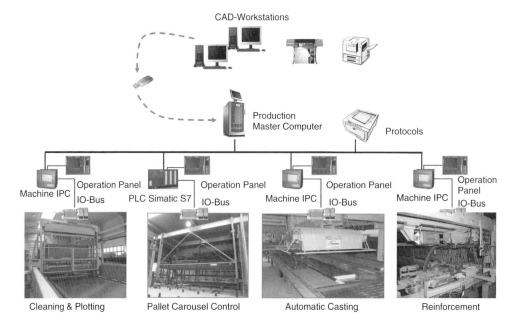

Figure 6.15   Schematic structure of Industry 3.0 integrated CAD-CAM structure.

with the Technical University of Vienna). In this plant a complete first CAD-CAM system for the prefabrication concrete industry was installed and operational as shown in Figure 6.15 as follows:

- CAD system to support the elementation for precast floors, reinforcement calculation, organization of transport stacks from the CAD's floor plan.
- Master Computer System (MES) to plan and organize the production of prefabricated concrete-elements onto production pallet, NC Data Handling for production machines, interfacing the production machines, visualization.
- Control system for a cleaning, plotting and shuttering machine with interface to MES.
- Control system for reinforcement straightening, cutting, bending and placing system with robotic arms with interface to MES.
- Control system for truss girder welding and cutting machine with interface to MES.
- Control system for automatic spreading of concrete into the forms with interface to MES.
- Control system for pallet rotation plant and curing chamber with interface to MES.
- Control system for batching plant with interface to MES.

Within the following years the master computer systems on the market were extended more and more to support the work preparation system by optimized pallet nesting systems, presentation and integration of information and intelligent NC Data Generators for flexible production of precast elements directly from the CAD Information. Furthermore, logistic functions, where supported by MES System and network connections, leads us to a so-called Industry 4.0 System where machine control systems are linked to master computers and to each other for optimizing productivity. Because these prefabrication concrete production systems are today completely integrated with the ERP (Enterprise Resource Planning) and BIM (Building Information Modelling) Structure this can be referred to as a real Computer Integrated Manufacturing System as shown in Figures 6.16 and 6.17.

Nowadays, due to utilization of industrial networks all parts of the production are connected and can exchange data for:

- Production planning and delivery – so called "Just In Time" (JIT) Production
- Geometrical CAD Data and NC Data for several production machines
- Productivity values including errors, alarms and maintenance messages as well as OEE – Overall Equipment Efficiency
- Graphical Data for workshop display and visualization
- Industrial I/Os – Inputs/Outputs of the control system, like sensors, actors, activation and operation

Because of open communications structures we are today very close to an implementation of Industry 4.0 in this branch. Of course some lower developed markets still use Industry 2.0 or 3.0 Systems to support their production process.

The following sections discuss software, hardware, production machines and products for modern mechanized, industrialized and the automated precast concrete industry.

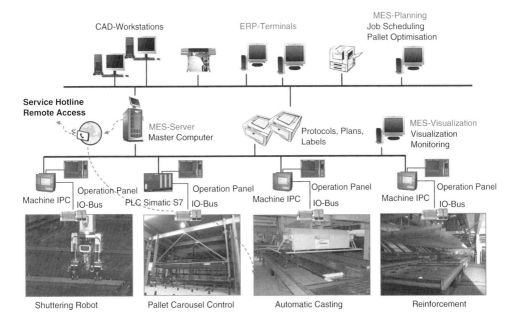

Figure 6.16   Modern integrated CIM precast production.

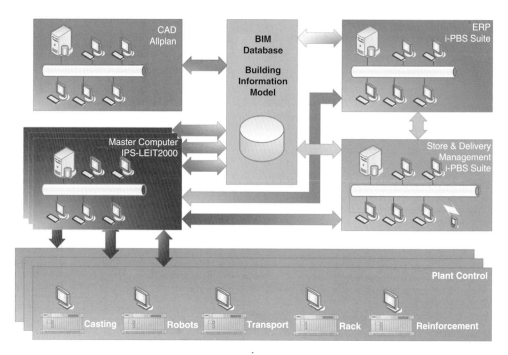

Figure 6.17 BIM Integration for automatic prefabrication concrete plants.

6.2 CAD-CAM BIM from Industry 2.0 to 4.0

According to the detailed discussions of Chapter 5 about CAD and BIM, this chapter focuses on the special needs and requirements to use this information for automatic or at least industrialized production of prefabricated elements. The majority of the pre-fabrication industry is presently situated, according to the illustration in Figure 6.18, between Industry 2.0 and 3.0, whilst the aim of new and major players lies between Industry 3.0 and 4.0. Of course the following needs are not only feasible for construction elements out of concrete but might also apply to other materials too.

CAM (Computer Aided Manufacturing) is a technology to transfer numerical data of a CAD Drawing to certain production machines and cells. The main goal is, due to the technical work in the CAD-system, the production process is automatically controlled by this information, with less user-interaction as possible.

6.2.1 Production of non-variable parts versus production in lot size one

The prefabrication concrete industry is very well developed for concrete elements with fixed geometrical form and functionality. Metal or wooden forms are built up according to the paper drawings of the consultant or in-house production engineer and the forms are used for the production of multiple repetitions of the element until the form must be refurbished.

Figure 6.18   Industry 2.0 to 4.0 (Courtesy RIB AG).

This method is still in use for batch production of prefabrication parts but usually there is no connection between the CAD system or the BIM Model and the production machines, except using paper drawings. All necessary work is done manually, automation is not possible.

Because building requirements were rapidly changing in the past, even for precast concrete buildings to be more beautiful and more flexible, the requirement of CAD-CAM systems raised to hand over the numerical data of an element's CAD drawing to several machines for production.

For a CAD-CAM systems using the Industry 2.0-method of stand-alone automated machines, it is sufficient that the CAD system is forcing the construction engineer to meet several designing rules and to draw objects, like walls, floors, reinforcement rods, precast embedments, and so on, rather than only drawing lines in the CAD system. Still until today, this is the first, very important step to consider, before facing an industrial and automated production. Very often there is the misunderstanding that a line drawing in AutoCAD could still be sufficient. An AutoCAD drawing meeting a certain kind of layered structure exported into DXF or HPGL Format could be the first step, as shown in Figure 6.19, for example to control a big plotting devices (DXF is the Data Exchange Format of AutoCAD and HPGL is HP Graphics Language to control plotters).

To enhance the production of precast concrete elements special Prefabrication CAD system are implementing object-oriented CAD models to meet the Industry 3.0 requirements. In the first stages, the architectural drawing must be manually transferred into the CAD system by the consultant or in-house production engineer, as shown in Figure 6.20. Nowadays, it is possible to import standard drawing formats like DWG, DXF and many others. After this task is finished, floors and walls as well as the connections between them can be virtually erected and designed in usually 2.5D or 3D level by level. The slab thickness as well as the necessary reinforcement according to the structural design values is automatically calculated. This leads to virtual floor and wall objects containing reinforcement rods, meshes, stirrups and connection parts. Furthermore it is possible to implement different precast embedments like tubes, connector boxes into the walls and floors.

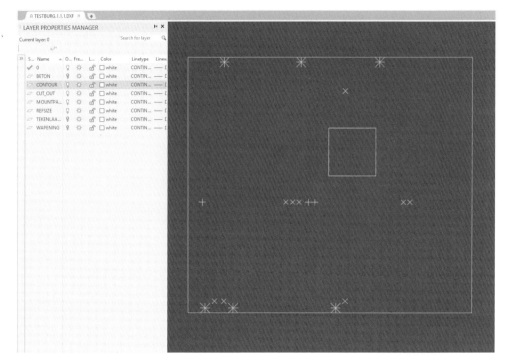

Figure 6.19 DXF drawing with layer structure suitable for primitive CAM startup.

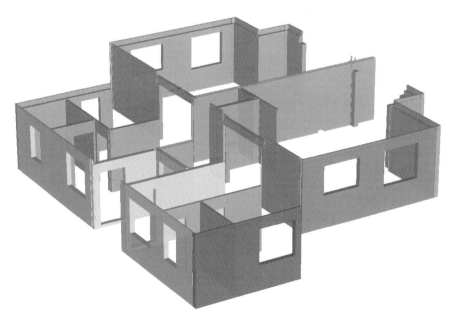

Figure 6.20 Modern virtual 3D-precast elements for automation. (Courtesy Precast Software Engineering GmbH.)

Having finished the object structure of the walls, floors and other prefabricated elements, the program can generate workshop drawings for these elements and technical production data could be exported in certain specialized formats. These could be transferred either directly to the intelligent machine – for slab by slab production – or to a production master computer which is doing the detailed calculation for the numerical data. Details about that will be explained in Section 6.2.4 about MES Systems.

Usual standardized and well-known formats are:

- IF-GENInterface
- CAD-CAMInterface Unitechnik (several versions)
- PXML
- UNIXML

To meet further requirements of Industry 4.0, the CAD systems must be based on real BIM Structures and allow the transfer of information from CAD to CAM and back from CAM to CAD as mentioned in Chapter 5. For further needs in the BIM Process the following data such as concrete receipt, quality data for concrete charges and reinforcement, production schedule as well as documentation data for the quality of the element could be passed back into the model. Unfortunately, the above-mentioned interfaces are not anymore suitable and modern, perhaps IFC-based information is required. This kind of bi-directional integrated CAD-CAM communication will lead this sector of industry to an Industry 4.0 world. Anyway a minimum of CAD-CAM is required to enable automation and industrialization of prefabricated concrete industry.

6.2.2   IBS – suitable prefabricated products for mechanization and automation

Intelligent Building Systems (not the same as *IBS* in Chapters 2 and 3) are a precondition for a modern, mechanized or automated prefabricated concrete industry.

Element-to-element connections, floor-to-wall connections, reinforcement and parts to be cast in, as well as transport and mounting devices for the construction site, must ideally meet the requirements of industrial and automatic production. Thus several prefabricated concrete construction systems were established on the market.

6.2.2.1   Intelligent building systems – the base for automation

Because most of these suitable Buildings Systems are covered in principle in Chapters 1 and 2, here is a list of systems widely used for automatic production, mentioning some important limits, which should be considered for automation:

- Precast reinforced floor slab: are easy for mechanization and automation in different factory types, and done on long bed as well as on pallet rotation plants according to productivity needs. Reinforcement is possible with automatic systems, reinforcement overlap must meet the shuttering system for automation, bridged girders are usual. If the form of the floor slab is curved or too complex it will need manual work in the factory.

- Prestressed floor slab: are mostly produced on long bed plants because of the necessary prestressing equipment but some are available in pallet rotation plants. Formwork in the cross direction to the prestressing strands is difficult for automation, and it is necessary to limit the prestressing patterns, typically to between 4 and 8 per unit depth. There are several automation methods for prestressed systems available.
- Double wall: are easy for mechanization, could be produced in the same plants and with the same methods like the precast reinforced floor slab.
- Hollow core: are easy for mechanization and automation, but only possible on long-bed production systems.
- Solid walls: if the thickness requires multi-reinforcement layers, reinforcement preparation will not be completely automatic or machinery for automation will be cost intensive. Shuttering system could be complex depending on the connection systems between the elements, thus the system and the different wall thickness must be standardized as possible to enable efficient automation. It must be considered that in production precast embedments will be much more difficult to mount and there is the need to smoothen one side of the element if produced on tables.
- Solid floors: more or less the same as for solid walls, in comparison with pre-floor systems all necessary pipe and electric installation must be completely done in the prefabrication factory and is usually manual work within automatic machines.
- Stairs: no real automation possible, except that industrialization with special, flexible (adjustable waist, rise and going) forms are possible, straight stair flights are sometimes produced on pallet rotation systems. The reinforcement cages can be pre-produced in automatic mesh welding plants.
- Beams and columns: mainly produced in fixed forms usually with prestressing equipment for most beams and some columns. There were some trials of mechanization and automation, but until now these trials were not successful.
- Sandwich walls: if the wall design is to have suitable and easy wall connection systems, limiting the different possibilities, this product is perfectly supported by automatic production systems. All other conditions for solid walls and floors – wall thickness, surface smoothening, effort for mounting precast embedments – have to be considered too.
- 3D-Prefabrication (volumetric) elements: complex prefabricated elements out of concrete are usually difficult to automate. Industrialization is possible for preparation of reinforcement and the production in fixed forms, moving around a circulation plant. The fixed forms must be manually prepared by timber or metal and usually will be reused several times, with tiny adaptations. If automation is requested, perfect prefabrication CAD data, including detailed reinforcement information and layered information for concrete volume, is required.
- Different concrete for different products: in general, the possibility to automate any kind of prefabrication product depends on complete object oriented CAD-Date in certain formats. Anyway, the detailed information about necessary concrete must be contained. Usually these data consist of:

  "concrete grade", for example, C25/30 specifying the type of concrete (e.g., LC for lightweight concrete), minimum compressive strength and many other values for cylinders and cubes

"exposure class", for example, XC2 specifying the resistance against environmental influences information about colour and several other concrete data
to be considered when mixing the concrete in automatic batching plants.

6.2.2.2   Design – requirements to support automation

As mentioned above, a suitable design of a concrete building structure is essential
for industrialization and automation of prefabricated concrete elements. Although
this chapter is about automation, it is very important to keep an eye on this topic
too, because professional design is highly important for a functioning and reliable
automation.

First of all it is necessary to use a professional CAD system equipped with precast
concrete extensions, such as mentioned in detail in Chapter 5. If the prefabrication
engineer must consider all requirements manually when drawing the element in a
standard CAD Software, this is very complex and will finally lead to data errors which
will hinder or stop the production later on.

Modern concrete structures are usually not matching to standard elements out of a
batch production, so each panel will be a little bit different from the other, in content
and structure. Thus, a special support for all necessary design requirements from the
CAD Software is very important.

Furthermore, these design requirements are sometimes in conflict with usual or
certified construction methods at the target market. To be cost effective and finally
competitive with the prefabricated concrete structure, it is sometimes necessary to
change some paradigms of construction or at least adapt used forms and method
appropriate to efficient precast structures. The necessary steps could be supported by
professional consultants, having experience in different markets.

Besides the geometrical precision and the complete volumetric content, the following requirements should at least be considered:

- Panelization

The suitable panelization of prefabricated concrete structures is a very important
design requirement for industrialization and automation. It must meet the possibilities of production tables or forms to finally fit into the production facility
including the necessary formwork. Very often this factor limits the exchangeability
between different productions because it was not taken in account. If the length of
a single prefabricated element is ideally designed to the size of pallets and forms,
it will be much more better for the productivity of the plants and installation to
ideally combine on forms, for example, if average wall lengths are suitable to be
combined onto one production table, the utilization of the production pallet would
be optimized and the productivity of the plant will increase.

That is why the prefabrication construction designer is a very important factor for
optimized, cost-effective and successful prefabricated concrete production.

- Side forms

The side forms of concrete elements influences the effort of formwork in the production very much. If it is possible to design straight wall contours rather than complex

Figure 6.21   Characteristic of edges for double wall and sandwich wall.

edge structures, as shown in Figure 6.21, this is essential for automation in prefabrication plants.

Comparing for example the edges of a double wall against the complex side shapes of a sandwich wall, this will influence the necessary side forms within the plant. To be effective and fast in shuttering work, the side form should be as simple as possible and therefore the designer must try to simplify the edge contours as much as possible. If it is not possible to use easy structures, like in the production of sandwich panels (Figure 6.22), the amount of different formwork heights as well as the forms/structures of the edges should be under control and limited. Usually this is one of the major tasks to be cleared between the architect, the structural engineer and the prefabrication designer. Not considering this topic during the design process will lead to expensive prefabrication construction, where automation is difficult, costly or sometimes impossible.

If limiting the flexibility of wall structures and thickness, it is possible to design a perfect formwork system to be used in automatic plants. The same applies to the forms for beams and columns, but as the automation possibilities in general are very low for these products, it is only influencing the cost of additional formwork.

Overlaying connection reinforcement would require special formwork with holes to enable the connection rods. There are still new developments for new formwork systems to enable overlaying reinforcement, but it would be a better idea to use a prefabrication connection system instead of this connection system if it is realizable in the target market for these elements.

- Reinforcement

Besides the connection reinforcement between the elements, the rest of the reinforcement has to be considered to enable successful automation. Single layered reinforcement like in semi-prefabricated floors, double walls and thin solid walls is much easier to handle than multi-layered reinforcement. If complex reinforcement is necessary, it should be designed as flat or bent mesh if possible.

Thanks to modern reinforcement plants for online production of bent rods and meshes a usual complex cage could be designed using lower and upper mesh, with or without bending, connection cages and if necessary additional rods for structural

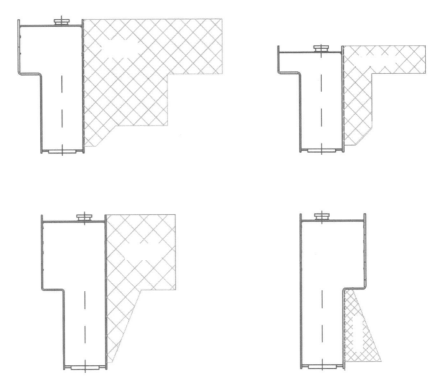

Figure 6.22   Flexible formwork for sandwich walls.

requirements. Doing so, the meshes could be placed into the forms by automatic handling, interrupted by manual mounting process for connection cages and/or additional structural rods.

Furthermore, it is important, to consider geometric collision between, rods, meshes, truss girders and precast mounting parts within the precast panel. Usually the prefabrication CAD system should support this task, but of course the precast designer should also care about it. Reinforcement will have to change, for example, diameter, spacing, additional bars, when mounting parts are built in or shifted.

Mesh welding plant for tailor-made mesh (Figure 6.23) as well as cut and bending machines will have certain limitations, like the diameter of the bending roller, a minimum distance between rods, a grid dimension, minimum rod length and many others, which must be met by the precast designer to enable reliable automatic production. This control mechanism is only done by certain modules of the precast-CAD software, which usually calculates the mesh on demand and reacts immediately if the designer has to change something. All manual methods will not be successful for automation.

Reinforcement design for prestressed elements should fit to reinforcement patterns and possibilities of the stressing devices in the production plant, there will be limits for the grid and forces.

The more the reinforcement will be designed for online automatic production the higher and better will be the productivity of the industrial prefabrication plant. There will, however, be structural needs to support the reinforcement manufacturing by the plants staff. If the reinforcement is too complex, pre-production of these cages

Figure 6.23   Bent, tailor-made mesh from the automatic mesh welding plant.

is necessary, perhaps a shift in advance in order to keep the flow running during processing the prefabricated element.

- Connectors

As two special mounting parts, the following are examples for the professional connection of precast elements (Figures 6.24 and 6.25). Doing this with overlaying reinforcement rods will spoil the effectiveness of automation and industrial production, because the reinforcement and the necessary formwork will be too complex. For example, it should be possible to use connection devices to connect walls and floors to each other. These tools could easily be mounted onto the formwork and only must be supported by additional reinforcement rods within the base reinforcement.

Another example is the connection of precast elements without having thermal bridges, or for fire insulation. There are many suppliers supporting precast construction with these structural mounting parts. Nevertheless, it is necessary to include them into the designing process unless they have to be considered in the reinforcement, formwork and mounting process.

- Precast mounting parts

Because it is very difficult to integrate parts into a hardened concrete wall, all necessary mounting parts must be included in the precast design right from the beginning. It would go too far to mention all the necessary and possible precast embedments within this chapter. There are many suppliers having specialized parts for openings (doors, windows, other openings), electricity, piping, transport and mountings, for example a lamp mounting shown in Figure 6.26. All these pieces must be integrated into the construction catalog of the prefabrication CAD system in order to be transferred to the automatic production in the correct form.

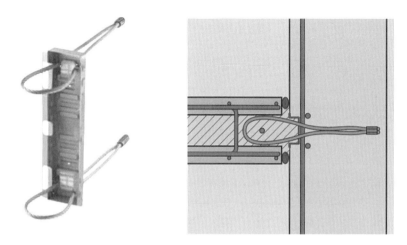

Figure 6.24   Pfeifer FS box wall connector system. (Courtesy Pfeifer Seil- und Hebetechnik GmbH.)

Figure 6.25   Schöck Isokorb® Type KXT as an example for thermo-insulating structural connection. (Courtesy Schöck Bauteile GmbH.)

Figure 6.26   Kaiser HaloX P for lamp mounting including transformer tunnel.

- Concrete type

Concrete type must be considered in the design system for automation. Not every concrete type is suitable for every machine in an automated production; for example, highly flowable concrete, such as self-compacting concrete, may not be suitable to certain types of automated plant, and does not require vibration, finishing.

The concrete bucket or concrete spreader should be able to spread the type of concrete into the form as well as the concrete mixer should be able to mix the appropriated quality and quantity. There will be some more details about in sections concerning the casting machines.

Depending on the type of prefabricated concrete element there might be some other requirements, which are usually mentioned and taken into account during the planning of an automation system. If the requested elements for a designed production are not industrialized at the moment of planning, it might be a good idea to consider professional consultants helping to adapt or redesign the pieces for a successful automatic production.

6.2.3   Just-in-time planning and production using ERP systems

Many of the concrete prefabrication plants are extremely depending on the progress and timing at the construction site, because prefabricated concrete elements shows their strong points, integrated into a fast, "Just-In-Time" construction system. That is why the production of precast concrete elements usually is directly connected to the transport and the erection process at the construction site.

Having a given delivery date and erection sequence, it is necessary to pre-plan the production in the factory in a detailed manner, sometimes including the whole prefabrication design process too. Because the precast elements are usually cured into the factory to their strength necessary for erection, it is possible to reduce the number of waiting elements in the stockyard.

The following steps and milestones, as summarised in Figure 6.27, are typically for a JIT planning and production in an automated precast plant:

- Sales: CRM (Customer relation management), prefabrication calculation, offer, price list, long-term load planning
- Utilization planning: Construction planning and time scheduling, capacity of plants, shift planning, load balancing between plants
- Engineering: Integrating JIT design for the precast designer, working lists, article matching between CAD and ERP System (article catalogue), purchasing of parts for the production, production dates and delivery, technical data collection
- Production: time and form scheduling for precast elements, production requests for automatic productions, supervision of production progress and delivery dates, production staff
- Inventory: intermediate storage of precast elements, monitoring automatic stockyard systems
- Shipping: Pick Pack Ship, truck monitoring, supplementary material, additional on site reinforcement, construction site accessories, delivery notes
- Mounting: supervision of the mounting process at the construction site, construction staff, building construction machines, product call-up for JIT Integration

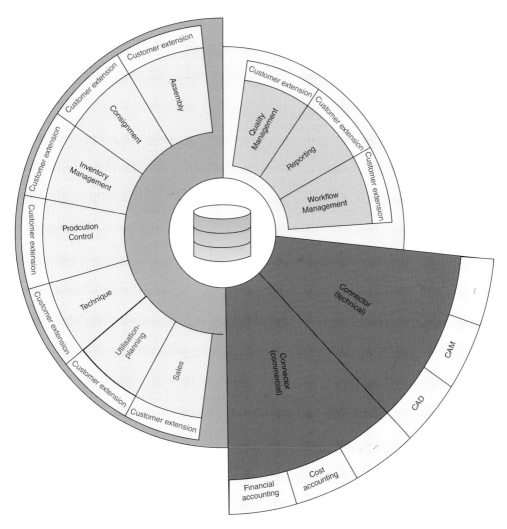

Figure 6.27   i-PBS enterprise suite, integration of data.

In former times, these tasks and steps were all monitored using list and eventually spread sheet programs. Today, having a very high productivity and output from the automated prefabrication plants, and being more and more forced by the market to a Just-In-Time planning, production and delivery, these tasks are usually performed using a specialized prefabrication ERP system.

Although there is a wide range of ERP Systems on the world market, they are usually not very well prepared for this kind of planning, production and delivery of construction elements in lot size one. If using standardized elements for a batch production these systems might work equally well.

Thus several specialized software systems have been developed and launched since the 1990s to meet these requirements and handled with the special data used in this branch. Examples are shown in the screenshots in Figures 6.28 and 6.29. Most of these Prefabrication ERP Systems are supporting the whole, above-mentioned

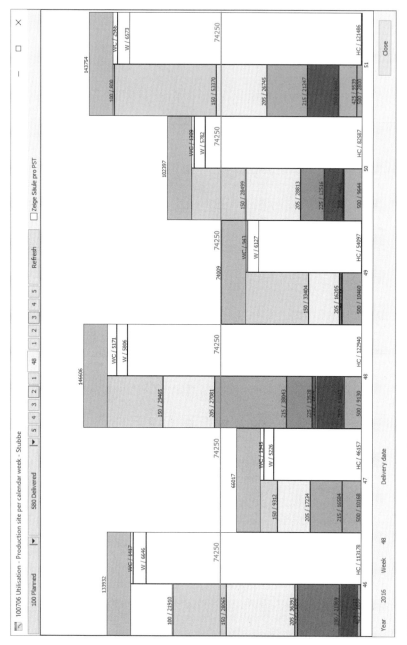

Figure 6.28 i-PBS enterprise suite, utilization planning overview.

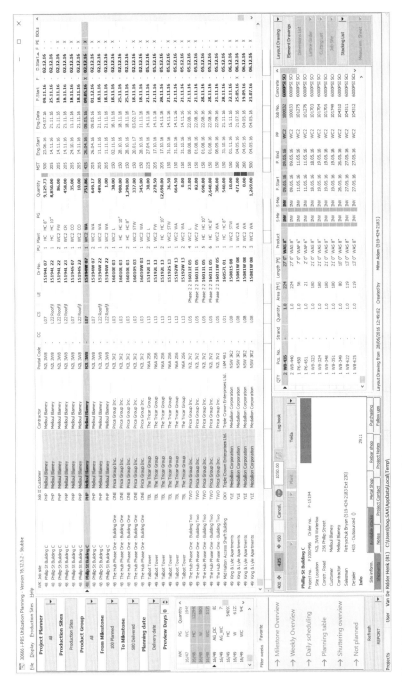

Figure 6.29 i-PBS enterprise suite, engineering modules.

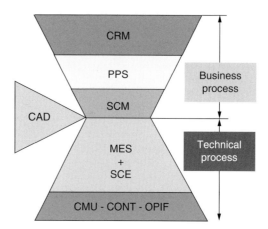

Figure 6.30   SAA-automation pyramid with CAD and ERP details.

process, including online interfaces to CAD systems, BIM Systems and MES Systems, leading to a totally integrated planning and production system. Bookkeeping as well as merchandise planning and control management is usually left beside and other professional ERP Systems are connected via interfaces for this reason.

This kind information management system is today essential for a successful JIT Production of prefabricated elements, especially when mixing several different prefabricated products to be integrated into one building at the construction site. This program is used by many different employees within the construction or prefabrication company and must be equipped with an effective client/server system to be operated from fixed workstations as well as from mobile ones. Working on the same BIM Data as mentioned in Chapter 5, a Prefabrication ERP System is also an important connection system between the BIM Partners. In some applications it is the operator of a Closed BIM System.

Therefore, for modern prefabrication plants it is important to design and implement a tailor made ERP System.

6.2.4   MES systems for mechanization and automation

MES (Manufacturing Execution System) is essential for mechanization and automation of production systems for prefabricated concrete elements. The MES System is one member of the Automation Pyramid, sitting between the ERP and the Control Layer, as shown in Figure 6.30. It is more than the so-called "Master Computer System" but it includes process management and SCADA System (Supervisory Control and Data Acquisition). Anyway in the area of precast concrete automation the term "Master Computer System" is very often used, because not every supplier is integrating all functions of a MES System into this software package (Figure 6.31).

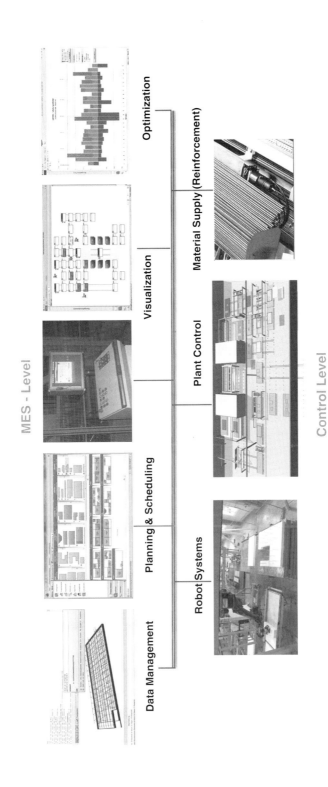

Figure 6.31 "Master computer system" connecting the production.

The figure above contains following technical terms:

- ERP – Enterprise Resource Planning
  Complete business process
- CRM – Customer Relationship Management
  Customers, Commissions
- PPS – Production planning and scheduling
  Planning, supervising, reporting
- SCM – Supply Chain Management
- CAD – Computer Aided Design
- MES – Manufacturing Execution System
  Short-time planning, data management
- SCE – Supply Chain Execution
  Storage, Pick Pack Ship Handling

According to the "Digital Factory Concept based on BIM" in Chapter 5, the MES System is the online connection to the production plant or to the production staff if we do not have automatic production machines for either or all steps of the fabrication process.

The following functions are usually supported by a professional Prefabrication Concrete MES System:

- Analysis and import of Prefabrication CAD Data either from the ERP System, directly from the CAD system or from the BIM Structure of the building object
- Handling of the production catalog and the necessary values for reinforcement, concrete, and precast mounting parts – these data could be different for the same part for different plants
- Time-based short-term planning of the production
- Automatic or semi-automatic pallet nesting (Figure 6.32), to place the available precast elements onto the production tables (long beds) and optimize the production according to lots of parameters and the order of delivery
- Automatic planning of the production order of production tables according to plant specific parameters and online reaction on plant states (e.g., available forms and pallets)
- Simulation and pre-optimization of the production sequence
- Online creation of production drawings, workshop drawings, preparation lists, labels and other documents and information necessary during the production
- Generation of production data for the machines and robots from the imported BIM Data including specialized parameter for the concerning machine and values from the part catalog. For example, it is necessary to interrupt the formwork whenever an upturned beam is designed to be the shutter instead.
- Transfer of the machine data to the different robots, computerized numeric control (CNC) and other production machines according to specialized interface definitions. Modern interfaces are and should be designed as bi-directional online transfer via computer network (industrial ethernet). With these interfaces the MES Systems gets back production states, values, material consumption and alter messages as well as necessary quality data to be sent back to the BIM Structure

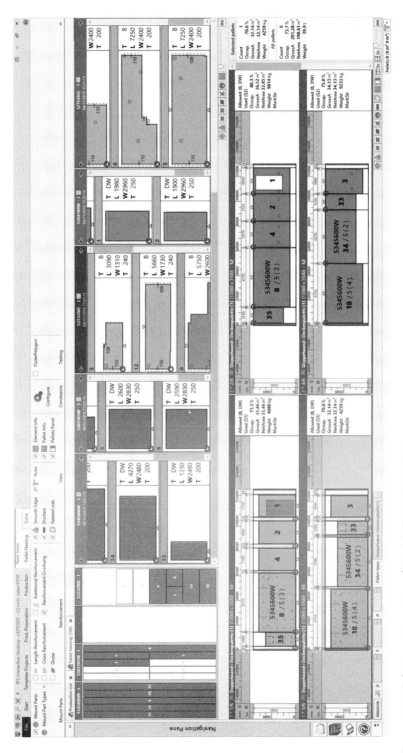

Figure 6.32   Palette nesting system of IPS-LEIT2000 (SAA-MES).

- Logistic control of the production plant using working plans (production recipe) for different precast products as well as resource monitoring and balancing and necessary pre-production (reinforcement, formwork, etc.)
- Curing control and control of the heating system
- Supervision and monitoring of the plants states and alert messages
- Visualization of the plant and plant components
- Defining, controlling and monitoring of the production staff
- Production data acquisition for detailed production times, plant movement, alert messages, material consumption and working times of the staff
- Online plant display, displays for CMUs (Cooperative Manufacturing Units) and HMI Displays (Human Machine Interface)
- Statistics, reports, histograms and management tools for supervising and controlling the production and productivity of the plant. For optimization, it is often necessary to watch the historic movement and production to find optimization steps for the future.
- Quality Management and quality data analysis which will lead to an OEE System (Overall Equipment Efficiency)
- Maintenance management system and control
- Report interfaces to ERP and stockyard management system

The special differences to the usual master computer systems or other MES are that the prefabrication concrete master computer has to work with changing production data because nearly no precast element is the same as another. Tables, and long beds as well as special forms must be utilized best, depending on many different parameters, which are sometimes really depending onto the plant layout. Furthermore, the integration of manual work between automatic machines is important to lead to a smart factory in future.

Therefore, Standard MES Systems have not been used for the precast concrete plant and specialized products were developed since 1987. Refer to Section 6.4 for more details on modern integrated prefabrication production process.

6.3   Automation methods

This section focuses on the automation methods for the prefabricated concrete industry. There are many other production methods for precast elements but the automation possibilities are very limited except the preparatory work.

Flexible automation in the concrete prefabrication industry is very well developed in Europe because of usual high labour costs, particularly in Scandinavia. Quality and precision demands for prefabricated concrete elements are sometimes very high and professional craftspeople for concrete prefabrication are getting more and more seldom.

That is why automation in this branch is still developing. Although there are lots of devices, machines and robot cells to do formwork, reinforcement, casting and many other task of the production process, manual work is still necessary for many intermediate and finalizing steps to get a perfect precast concrete element. The industry is still trying hard to automate more and more working steps, but unconditional full automation will never be achieved at this moment, neither in Europe nor elsewhere in the world.

Accepting this, necessary employees will have to be grouped into two sectors:

i.   few high-skilled persons for designing, managing, maintaining and supervising the industrial production and on the other side back staff to fulfill trained standard tasks with high speed, reliability and quality

ii. machines and processes. The manual production tasks will be ideally supported by the MES Solution in order to support quality and precession of the end product.

Additionally, it is important to mention that industrial and automated plants must have a skilled maintenance staff able to perform or organize all necessary mechanical and electrical maintenance work as well as supporting the elimination of faults and standstills in the plant. This person should know to communicate with the hotlines of several suppliers, search mechanical and electrical errors, find reasons and perform remedy. All professional suppliers of machinery and control systems are providing hotline service including remote access to the device to support the local staff. Knowledge to communicate in English by telephone, mail or any other media should be a precondition for this job in the plant.

6.3.1   *From simple to the highly sophisticated*

It is still proven, at least in the European market for prefabricated concrete elements, that automation will raise productivity, quality and precision of these construction elements. Furthermore, "Just-In-Time" (JIT) production is only possible with some amount of automation and/or industrialization.

Approaching regions with low labour costs, unfamiliar with industrial production using highly automated plants should be done very carefully. A step-by-step approach is often a better idea and will finally lead to a greater acceptance, satisfaction and better further development for the end products. The following steps could be of advantage:

- Introduction and application of a modern construction system with prefabricated concrete elements, including authorization and first successful tests
- Industrialization of the prefabrication concrete production in factory, using easy to use machines and methods if labour cost for the rest of the tasks is affordable for the product
- Organization of sales, engineering, delivery and mounting.
- Automation to extend the productivity and quality, reducing personal manufacturing work

6.3.2   *Automation methods*

The following subchapters describe the possible automation methods in precast concrete factories. Some of these features have been in use for several decades, whilst others have been running for a very short time. All of this automation might be combined according to the needs and possibility of respective prefabricated concrete products. The machines and automation cells based on the production of precast floors and walls, might also be applicable and meaningful for the production of bar-shaped or volumetric concrete prefabricated elements too.

6.3.2.1   Circulation plant and long-bed production

The basic task for industrialization and a precondition for automation is the organization of "work pieces" onto "work piece holders" in general. Doing so, the production of single elements will be grouped together and all preparatory and following tasks will be based on these working units. These could be tracked and supervised and will lead to a reliable JIT Production.

Usual "work piece holders" are:

- Tables, tilting tables, flapping pallets
- Long bed tables
- Beam and stair forms
- Pallets for pallet circulation systems

Within a typical pallet circulation system, illustrated in Figure 6.33, the "work piece holders" are bringing the elements step by step to working places to be more and more completed. The empty pallet usually starts with a cleaning process (Figure 6.34), continues to the formwork stations, followed by stations for mounting parts, reinforcement, inspection and finally casting. Curing will follow and dependent on the complexity of the precast part, it might be re-sent into further production for the next steps. After the final curing step, the readymade element will be lifted off the form at a demoulding station and usually directly put into/onto a delivery cage/package. The steel-formwork will be removed, cleaned and reused for the production of the following element.

In a plant like this, the work is automatically sent to the station where tools, work equipment and materials are available and the workers do not have to move from station to station. If the plant is correctly designed, this method is the most efficient way for production of flexible wall and floor elements. Rather seldom this production method is used for prestressed elements and volumetric pieces, but still there could be some advantage in work organization doing so.

Another form for automation are long-bed productions. In these cases, the automation machines must move over the long beds, the material and tools must be transported too and workers must change their working place very often. There are some products which more or less require this kind of production, such as prestressed precast products or hollow core floors (Figures 6.35 and 6.36) and for certain kind of products this system could be much more efficient than circulation plants.

If it is necessary and feasible to produce wall element onto a kind of long bed structure, which could be meaningful especially for small productivity and in a startup phase, a special system is successfully used. Kinds of tilting tables are aligned in lines, nearby each other, building a production line. A movable trolley (shuttle) runs under these tables (Figure 6.37), having the tilting device and a compaction device onboard. When it is necessary to compact or tilt, the trolley moves up to a certain table and does the job. Thus a lot of money is saved in comparison with tilting tables, reusing the same device for all of the tables and it is possible to use more or less all machines for plotting, cleaning, casting and trowelling, which are described later.

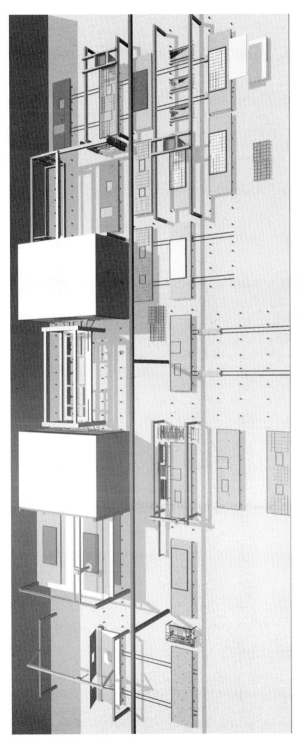

Figure 6.33   Sample of modern pallet circulation plant. (Courtesy Sommer Anlagentechnik GmbH.)

Figure 6.34   Cleaning the pallet.

Figure 6.35   Prestressed long-bed production for floors, with moving machines.

The work piece carriers are usually supervised by a master computer system (MES) and from the moment after cleaning the pallet until the moment of the liftoff and demoulding process the systems exactly knows what is on the table, what are the working steps to be done and by whom these working steps are achieved. All data for the automation machines are mostly arranged by table or long bed and industrial data capture is done on this basis too.

Figure 6.36   Hollow core production lines.

Figure 6.37   SPP (shuttle processing plant). (Courtesy Sommer Anlagentechnik.)

6.3.2.2   Formwork

The usual first step in the production of prefabrication concrete elements is the fabrication of the formwork. As we still have a table – a pallet or a long bed – the "lower side" of the formwork is still established. In the production of lot size one or very low batch sizes, it must be considered that the formwork is changed every time.

After lifting off the steel formwork, the shutter pieces are usually cleaned and oiled by an automatic device (Figure 6.38). If the magnets and magnet boxes for manual mounting of formwork for recesses are not integrated into the formwork system, the must be cleaned and oiled too.

The first machine to help the shuttering process on tables is a big plotting device. Depending on the size of the tables, the precision of the pallet locking device and some other environmental factors, the precision of the plotting line is about ±2 mm. The plotter moves over the table and draws a picture of the:

- Element contour to be formed, including formwork marker, chamfer information and information about overlaying reinforcement
- Recesses within the element to be shuttered manually
- Position and symbols for precast mounting parts, to mounted onto the pallet as well if applicable elevated (solid walls)
- Special markers for reinforcement (position of lattice girders)
- Other important information for the following manual work.

These plotting devices are often combined with a pallet cleaning device and eventually including oiling of the shutters and pallet and the positioning of the cross shutter. The MRP Machine – a cleaning, plotting and X-shutter machine shown in Figure 6.39,

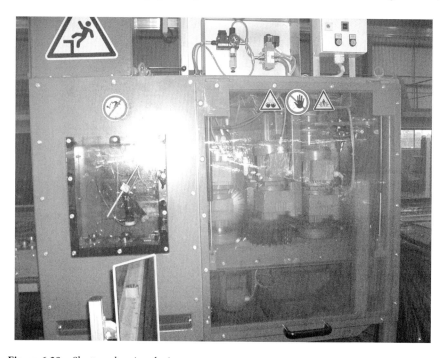

Figure 6.38   Shutter-cleaning device.

Figure 6.39   MRP-machine – cleaning, plotting and X-shuttering. (Courtesy Weckenmann.)

was one of the first automation systems in the prefabricated concrete factory and this device is still used in new factories too, sometimes without placing the X-shutters.

These combined machines have several well-known suppliers with dependent names on the market, such as MRP, RPE, RPÖ, QRP, RES, … and working on one pallet station secured by fences or other safety devices. These types of machines are often used in long-bed production too (Figure 6.40), where the device moves over the whole line, being shifted by a cross shifting device at the end of the line, to work on the next line. Due to safety bumpers and special distance measuring system, these devices are also running in automatic mode.

Because the formwork parts, the metallic shutters are sometimes very heavy, and the precision requirements are increasing in the past, for example, ±1 mm. The next step of automation is the shuttering robot shown in Figure 6.41.

The shuttering robot is mostly doing the following functions, but of course these functions and whether they are performed or not depends on the requirements of the concrete products and the requests of the end customer:

- Surveying the pallet to enhance the precision
- Placing standard shutters around the elements' contours according to CAD-CAM Data, closing the gap between the shutters and activating the integrated magnets. The shutters are automatically calculated to make the tiniest gap possible based on the existing shutter system and the available shutters in the shutter magazine. Haven sufficient different shutter lengths the residual gap to the next shutter may be that small, that there is no need of manual formwork extension. If necessary some edges will be left over without shuttering, plotted only to hold later on upturn beams or other shuttering elements

Figure 6.40   Special Sommer RES-machine for long-bed production.

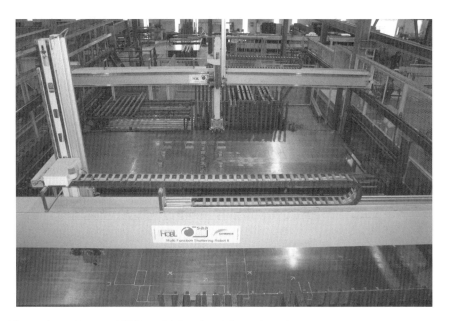

Figure 6.41   Sommer MFSR multi-functional shuttering robot.

- Calculation and placing of magnets or magnet boxes to later on fix formwork for windows, doors or other recesses. Sometimes recesses are shuttered with standard formwork elements too
- Plotting of all necessary lines and symbols to complete the formwork and place cast in mounting parts
- Applying mould release agent (oil) under the shutters and on all surface parts where concrete is poured
- Placing marker stickers for mounting parts
- Placing special magnets to center mounting parts later on
- In future: cutting and placing shutter additions out of other materials (styrofoam, …)
- Storing shuttering parts between active magazine (directly feed after the shutter cleaning device) and passive magazine (fixed position nearby the pallet). According to productivity requirements, this task will be sometimes done by a separate handling robot.

This robot cell is usually protected by fences (e.g., steel mesh) and is running in an unattended full automatic mode. Because the shuttering system might be more complex than in Figure 6.42, for example, for sandwich shutters and so called SMS (Shutter Module System) shutters to meet different formwork heights. The shuttering robot, or its handling robot, will in some applications dismantle and assemble the necessary shutters from the base types and top shutters. The environment for this modular shutter system is also mentioned in Section 6.2.2.2.

All shutters are registered and handled in automatic mode decoding the shutter length and height, sometimes RFID tags are used to identify the structure of shutter combinations. Of course these shutters could be equipped with forms for chamfer and for connecting grooves or other cross-sections.

Figure 6.42   Handling the shutter module system, with forms for connecting grooves (Courtesy Sommer Anlagentechnik).

After demoulding of the element, the de-shuttering process can also be done by robot. Depending on the required cycle time of the pallet, the de-shuttering robot is a separate device or the shuttering robot is working as a combined shuttering/de-shuttering robot.

The basic feature for this task is a scanning process of the pallet to identify the different shutters on the pallet regardless of the waste lying around. Because the shutters may have to be moved for the liftoff handling of the element just before, and because of the environmental conditions in a concrete prefabrication factory and the low contrast, this is usually done by special laser image processing systems to achieve the process safety (Figure 6.43).

After this the de-shuttering function automatically opens the magnets, which are still closed, and removes all shutters to the conveyor or active magazines (Figure 6.44) leading them to the shutter-cleaning device (Figure 6.38). If the pallet is treated by a combined shuttering /de-shuttering robot, a moving pallet cleaning machine must work at the same station under the multi-functional robot.

Today multifunctional shuttering and de-shuttering robots may work with a productivity from 4 up to 10 pallets per hour, but every time with respect to the amount of tasks, the amount of average shutter pieces per pallet, the size of the pallet and the transport situation in and out the robot cell. Productivity of this device must be carefully designed in the project phase, to avoid unrealistic assumptions.

Additional automation support could be done for the preparation of supporting formwork and formwork for recesses. Being supported with data from the master computer – only with online Industry 4.0 communication – styrofoam pieces for supplemental formwork could be cut by automatic devices.

Figure 6.43   Scanning the pallet for shutter identification and image processing.

Figure 6.44   De-shuttering robot.

Formwork for window or door openings made out of plywood or other suitable material can be cut by CNC milling machines or water jet cutting devices. Again these machines are supported by flexible data from the master computer. The cutting machines will also be used for generating formwork for more complex prefabrication concrete elements, like stairs and volumetric pieces. In these cases, a formwork generator calculates all necessary boards for manufacturing the formwork for these objects.

All of these automatic machines to support the formwork process will considerably reduce the manual working time, enhance the precision and achieves a high flexibility of the element's contours without additional cost, because as long as the robot can meet the designed cycle time the formwork costs are more or less the same. However, depending on the complexity of element geometry, manual formwork will be necessary in nearly any production facility.

6.3.2.3 Positioning of mounting parts, support by laser projectors

Precise placed and cast-in mounting parts are one of the key features of precast concrete elements. Mechanical connections between walls, floor and an eventual necessary load-bearing structure, electrical connector and mounting boxes with pipe works, plumbing and supporting sockets for transport und mounting are only few of these (Figure 6.45).

The above-mentioned plotting device can mark the positions where to place this pieces in a certain range of accuracy. Depending on the plotting nozzle, color and plotting machine the precision could be around $\pm 1,5$mm to ± 3mm, excluding the problems of manual gluing the part onto the pallet.

Sometimes the coloured lines might make a tiny mark on the outside surface of the precast concrete elements and though some operators have to wipe away the coloured line shortly before placing the piece on the table. Anyway it is the advantage on plotted lines that they stay on the table as long they are not covered by concrete or any other layer of material, so they can be used at nearly any workstation in the plant.

If the coloured line is a problem, laser projector can do the job of marking out positions much better as shown by the bright lines on the edges of the steel shutters in Figure 6.46:

- Very thin lines and thus at least same precision at the plotter
- Filtered projections, for example, to display only part of the mounting parts like electrical pieces
- 2.5D or 3D support enables projection on different layers. If supported by the CAD Data and the master computer, different layers could be defined, for example, all formwork contours on the pallet surface, all formwork contour on top of the formwork to do quality check, all mounting parts which are glued to the pallet surface, all mounting parts glued to the pallet surface but elevated, all mounting parts above the insulation, all insulation parts, and others
- Projection of additional information with mirroring
- Switching the picture by remote control
- No costs for colour

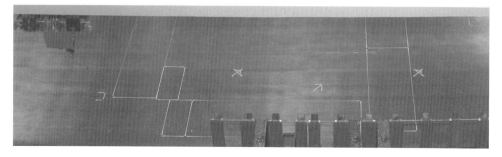

Figure 6.45   Plotting lines to mount electrical connector box and additional formwork.

Figure 6.46   Laser projection at the tables.

Of course there are some further issues to be considered:

- Precision and cost is depending on the height and mounting possibility over the respective station
- The "plotting" picture is only visible on the station where the projection system is mounted and only if the projector is on
- Due to electronic and laser equipment recalibration is necessary and regular maintenance should be taken in account
- The mounting of the projection systems should be considered. Mounting must be stable enough the environmental tools and machines do not move the structure. Otherwise the picture will move during projection and will induce sometimes

heave inaccuracy. Furthermore, there must be a locking system for the pallets at every laser station or any other alignment system for the laser picture, manual alignment might be very error prone

- If the count of lines – on one projection section – for one projection is too high and there are not enough overlapping laser projectors for one station, the picture could get glimmering and difficult to work with.

Because of these advantages and high flexibility, laser projection systems are getting more and more common in modern factories.

To avoid human inaccuracy and faults, some special mounting parts might be mounted or at least guided by robotic functions, mostly using the shuttering robot, as shown in Figure 6.47.

Mostly these specialized mounting parts are guided by position magnets, set by the robot, where the parts could be screwed on or snapped over. Of course, robots can set the mounting part directly, but the flexibility for different mounting parts will be limited by gripper and magazine and could be expensive if lots of pieces should be handled.

High productivity requirements often lead to mounting part robots to enhance productivity.

6.3.2.4   Reinforcement

Reinforcing the precast concrete elements is a very important task during the production process. In different countries it is necessary to meet a lot of different requirements for reinforcement, and their connection in between. That is why there are many different machines and devices to support this process:

Figure 6.47   Mounting magnets in matrix magazine for supporting screw holes, set by shuttering robot.

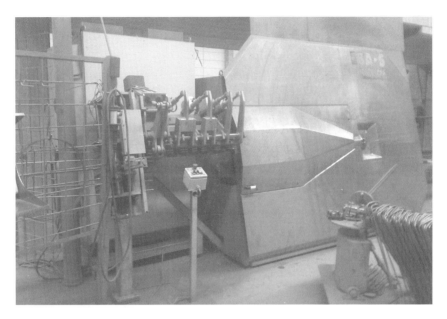

Figure 6.48   Rebar cutting and bending machine.

1. Bar cutting and bending devices: For low level industrialization, usually a reasonable amount of workers are required to cut single bars, bend them into shape and connect them to reinforcement modules and cages. Even in highly automated plants it might be necessary to have some of these manual tools, to prepare additional reinforcement rods with diameters (e.g., 20 to 40 mm) not suitable for automatic processing

2. Cut and bend machines: These automatic machines will support the production of prefabrication elements with cut and bent rods, straightened from steel coils for stirrups and connecting cages (Figure 6.48). Data might be sent directly from a master computer, to integrate this tool into the JIT production process, nevertheless, usually these reinforcement modules must be prepared some shifts in advance, because the single pieces must be connected or welded into a form of cages or other modules before.

3. Automatic reinforcement machines: Very often used within automatic precast factories especially for semi-prefabrication elements (Figure 6.49). These machines are usually used in long-line production, producing, just in time, the required rods for direct positioning within the reinforcing cage. For this kind of machine it is essential to provide perfect reinforcement data with multi-layer instructions by the master computer system; manual data handling is not convenient. These systems are usually equipped with following functions:

    a. Automatic and integrated straightening device (usually rotor systems) for different rod diameters

    b. Payoff from different coils

    c. Automatic change of diameter for any rod

    d. Automatic cutting to a certain length

    e. Application of spacer parts onto the rods to keep a certain, designed concrete cover from the pallet surface

Figure 6.49   Automatic reinforcement machine with robot.

    f.   Automatic bending of bend forms on one or both sides of the rod

    g.   Automatic positioning into the element by robot heads or other devices. Sometimes the readymade bars are placed manually into the form, depending on the possibility of labour costs and designed cycle time.

4. Lattice girder cutting and welding systems:

Because the lattice girder is still one of the usual reinforcement pieces used in semi-prefabrication concrete elements, several machines within this sector of reinforcement could support the professional production (Figure 6.50). Cutting and welding pieces of truss girders to fit the amount and lengths of an element is used in production worldwide. The machines usually receives data from a master computer system and organizes the cutting to length and supports at the welding process of rest pieces to a longer girder. In these systems the raw girders are delivered by the supply chain in different raw lengths, heights and truss forms. In the beginning of such automatic prefabrication plants the truss-girders were prepared by an inline welding machine, directly from the coil. Later these machines were used less until the inline automatic girder welding machines gained the ability to weld girders with different configurable girder-heights during the normal cycle-time without any manual back fitting of the machine. Both machine times are very often used today. These girders might be additionally placed onto the pallet by the reinforcement-robot.

5. Mesh welding plant:

The highest level of reinforcement is done by mesh welding plants (Figure 6.51). These plants – it is usually more than a machine – are more or less in line with the production and fabricates tailor made wire mesh exactly according to the

Figure 6.50   Automatic lattice girder welding and cutting.

CAD Data, transferred and optimized by the production master computer. The rods will be straightened and cut directly from coils similar to above-mentioned machines and welded into a suitable grid. Usually not all cross-sections have to be welded, because it is not necessary to manually cut the mesh anymore. Modern mesh welding plants might be equipped with integrated mesh bending devices so that these plants can prepare the mesh including the stirrups to receive the second layer mesh right after finishing the first. Doing so, it is today possible to prepare the complete reinforcement module with few manual intervention to connect the different layers. If bending devices are not integrated usually manual work is necessary to bend meshes and finalize the reinforcement module.

Totally integrated mesh welding plants are additionally equipped with an automatic mesh crane, which is placing the wire mesh or prepared reinforcement module layer on the correct place onto the production pallet without manual intervention. These plants are producing pallet per pallet, supporting JIT production. Because some of these plants have a productivity greater than for one inline pallet circulation plant, meshes for other parallel production (long bed, table) and for the construction site could be produced, when the intermediate mesh storage for pallet production is filled up. This makes a mesh plant much more valuable.

Of course it would be possible to externally buy reinforcement and reinforcement modules and include them into the supply chain management of the production. If a concrete prefabrication plant is optimized for JIT production, this could be very difficult, because the forerun will be difficult to control and it might finally lead to inefficient production. Of course some plants still work with standard mesh, bent and

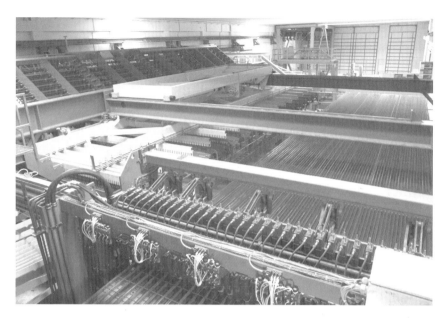

Figure 6.51   Automatic mesh welding plant with mesh crane.

cut into required forms within a local workshop. This is usually only the second option for real industrialization of the prefabrication production.

6.3.2.5   Batching concrete

Mixing of concrete seems to be the most standard task for prefabricated concrete elements. In order to industrialize this process and receive high-quality concrete elements with perfect surface it is usually not enough to work with standard concrete like available by usual ready-mix plants.

Batching receipts for precast concrete often varies from ready-mix concrete. It usually depends on the required surface paired with the mould release agent, required fluidity for automatic casting or extruding and curing time especially for semi-prefab and JIT delivery additionally depending on the available curing system. Sometimes there are up to 10 aggregates and chemicals to be combined to receive a perfect precast concrete and some experience of the batching master is necessary to fulfill these requirements.

That is why automatic batching plants are used in all industrial precast concrete systems (Figure 6.52). Automatic weighing systems as well as the measurement of the humidity of aggregates is a "must have" within this plants and the mixing machine as well as the control-system has to be equipped differently, according to the needs of the different concretes to be batches:

- Standard grey concrete with different strength
- Coloured concrete – needs a colour dosing system (liquid or powder) and sometimes if exposed aggregates are required lots of aggregate bins
- SCC – Self-Compacting Concrete: high-precision dosing and fast transport to meet fluidity requirements

Figure 6.52   Automatic batching plant for prefabricated concrete.

- Concrete with fibre reinforcement or special material
- Suspension concrete – could safe cement but usually needs a premixing device to get a suitable slurry.
- Foam Concrete – needs special premixing system with high speed mixer
- HPC (High-Performance Concrete) or UHPC Ultra HPC needs special material and perfect dosing.

Again, these machines could be connected to a master computer system receiving the batching requests with the precise amount and receipt for a pallet or on batch for a bigger concrete lot. Usually the flying bucket system to transport the concrete to the casting device will be controlled by the batching control system too and will receive the required drop position by the master computer (Figure 6.53). Thus the batching could be optimized according to JIT production especially if different type and strength of concrete is required.

6.3.2.6   Casting

Casting of prefabricated elements could be done with different devices, sometimes manual, especially for complex, volumetric precast forms, and sometimes full automatic in multiple layers. Automatic casting is usually controlled by a combination of weight measuring, motion control system and pouring control system. Professional control systems, leading to high-quality results and predictable layer thickness are using highly flexible fuzzy logic control systems to deal with different concrete types, fluidity and other problems of this material. For highly fluid concrete like SCC, special pouring algorithms are necessary to ensure an even casting.

Figure 6.53   Flying bucket system for concrete delivery to the casting machine.

- Simple Buckets:

  Manual casting buckets are used worldwide mostly for tables, battery moulds, stair and column forms. For the rest of mechanized production other systems will support better.

- Flap Concrete Spreader (Figure 6.54):

  Simple but effective casting system, having a bucket, several flaps to be opened separately and some spiked rollers to force the concrete pouring. According to the fluidity and their variability of the concrete, these machines are good for automation too.

- Concrete Spreaders with Augers (or cellwheel):

  These kind of spreaders are very similar to flap concrete spreaders except that there are augers to transport the correct amount of concrete out of the bucket as shown in Figure 6.55. Depending on the casting width of this machine there are usually 8 to 12 augers, which might be controlled separately. Sometimes the augers are closed by separate flaps to deal with highly liquid concrete – like SCC – too. This system is usually best for automation, because the amount of concrete is very good controllable, more or less regardless of the liquidity. Sometimes, for higher pouring speed, these casting machines are designed for the full width of the pallet or table, for pallet rotation systems the spreader is usually designed for the half width, which saves costs and is absolutely sufficient according to usual cycle times at concreting stations.

Figure 6.54   Flap concrete spreader.

Figure 6.55   Concrete spreader with augers.

- Concrete-Spreader with belt:
  This type of concrete distribution device is sometimes used to distribute a higher amount of concrete at a time (Figure 6.56). Furthermore special concrete types might be used with this spreader. The spreading belt has a fixed width but the spreading width might be smaller due to an adjustable guiding plate. Using a turning bucket and a remote control, the operator can fill large quantities in a short time. Unfortunately, this type of casting machine is not very precise and not designed for tiny concreting areas, thus it is normally not automated.
- Slip-formers:
  This casting machine is used mainly for hollow core slab production (Figure 2.21) or for manufacturing of special concrete elements with longitudinal hollows inside, such as hollow core walls. A relative dry concrete is necessary to use this machine and it is only possible to cast long stripes of the same concrete width in one piece and cut or reshape this long-bed strip later on. An integrated vibration device compacts the concrete immediately when it is flowing out of the orifice. The form of the orifice defines the height, form and width of the final long concrete element. See Section 2.4.2 for further details.
- Extruders:
  The extruder is again used for casting of hollow core products but differently to the slip-former, the casting and compaction processes are controlled and performed by special augers, again through an orifice as shown in Figures 2.23 and 6.57. Usually the result of extruders is of higher quality throughout the whole length of casted strip and the compaction is more homogenous. The technology of augers and orifices is more developed and costly than the slip-formers. See Sections 2.4.2 to 2.4.4 for further details.

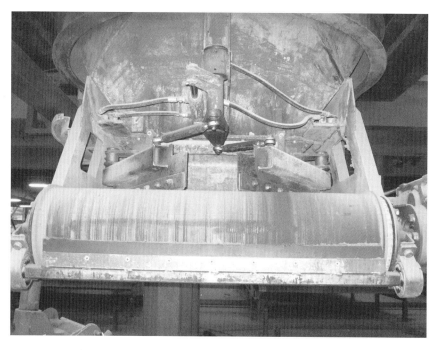

Figure 6.56 Belt-casting machine.

Figure 6.57   Extruder for hollow core slab production.

- Pumping technologies for concrete
  Concrete pumping is very often used in cast *insitu* environments but could be used for precast concrete technologies too. Forms, pumps and concrete must fit very well together to lead to high-quality products and only a few companies worldwide are providing professional technology for pumped precast elements. Usually the elements are cast in a complex volumetric form, to cast one prefabricated prefinished volumetric construction in one piece. Thus the forms are more complex and the necessary lot size to be successful must be reasonably high.

6.3.2.7   Turning devices for double-wall technology

The double-wall system, Figure 6.58, is an originally European construction system for semi-prefabricated wall elements, which are easily transportable because of their relatively low weight and being finally filled with concrete at the construction site, to form a homogenous and monolithic concrete wall/floor structure together with semi-prefabricated flooring systems. The production is done in two shells, first the pre-production of the first shell including necessary reinforcement and connection pieces, eventually inside insulation. After curing the first shell the second one will be concreted and joined together for the finished element.

The production of these prefabricated parts is described in Section 6.1.1 as a stationary production on tables. To effectively produce double walls within a pallet circulation plant or on long-bed systems, specialized turning devices are used.

- Suction Cup/Vacuum Cup–Turning Devices
  Using these machines, the first shell – pre-production – will be lifted off the pallet after curing and exactly positioned onto the vacuum turning device to fit into the second shell on the other pallet as shown in Figure 6.59. The more

Figure 6.58   Double wall.

Figure 6.59   Vacuum-turning device with fixed first shell elements over the second shell pallet, just before joining.

suction/vacuum pads are available on the device the easier is the positioning of both shells on their pallets, because even small strips near to door openings must be lifted by certain number of suction pads. Sometimes the position could be supported by laser projection too. After positioning all of the necessary elements, the suction cups are activated, the elements are lifted, turned and put back into the second, freshly concreted shells on the second pallet. After compaction, the second shell pallet leaves the turning device for curing.

This turning system is very flexible according to wall thickness, but more professional operators are required to precisely fix the elements onto the device. Additionally, the nesting of the elements must be done according to the possible positions of the suction cups and the weight of the single elements – high professional MES software is supporting this process.

- Pallet-turning devices
  With this system, the pallet of the pre-production – that is, first shell – is clamped into the turning device and turned as a complete piece over the pallet of the second shell as shown in Figure 6.60. To do so, the elements of the first shell must be fixed onto the first pallet, not to fall down or shift during the turning process. The fixing is done with bars or special magnetic boxes. After turning and joining the two pallets and their elements, the fixing bars or devices must be removed and the shells are again compacted together. Due to the thickness of fixing bars and devices, the possible minimum wall thickness is limited and double walls with zero gap cannot be produced with this device. The precision of this joining process is limited only by the formwork process and the tolerances of the turning device, and not influenced by human work.

Figure 6.60   Pallet-turning device with fixed first shell pallet over the second shell pallet.

After joining the two shells, the first pallet is lifted up again whilst the second pallet leaves for the curing chamber and the empty first shell pallet is sent for de-shuttering and cleaning to be used in the next cycle. There are lots of similar system like "big toasters" on the market doing more or less the same procedure either with elements or with pallets.

For long-bed systems, mentioned in Section 6.3.2.1, there is also a suitable turning device available to enable double-wall production, of course with lower productivity. These machines move lengthwise over or nearby the long bed tables and are based on suction/vacuum cup systems. The positioning of the elements on the turning device and in length direction must be done manually and handled with care to ensure a good quality of the double-wall product.

6.3.2.8   Treatment for solid parts

Usually the second surface – this side, which is not facing the pallet side of a solid or sandwich wall must be smoothened. During concreting with standard or lightweight concrete it will be necessary to place a little bit more of concrete into the form to meet the compaction needs, as with manual concreting this is sometimes done during the compaction process to be sure that certain wall thicknesses will be achieved. Even if automatic casting is done it might be meaningful to have little bit more concrete to enable proper screeding after compaction. Using self-compacting concrete, the upper surface will be as though it had been screeded when filling the formwork exactly up to the desired height.

- Screeding

Screeding takes place directly after casting more or less for every solid-wall product to ensure a smooth upper surface. Screeding might be done manually just with boards but should be done with mechanical devices moving over the pallet after concreting or directly mounted to the casting machine as shown in Figures 6.61(a) and (b).

Usually the screeding device, such as shown in Figure 6.62, is a combination of lateral moving bars and vibrators to receive a good result. Just screeding surfaces will never match the results of table or shutter surfaces. For lightweight concrete, smoothening rollers are available too.

- Trowelling

Because sometimes the solid-wall elements need to have two smoothened surfaces, the upper surface on the table needs to be troweled after a certain time of pre-curing. Trowelling in modern mechanized plants is done by trowelling devices (Figure 6.63) or seldom by hand trowelling devices too.

Normally this process is done by rotating plates or "helicopter" paddles, guided by a machine and moving over the concrete surface with a certain rotation and movement speed. The system of trowelling depends on the stiffness of the usually pre-cured concrete surface and it needs a certain kind of experience of the operator to reach a high quality result. To have the best surface, it will be necessary to do different trowelling steps. Up to three work steps might be good, starting with a helicopter trowelling and finally a smoothening with plates.

(a)

(b)

Figure 6.61   (a) Screeding device mounted on an automatic casting machine;
(b) Independent screeding device.

Figure 6.62   Roller smoothening machine.

Figure 6.63   Semi-automatic trowelling machine with "helicopter" paddles.

The first steps of trowelling could be automated with surface supervision by the operator. The last step should be done mechanized only to ensure perfect quality by the operator. Manipulation of these machines are done by remote control or cable oriented systems.

6.3.2.9   Curing

Depending if we are using rotating pallets, fixed tables or long beds, curing is mechanized differently. Having fixed installations, the curing process is only supported by heating systems and special coverings.

Using a pallet rotation system gives a wider range of curing control. After casting and compacting, the pallets are stored into a curing chamber, Figure 6.64, which is usually closed and sometimes heated. Climatic conditions within the curing chamber are essential for the final quality of concrete, and the strength of the precast concrete element. It is not only the temperature that should be controlled, furthermore the humidity should be held to a certain value throughout the chamber. Therefore, professional heating and ventilation systems are available on the market and must be taken into account when mechanizing the production. With certain admixtures and additives to the concrete and a controlled atmosphere in the curing system, quality could be raised and curing times might be reduced.

Curing chambers are usually filled and emptied by rack cranes or lifting devices, working in automatic or at least in semi-automatic mode to store the casted elements and fetch them out after a certain time span for demoulding or further treatment of the surfaces – trowelling, tiling, and so on. For easier products like semi-precast floors, it is possible to use stacking systems, Figure 6.65, where the pallets are only stacked over each other because this is cheaper than rack systems. Of course these stacking

Figure 6.64   Curing chamber with sectional doors and rack-lifting device.

Figure 6.65   Curing chamber with stacking and stacking crane.

systems are not that flexible for the process, because there is no random access to the pallets and there must be a specialized planning system to get the pallets out of the curing chamber in the desired order to build up transport units after the demoulding and lifting of.

All these kinds of curing chambers are without random access – there are rack systems with multi-pallet trays too –sometimes called "LIFO" curing chambers (Last In First Out) because the last pallet put onto a stack will be the first to get out, unless there is a restacking system implemented. Restacking might be either done by the stacking crane or special, heavy lifting devices, which enable the lowermost pallet to be removed from the pallet stack.

6.3.2.10   Cutting of prestressed wires and strands.

For production of prestressed concrete elements, usually for prestressed floor slabs and beams, it is necessary to use metallic strands and prestressing bulkheads on either long beds or special prestressing pallets. The prestressing strands are inserted in certain, pre-calculated patterns into all elements on a long bed and tightened to a certain force before casting the concrete.

To lift off the elements from the table after curing of these elements the prestressing strands must be cut at a certain position either exactly at the edge of a slab or with a certain defined reinforcement overlap outside the slab (Figure 6.66). Because there is a lot of work to be done, this cutting was automated with a system working on the base of long-bed production.

Before cutting, at least the cross shutters have to be removed – in such plants this is automatically done by another run of the cleaning plotting shuttering machine – before the strand-cutting device can do its job. Every time the strands are cut at any side of an element, all the elements will slightly move in the length direction on the table, because of the rest of strengthening force in the wires. Therefore, the cutting position is automatically scanned by suitable devices and recalculated every time. After this calculation, the cutting is done, for example, by a CNC-4-axis circular sawing system. Perfect distance measuring system for long range as well as optimized scanning technology for uneven concrete edges are key technologies to make this process work.

Figure 6.66   Automatic strand-cutting device for long-bed production.

Because this machine is running self controlled over the long bed with a high cutting speed, it will save a lot of manual work and lifting off of the slabs can immediately follow the machine.

This cutting system is until now only used for prestressed semi-prefabricated floors and solid prefabricated floors up to 100 mm depth, but could be implemented to all shuttered production systems.

In hollow core production there has been another slab-cutting system in use for a very long time, but this system is completely different. At HC production there are no cross shutters and the whole extruded or formed slab with a fixed cross-section will be cut by big sawing machines according to the requested length by the production planning. There are only straight cuts; the rest has to be done manually. At the cutting of prestressed floors, the slabs edges are formed by shutters, which might have different forms – curved too – and only the prestressing wires are cut by smaller cutters.

6.3.2.11   Insulation

Because of the stronger physical demands to concrete prefabricated elements, thermal insulation inside the concrete elements is becoming more and more important. As outside insulation can usually be applied only at the construction site by extensive manual work using a scaffolding structure and following plastering, insulation inside the concrete element makes it really pre-fabricated and not semi-prefabricated. Insulation materials of all kinds of polystyrene or rock wool are in use. To avoid thermal bridging between the concrete shells separated by the insulation material it is necessary to use either stainless steel connectors, like stainless steel girders or connector bars or other forms of fiberglass or similar materials, like Schöck ComBAR®, Thermomass®, Kappema©.

In some cases polyurethane and mineral foams are used for insulation but the application is not very easy because of the unpredictable expansion of these foam products depending on temperature, humidity and other factors of the concrete element and the environment.

Today thermal insulation is used within solid/sandwich wall and double-wall systems. Because the enhanced manual efforts of doing this kind of insulation there are automation systems for different steps of this procedure:

- Cutting/shaping and connecting insulation material
- Placing insulations parts into the precast form
- Mounting concrete shell connectors

The automatic cutting and shaping is done by different machines on the market, either using hot wire (for polystyrene only) or water jet cutting, for example, Figure 6.67. The raw data of the insulation elements must be transferred out of the precast CAD system and will be adapted and transferred by the master computer system. In fully automated insulation cutting machines, the machine is loaded with standard blocks of raw insulation material, transporting the pieces through the cutting and shaping process and either stacking them according to the element structure or directly feeding the blocks to the placing robot system.

At installations where the blocks are positioned manually into the forms, sometimes the raw material is manually placed onto the cutting table too but cut automatically. Systems to drill holes for the wall connector or other cutouts are partly included in the tailor made appliance. To handle the elements more easy they can be automatically joined to longer strips by gluing them together.

Combined automation systems, like the IPAR-System from Sommer Anlagentechnik/Germany, are cutting and shaping the insulation blocks, transporting them to the robot cell and placing the pieces with high precision onto the lower shell of poured and compacted concrete. Immediately after, the wall connectors are put in by use of robots (Figure 6.68) until each single rod is in the right position in the lower still wet concrete shell. It will be necessary to re-compact the concrete again so that the wall connectors are fixed. Depending on the precast product, the insulated element will be put into the curing chamber to be turned into the other shell of a double-wall or the next level of reinforcement and concrete will be poured on top of the insulation to produce a sandwich element.

6.3.2.12   Application onto the surface

To meet certain requirements of a more durable building surface or to have an optical covering of the concrete surface, for example, in Figure 6.69, it is possible to apply stone- or tile-based parts onto the concrete element.

In principle, there are two methods used on the market:

- Gluing tiles or clickers onto the hardened concrete surface
- Placing tiles into the forms and pour wet concrete on it.

Figure 6.67   Stand-alone insulation cutting and joining machine. (Courtesy Sommer Anlagentechnik, Germany.)

The first system has been used for several years and because the clinker stones are very small it is done efficiently by robots. If the plant is designed for, this task can be done by a separate robot cell or by using the shuttering robot with an exchangeable gripper tool in the second shift.

A more developed system is the JFI-Method (patented by Sommer Anlagentechnik GmbH, Germany) shown in Figure 6.70(a). This is based on bringing the tiles in before concreting the lower level of the precast element. Doing so, the tiles are integrated into

Figure 6.68   Sommer IPAR-system mounting insulation and wall connectors.

Figure 6.69   Sandwich element with tiles on the outside surface.

(a)

(b)

Figure 6.70   (a) Robot is placing tiles into the grid of joint filler; (b) Element surface with surface tiles applied with JFI method.

the stable concrete structure of the wall. Different forms and colours are possible, if the data are professionally prepared by precast CAD and MES-System.

There are the following steps to be automatically performed by robot and handling systems or manually in a later stage:

- Applying the special joint filler to have suitable forms in between the tiles/plates after demoulding (Figure 6.70(b)), such that the surface is not directly connecting with the joining concrete. The joint filler pattern and the amount of filler material is exactly calculated by the master computer.
- Selecting, handling, cutting and queuing the tiles, plates or clinker stones.
- Positioning the pieces into the still flexible joint filler to have a precise gap form in between the plates.
- Normal production of solid-wall or sandwich element
- Manual removal of the hardened, but still flexible, joint filler after lifting of the element from the pallet.

Instead of the JFI technology, other jig grids are used but in these cases the preparation is very sumptuous and only will be economical with high repetition which would be against the flexibility of automated precast concrete plants.

For all of these processes it is very important to feed the tiles/plates in a suitable way into the robot cells. Sometimes this is directly done by robots from well-defined stacks and cages, coming directly out of the tile factory, sometimes manual workers have to queue them up into jigs eventually supported by laser projection.

Of course there might be several variants of these application methods with regards to the patented systems. In future, even inside walls could be equipped with tiles for example in using them for prefabricated bathroom units.

6.3.2.13   Treatment of concrete surfaces

Another way to make the concrete surface of precast concrete elements more fancy, sometimes really hiding the concrete surface, are kinds of surface treatment. Mostly this is a combination of different methods, which are presented here only briefly. Of course it is partly possible to support this work with automatic systems too.

- Form liners

Form liners are used to give the concrete surface at certain structure, in contrast to the flat and smooth surface of a steel table or a steel mould where precast concrete is usually produced on/in. Thousands of different surface shapes can be produced using a structured plastic or rubber mould, placed onto the pallet of a precast form as shown in Figure 6.71. Brickwork, bamboo style, wave lines, stones and many other can be optically simulated and numerous vendors of these form liner systems are well known on the market.

To integrate form liners in modern precast concrete production the following environmental factors are considered:

- Usually, form liners are designed to be used more than once, because of their costs, so they are best for batch production of elements. When using them for flexible

Figure 6.71   Sample pieces for form liner surfaces.

element size and contours, it is necessary to have at least a grid of length and width
a possibly a suitable stockyard system for these form liners
- To prepare a beautiful surface it might be necessary to have special concrete
too, for example, self-compacting concrete with well-graded and small-sized
aggregates.
- The circulation plant should consider stations to mount the form liners into the
form and to remove them again for cleaning, as well as perhaps considering a
multi-layered casting procedure to pour a thin layer of "beauty concrete" followed
by the load bearing normal concrete.
- Form liners are often used together with certain coloured concrete or different
aggregates, bit the batching plant has to have the necessary ability.
- Last but not least, the CAD Data and calculation methods of master computer
systems are again very important, because the form liners will have a different
thickness and might have a very deep structure making it challenging to calculate
the right amount of layered concrete and the suitable element thickness to set
the suitable shutters, even by a shuttering robot. In most cases special shuttering
systems are required.

Nowadays, due to 3D printing and milling technologies it is possible to make the form
liner structures according to specialized CAD Data too, by use of a automatically produced negative form to produce one's own form liners. Due to the cost of the form
liner material, batch production should be considered for similar elements.

To ideally store the unused form liners, and to reuse them whenever needed for production, an automatic storage system is required, as shown in Figure 6.72. The form
liners are laminated onto carrying frame boards stored in a rail-based system. When

Figure 6.72   Automatic form liner storage system.

the master computer recognizes that a certain form liner is necessary for production it will search within the rail storage and the system is handing it over into the form.

After lifting of the element the form liners are reinserted into the rail system waiting for the next utilization.

• Retarding and washing exposed aggregates

By use of certain chemicals to retard the hardening of concrete, it is possible to produce concrete elements with exposed aggregates. First it is necessary to paint the steel form with a retarding chemical available on the market. The concrete must be mixed using the type of aggregates to be exposed later on. When the concrete is casted onto the pallet the surface gets in contact with the retarder and after hardening and lifting of the element, it is possible to wash the surface with high-pressure water to get the exposed aggregates. The wastewater should be carefully treated because of different chemicals.

Because thickness and type of washing result depends on the retarder and the application, sometimes automatic retarder spraying machines and automatic surface washing systems are used to ensure perfect quality.

Using retarders with different intensity and applied onto a release paper in a form of a negative picture, will lead to photo concrete after washing the surface. Combined with a polishing finish, this will absolutely lead to a beautiful concrete element, for example, Figure 6.73.

Furthermore, concrete could be treated with several other chemicals, for example to receive an etched surface.

• Grinding and polishing

Grinding and polishing is usually used together with coloured concrete and special aggregates giving the concrete a marble like surface. This can be achieved either by

Figure 6.73   Beautiful concrete element with polished picture concrete.

hand or using larger bridge polishers (Figure 6.74). If used together with a form liner structure and picture concrete retarding every outdoor façade or indoor surface is possible.

Concrete grinding and polishing machines are mostly automatic. Having enough space, the surface finishing for numerous concrete elements could be prepared and the machine is running unattended in a separate production shift, which could take some time. Usually this process is not in line in a pallet circulation plant because, the element needs to be lifted off first and polishing is time consuming. Treatment of environmental dust and wastewater is an important side issue too.

6.3.2.14   Lifting off the precast elements

After hardening of the precast elements, the lift off or demoulding process usually signals the end of the precast production, of course before doing any surface treatment or cosmetic operations.

Depending on the shape of the precast concrete element and the shape of edges, it might be necessary to open or shift the formwork away to get the piece out. De-shuttering robots usually have the ability to move the formwork parts in the right way, without damaging the edges. Using thin-layered semi-precast elements like filigree floors or double-walls and a shuttering system with integrated magnets, the formwork can stay on the pallet and it is possible to lift it out the slab. Especially at the floor production, using standard pallet width, it is necessary to consider a suitable deforming slant at the fixed side form or some other systems (mould venting, mould bending) to get the elements out.

Figure 6.74   Grinding and polishing machine in a large production hall.

To ideally assist the demoulding process for wall and floor elements, the following tools and equipment are used:

- Tilting appliance for pallet system (Figure 6.75), to lift of wall element in a vertical way and to avoid damaging the edges of finished element.
- Movable supporting beams at the tilting side of tilting devices to support heavier wall elements and to prevent them from sliding over the pallet.
- Lift off cranes supporting the effective demoulding of filigree slabs (gripping the top flange of the truss girders).
- Balancing traverse for equable force transmission into the lifting hooks.
- Self-moving ladders to safely hook on the wall elements.

One very important point for damage free demoulding of concrete elements are the various kinds of lifting anchor systems. Many vendors on the market are providing different mounting parts to be concreted in, transferring the lifting forces into the concrete and the reinforcement structure. Sometimes different anchor systems might be necessary for demoulding and mounting the elements at the construction site. Nowadays, precast CAD-systems are able to calculate the correct position to place these tools.

To ensure a minimum necessary concrete strength for lifting the element without cracking the hardening time in the curing chamber or the heat introduction into the element should be controlled. Usually it is enough to check the curing time according to experience of the concrete type and the volume of the element. To be more accurate, as radio connected devices are now available to measure the hardness of the concrete element within the curing chamber or by use of temperature measurement, the heat input could be calculated and controlled. Furthermore, it is possible to shorten the curing time before lifting by means of concrete additives.

Figure 6.75   Tilting appliance in pallet system and lift-off crane.

6.3.2.15   Intermediate storage of precast elements

Even if the industrialized and automatic production of precast concrete elements is highly planned, computer organized and working quite near to a "just-in-time" production, it is necessary to at least do an intermediate storage of the elements. Because precast concrete products have often to stay in the stockyard for a certain time to gain a pre-defined required lifting strength, the stockyard management for different products must be flexible and sufficient.

There are two common strategies for intermediate storage:

- Storing the elements in pre-planned transport units usually called stacks or cages (Figure 6.76(a–d)). This is a very efficient method and used by the majority by avoiding double handling of every single element. Unfortunately, it is necessary to fix the element order of mounting and the grouping of the delivered units immediately before production. Modern production master computers are taking in account this exact mounting sequence, optimizing pallet nesting and other logistics to the circumstances. The saving in handling effort is huge.
- Storing single elements and commissioning them to delivery units immediately before transporting them to the construction site: This method is much more easy and flexible for the site. It is necessary to have some supporting structure for holding single elements and, because the elements are not pre-grouped, a stockyard logistic system and fast stockyard cranes are highly recommended to quickly find the elements at the pick-pack-ship process and load them to trucks. This method is often used for heavier solid or standardized precast concrete elements.

(a)

(b)

Figure 6.76   (a) Inloader cages: needs special trucks, but easy to load/unload; (b) Transport stack, consider last-in-first-out at the site; (c) A-frame, consider last-in-first-out at the site; (d) U-frame, flexible loading and unloading.

The pre-defining of the delivery and mounting order must be done by software linked to the BIM Model of the building so that it is able to organize the mounting process according to the agreement with the construction site. Doing it that way is an important point to minimize construction time at the site and to make use of the big advantages of prefabricated construction.

(c)

(d)

Figure 6.76   (*Continued*)

Of course, both of these stockyard strategies might be mixed at one production site according to different products. For intermediate storage it is common to use any kind of transport cage, to easily handle the elements and to avoid damages during transport. The following systems are widely used:

Although most of the precast producers are working with giant fork-lift trucks or gantry cranes in their stockyards, automatic stockyard systems (Figure 6.77) are

Figure 6.77   Automatic high rack stockyard system with automatic loading bays.

gaining market because high flexibility and unattended actions during night and loading at peak hours.

After the production and finishing of the elements, they are stored into an automatic high rack or container stack storage, controlled by SCE – Supply Chain Execution – software and stockyard master computer systems. Because these stockyards are usually outdoor, handling giant weights including optimization of stockyard space, perfect sensor techniques are necessary to avoid any problems during the handling of these pieces. According to the planned delivery order, the transport units will be directly loaded into loading bays for manual commissioning of the trucks or transported into automatic loading stations, where the truck driver can load their trucks themselves, arriving shortly before start of delivery.

Doing it that way the prefabricated concrete industry is really approaching a so-called "Industry 4.0" smart production and delivery process, very similar to automotive industry, but producing a real lot-size single product.

6.4   Integrated and automated prefabricated production process

As described at the beginning of this chapter, it is not enough to have automatic machines and transport system to support an optimized production process. Today's successful automated prefabricated production processes are have to be highly integrated into the whole planning, production and delivery tasks as well as the information of the worker in the factory must be carefully considered. Industry 4.0, Smart Factory and BIM are of course important keywords. But not knowing such words in the early days of automated precast concrete automation some of the functions were still considered in the 1980s.

The more the prefabricated production was requesting using JIT – Just-In-Time – production and delivery, the more it was necessary to have a continuous flow of information and data in all directions.

Every part of the plant must be informed exactly about the necessary data, not too much and not too less, to do automatic production by machines or making manual decisions by staff.

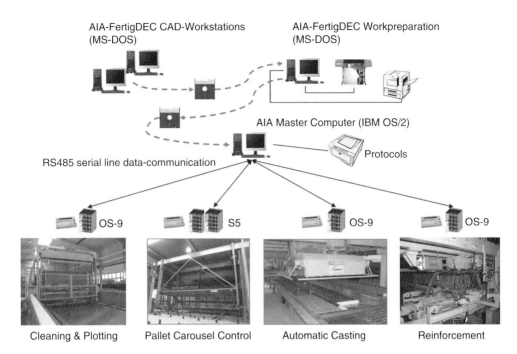

Figure 6.78   Structure of c. 1987 CAD-CAM production.

6.4.1   Structures

Former integrated structures were usually designed from top down, more or less this was Industry 3.0 CAD-CAM systems prepared production data out of an early pre-cast model, handed over as files to a first version of production master computer, and transferred via serial lines to the different production machines (Figure 6.78). Nevertheless the pioneers of CAD-CAM production in the precast concrete industry still had a two way communication between master computer and machines, reporting back the production time, cycle times of the transport systems as well as state messages and alert messages. So this was still the first appearance of Industry 4.0 because machines were talking to each other, relying on this information for their well-programmed decisions.

Today this structure is more a mesh of informational paths connecting ERP-Systems, PPS-Systems, MES-Systems, production machines and staff, working on different levels of these factories, as defined in Figure 6.79.

The list of information transferred within these levels of the automation pyramid is usually incomplete and growing day by day, finding out about new necessary information:

- Planned and real production and delivery dates.
- Complete BIM Data of the building as well as the detail for each prefabricated element
- Catalogue information about mounted parts, reinforcement, concrete and any other material necessary for the production

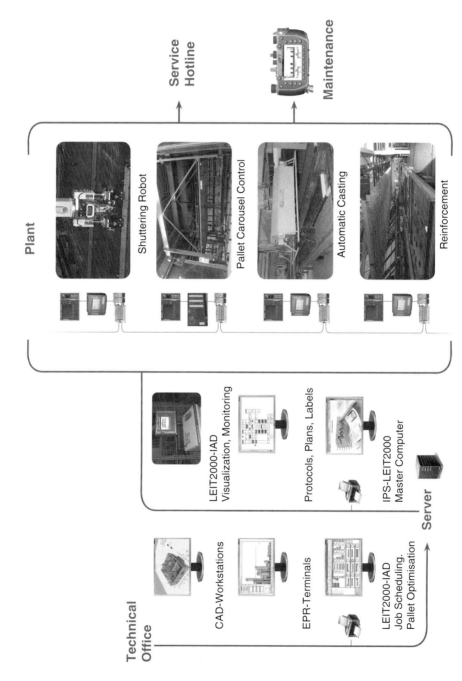

Figure 6.79   Modern integration structure of prefabrication plant.

- Data of specimens defining and ensuring the quality of the product
- Graphical, geometrical information for the workers in the plant
- Production output and key figures, like OEE (Overall Equipment Efficiency)
- State and error messages of all automatic machines for predictive maintenance and plant supervision
- Real live NC-Data or CNC-Data for the production machines
- Interlocking information between different machines able to interact with each other
- Information Screens
- Drawing Monitors with production details
- Mobile Terminals
- HMI for machines
- RFID Readers/Writers for IoT Methods
- QM Information

6.4.2   ERP, CAD, MES, PROD machines, HMI

Because there are so many participants within this communication system, modern and powerful communication methods are essential. Profibus, Interbus and industrial ethernet – like EtherCAT, ProfiNET is widely used within the factory's automation structure, where standard ethernet, hopefully TCP/IP or Web Service Interfaces are at the planning and manufacturing execution level. Of course some standard machines are still connected only by file transfer or serial line connection, which could be very stable too, if supported by professional MES.

Network and IT security is today very important for fully integrated prefabrication factories to ensure reliability of the whole system. Therefore, industrial networks must be technically separated from office networks. The access to the internet must be controlled and as secured as possible, to grant remote access for maintenance as well as to avoid viruses coming into the systems.

6.4.3   HMI – Integrating staff into the process

Real labour-free, fully automatic production will not be realizable, or at least not be affordable, even in modern precast factories. Human staff are not only necessary in all areas of data generation, planning and organization, but the production staff must be integrated too. Human Machines Interfaces must be easy to use even for personnel usually doing manual work.

The following systems could help in the office environment:

- Multi-screen applications for work preparation and planning workstations
- Big screens and if possible touch-screen–enabled applications
- Mobile integration to smartphones, Figure 6.80, for the staff and management to be informed about the actual plant state and productivity
- Software systems
- All tools must be online and automatic recovering from connection problems.

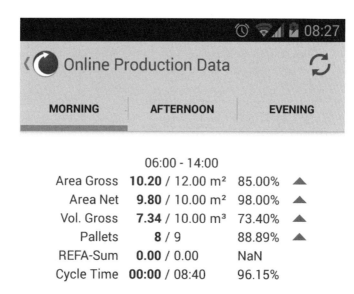

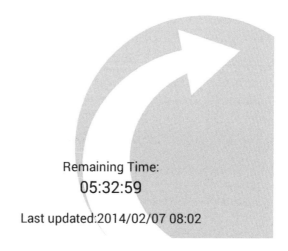

Figure 6.80   RIB SAA mobile app for the concrete prefabrication industry.

Within the factory robust tools for the rough environment of these kinds of factories is a must. It is especially necessary to have easy to use visualization and all visualization systems within one plant following one style guide to enable the plant staff to interchange their workstations (Figure 6.81). Graphical visualization with interaction of the user are very comfortable to us and extremely reduces standstill of plants and machines even in cases of seldom electrical or mechanical errors.

Paperless factories are developing fast giving much better information to the worker and saving an unbelievable amount of paper and toner material. Furthermore, the

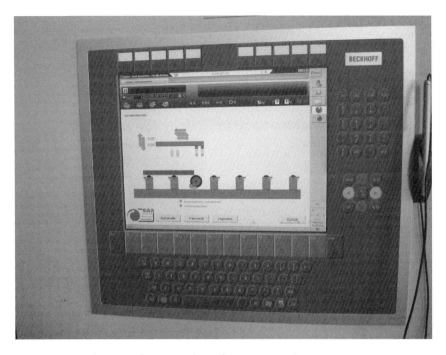

Figure 6.81   Machine visualization with touch-screen operation.

production is getting more flexible because production documents are not pre-printed and so the information can easily be changed just before the production without wasting paper. Of course to integrate these HMI directly into the production, as shown in Figure 6.82, stable and easy to use mobile or fixed terminals need to be connected to either the master computer or the control systems.

Even today one missing link is still not solved. Mobile operation by the use of radio network technology is still not (rarely) certified and allowed. As long as the terminal is cabled, a pre-defined area cannot have the machine under optical supervision, as shown in Figure 6.83. To drive machines in a wireless and remote way it must be 100% secure, so that the distance to this appliance is not too large, and to watch the operation effects. Still until today only very few industrial terminal systems are certified for wireless operation all the others still needs a cable for acceptance operations.

Last but not least the development will lead to augmented reality information systems. The worker will have to wear very lightweight glasses and will see the necessary information as a kind of head-up display during his work. Because the position identification could be achieved very easily, even geometrical as well as logistical information could be presented with the right position within the viewpoint of the worker. Systems like that are still existing on the market, unfortunately, the devices are still not light enough and not prepared for the rough use in a concrete prefabrication factory.

6.4.4   Smart factory, industry 4.0 – integration into BIM

An integration of all physical and virtual members is the target of Industry 4.0. If the virtual system, for example, a connection of ERP and High Level MES of the whole

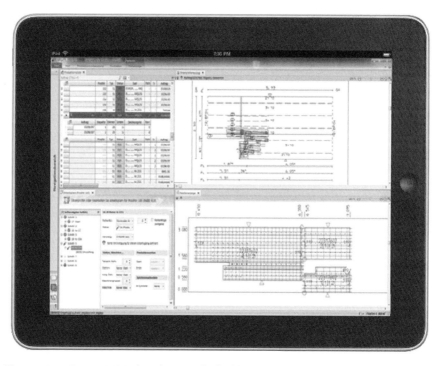

Figure 6.82   Factory HMIs based on standard tablet.

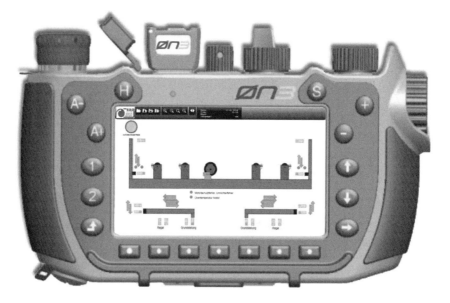

Figure 6.83   Visualization HMI with wireless control. (Courtesy EXOR GmbH.)

production plant is well designed, it is possible to pre-simulate the whole production process in advance, providing perfect information for the planning process. If this virtual system is connected and continuously updating its information out of the real factory and the machines, the process of planning can be adaptive to the reality of production and so optimized in the best way. And more or less in all prefabrication factories we would have to consider humans as a hardly plannable factor. As we are including intermediate and end products in our observation structure, and automatically deciding actions for the production optimization, this will lead us to a real smart factory.

The prefabrication construction industry has to consider the BIM Model of the final building. In Chapter 5 the topic of BIM was covered in detail. If all parts of the process know about the pieces, the elements to be produced, their contents, states and material as well as the position within the final building, even manual decisions for the construction site could be made much easier and be more efficient for a cheaper, professional and just-in-time production and erection of a building.

The information out of the production reality – that means real used material, energy consumption, real production time, quality information, charge information and much more, could be handed back from the production via MES System into the real BIM model. This makes it very easy to cross check for example the LCA (Live Cycle Analysis), to document the quality and last but not least to learn lessons for next construction projects or the actual production

Another integration matter of smart factories following this concept is the possibility to analyze the production. Getting values for productivity, reliability and quality of the prefabricated production leads to OEE (Overall Equipment Efficiency), Figure 6.84. This value shows better the real success of an automated production or the need to do some more adaptations.

Of course there are many other values for plant managers to consider. Cycle time and work plan statistics could help to optimize the plant if the product range is heavily changing, at the startup phase of a factory as well for all later adaptations. Analyzing information, alert and error messages out of the plant can help to install a preventive maintenance system (Figure 6.85). This is reducing or hopefully even avoiding standstills or production interruption of the plant, which is usually very costly. It is necessary to have all machines and devices integrated into this alert message system and the machines should report alerts even if they could recover them for the moment themselves. In case of real errors to be handled in the moment, the error information should be as detailed as necessary to quickly find the reason and perform counter actions.

6.4.5   QM included

Today the quality management within modern concrete prefabrication factories gets more and more important. Only perfect concrete elements allows JIT delivery and a continuous erection process at the construction site. That is why quality problems must be recognized as early as possible, to avoid any problems or in the worst-case standstill at the site. It is an advantage of precast elements that it is possible to avoid bad quality by usage of automation and an early QC system. Depending on the finishing

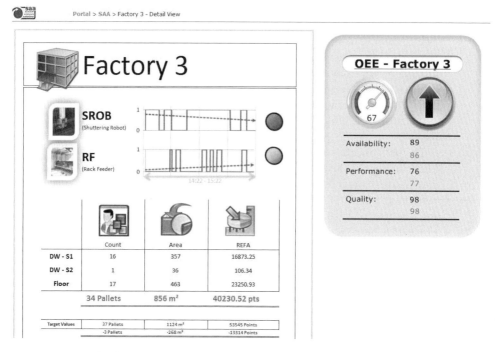

Figure 6.84   OEE display of concrete prefabrication plant.

process and a perfect material experience, a meaningful quality check could be done before curing and at least 10 hours before delivering. Recognizing defects at this stage enables repair or JIT reproduction as well as possible. Damages at demoulding and during transport are often harder to avoid, but if well planned before delivery this issue could be solved too.

Some systems of integrated quality management are still implemented into prefabricated concrete production; some of them are in elaboration.

- Supporting QC – staff by mobile computers, integrated into the MES system, using a predefined questionnaire and a digital signature as well as a quality error tracking with follow-ups, repair or reproduction. Sometimes there is even an interface to ERP systems to automatically inform about possible delays, repair costs or reproduction.
- Cameras and image processing – using cameras and image processing is sometimes very challenging in the grey, greasy and dirty environment of a concrete prefabrication factory (Figure 6.86). However it is realized to take photographs before and after concreting by use of high-resolution, industrial cameras and perhaps using rectifying systems for these pictures to document the state for each element in the production. This is done automatically. Unfortunately the pallets carrying the elements are large and the contrast is small, so by use of today's resolution and cameras it is not possible to get a mm precision on measuring these photographs.

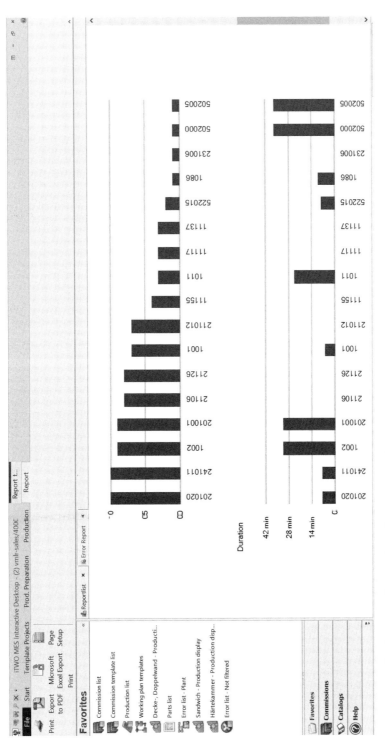

Figure 6.85   Graphical error analysis to show the most occurring messages.

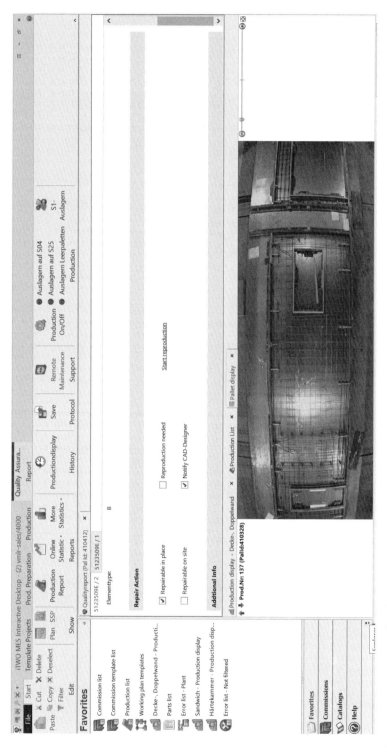

Figure 6.86 Automatically taken QC photograph of a pallet, without rectification system within an MES QM system.

Modern image processing systems can measure the content of an element much more in detail using a combination of laser scanner, camera and high precision image processing. Because of having to scan the whole pallet and using a highly developed system costs are very high and not proven for final return on investment, especially if the production is designed to make very few errors by itself due to high degree of automation.

- Because in most countries on the world the supervision of material quality is done in-house as well as by external third-party surveillance, the registration of material specimen as well as the charge numbers for example for reinforcement steel and concrete hast to be automatically integrated into the MES QS system to report this information back into the integrated BIM Model for QS. The amount of any material actually utilised could be registered too, for later, on LCA cross-check.

- Integrated production systems records many necessary and useful data for each precast concrete element. To retrieve the data at a later stage, the elements should be marked. Until the mounting process at the construction site it would be enough and very cheap to mark the elements with any kind of barcode labels. When the barcode is scanned with the mobile MES QM software, staff can find all information necessary to treat the QC case. If it is necessary to check these values after the construction is finished, only RFID chips concreted into the precast element will help. For reinforced concrete you have to use special tags and mostly agree about a rough position within these elements to find them again. Modern development of RFID – leading to a real IoT – Internet of Things – system for precast concrete will make reading distance and amount of possible information much bigger (Figure 6.87). Anyway the costs of concreting in RFID tags are still slightly high, so they are used very seldom at the moment

Figure 6.87   IoT for concrete prefabricated elements: RFID concreted into the structure.

6.5   Limits of automation

The possibilities of automation are various today and it seems there is nothing that cannot be automated. Of course automation brings a lot of predictability to an industrial production process where working staff in between automation cells might bring some incalculability. Besides arguments from labour policy and industrial psychology there are some other issues sometimes limiting automation in the precast concrete industry.

6.5.1   *Labour cost versus automation*

One factor, which has to be considered at any automation project, is the comparison of costs for automation versus the cost of manual working staff.

Where in Europe the cost of manual work and all associated taxes are very high and so automation is highly requested to reduce low skilled labour in prefabrication factories, in other countries in the world labour cost are so small that automation is not economic. Even in countries with low labour cost it might be meaning full to do a certain kind of automation or at least mechanization because prefabricated concrete elements have to meet certain quality requirements to be successfully placed on the market, for example, hollow core floor slabs are manufactured throughout south east Asia. If sometimes these elements are manually produced with poor precision, this highly effective, fast and durable construction system could get a bad reputation in the market, which will again lead to much more cast *insitu* construction, wasting all the advantages of prefabricated construction.

Labour cost of mounting the prefabricated elements on site must also be considered, as well as considering cost savings due to faster erection, better quality, independence on weather and the possibility to use the building earlier in order to start return on investment.

6.5.2   *Costs, necessary skills and ROI*

Of course it is not easy to modernize, industrialize and automate the construction industry and there should be some environmental consideration too. Last but not least on every local market in the world, prefabricated construction industry was introduced and is slowly gaining shares. For a precast concrete company to be successful it will be useful to either design and build in-house or have a good network of builders supporting with high-quality concrete elements.

To gain a safe return on investment not too far in the future it is necessary to consider all environmental conditions in the market. For automation issues the following should be considered:

6.5.2.1   *Step-wise automation*

Depending on the planned factory output it might be meaningful to start with a smart automation, especially in countries where the concrete prefabricated technology is not so well known, unlike in Europe and North America, and where labour costs are low.

Using a plant layout, which could be easily extended and prepared for this, as well as having simpler automation just to ensure high quality – plotting devices, high quality pallets, concrete spreaders, batching plants and minimum of MES – could ease up the startup phase if the necessary production volume is not too high. After some years, if the market share has grown, an extension is possible.

If the initial required productivity seems to be very high, it even could make sense to start with smaller plants and extend them for a second separate plant, than to build huge prefabrication factories. Huge size plants are sometimes very hard to operate because wide experience, professionalism and IT/ERP/MES equipment is necessary to keep track and organize all of these issues including supply chain management and deliver, transport and mounting.

Sometimes super prefabrication plants are not gaining the designed output capacity within a short range, and the return on investment would be delayed because of environmental conditions of the local construction market. In any of these cases, these factories themselves may receive their technical and output acceptance but have to wait for the ROI due to over-dimensioning or over-automation.

6.5.2.2   Technical employees and their necessary skills

It is necessary for industrial and automated factories to have a certain number of well-skilled employees. The skills are sometimes very special and the plant management should consider to take experienced staff for at least the first years until its own staff has grown its industrial precast concrete knowledge.

- Draftsmen for industrial prefabrication construction: CAD technicians are not architects, they must be construction engineers or at least draftsmen with knowledge to consider a lot of topics included in this book. They will by highly supported by CAD systems but finally the digital model of the building, constructed out of prefabricated concrete must be error free, otherwise the factory will produce errors and have costly repairs on site.
- The best industrial factory must have a continuous preventive maintenance, even if the plant is brand new. Automation is sometimes very technical and relies on this techniques, so mechanical and electrical staff is necessary to keep it successful running for decades. Even in the hopefully rare cases of standstills, these staff members have to be fast in analyzing the reasons (supported by the MES) system, effectively retrieving remote assistance to finally solve the problem in a very short time.
- A concrete prefabrication plant is only efficient and successful if the delivery transport and the erection logistics is perfect. That is why some smart staff members must be located in the logistics office organizing the information given by the integrated IT solution. These staff are ideally bringing together the unpredictability of the construction site with the high JIT predictability of prefabrication production. Managers for JIT production are the internal partner to deal with and to give the productivity targets into the plant.

6.5.2.3   Permissions and authorizations

One more topic should be considered as a limit of automation. Construction systems in the same way that precast concrete construction systems are usually authorized by government or any other organization responsible for the government.

If the construction regulations are difficult for automation, like the above-mentioned connection technologies, it is necessary to consider the time for the certification of the prefabricated variation to enable automated production in a certain plant. If it is slow and difficult to develop local construction rules a "back door" must be considered to produce the actual certified elements for the first period and change to more efficient methods after the certification is done.

6.6   Summary and outlook

Mechanization, industrialization and automation is for sure a key feature of the precast concrete industry. It might be the most durable system to construct buildings out of concrete elements, which could be very flexible in shape and equipment, as well as extremely durable. So flexible that even architects could be satisfied in their ambition of modern buildings.

For automation of this industry, it is not enough to have many automatic machines. It is necessary to have either a good IT solution to really make use of an industrialized JIT production plant as well as having highly motivated and skilled staff to achieve industrialized production and delivery. Moreover, smart management and sales should identify the market to make maximum use of such a plant.

Software for integration will be fast developing, not only considering the production process, but rather integrating real estate planning, construction, planning, BIM, production and delivery.

Machinery will further develop for different new prefabricated construction methods, like PPVC (Prefinished Prefabricated Volumetric Construction) systems which could lead to the productivity, quality and reliability of an automotive factory with the only difference that prefabricated concrete elements could be extremely heavy and often produced in lot sizes of one versus batch production.

Due to the worldwide increase of labour costs and the demand on quality and durability of modern precast structures, the market will anyway develop in this direction – in some places very slow, but in others very fast.

Part 3

Industrialisation of Concrete Structures

Chapter 7

Lean Construction – Industrialisation of On-site Production Processes

Part 1. Construction Production Process Planning

Gerhard Girmscheid

ETH, Swiss Federal Institute of Technology, Zurich, Switzerland

This chapter introduces the importance of the work planning process and the procedures to ensure production planning is industrialized in a lean management way. To enable industrialized lean management processes on site, the planning of the execution process for construction sites requires systematic incorporation of the environmental conditions in the work preparation and logistical planning processes. Particular attention must be paid to "what" is being built, "where" it is being built, and "which" boundary and environmental conditions need to be considered in the construction production planning. These three "Ws" influence the choice of construction methods and logistics on a construction site. Building on these three "Ws", the construction task itself must be systematically broken down into main, module and elementary processes, known as the "work breakdown structure". Once the processes have been broken down for the individual construction phases, components and manufacturing steps, the construction method can be selected. In doing so, the importance of a systematic selection procedure for identifying the best possible construction method, given the specific boundary conditions of the project, is highlighted. The methodical procedure for assessing whether the methods are fit for purpose, and for performing both qualitative and quantitative comparison of the methods, is described. The selection of construction methods and manufacturing workflows must comply with, or undercut, the specification of the target work execution estimation. The work execution estimation then provides the key performance factors – in terms of both work hours and cost – for the site, in reference to the production process for each construction element. Once the methods of construction have been selected, the schedule and resources for compiling the construction program are planned, taking account of the temporal and spatial interdependencies of the various production processes, as is the final work execution

Modernisation, Mechanisation and Industrialisation of Concrete Structures, First Edition.
Edited by Kim S. Elliott and Zuhairi Abd. Hamid.
© 2017 John Wiley & Sons Ltd. Published 2017 by John Wiley & Sons Ltd.

estimation containing the final work specification and key performance factors for the site management. Among the execution planning processes, logistics planning plays a key interface role, interactively linking the other work planning areas. The key to successful, value adding cycle and flow production on a construction site is to coordinate the logistics planning processes with the demands and needs dictated by the construction workflows. Based on these principles, the cycle and flow work processes can be planned in detail *top down*, and executed and refined in detail by the site management using a *bottom-up* process.

7.1   Work process planning (WPP)

7.1.1   *Construction production planning process – introduction*

More and more construction companies are unfortunately experiencing customers who are dissatisfied either with their adherence to schedule and/or their quality. "Defect management" has become a common phrase in the construction industry, whilst "Project success" poses a virulent challenge for companies. Far too many projects that started off with a promising estimation ultimately ended in loss, despite project and site managers initially forecasting a profit. The deficit grows from one month to the next as execution progresses. Is that necessary, even inevitable? It raises the question as to whether the estimation department is making mistakes or the site manager does not have his site under control.

Commercially successful construction execution must satisfy two fundamental requirements:

- the customer must be satisfied
- the project must produce a profit for the company

If these requirements are to be met, the construction and interior finishes companies must plan and coordinate their fabrication workflows as cycle and flow processes, which they then implement systematically on the construction site, while controlling them on a weekly basis and improving them continuously.

"Workflow process" or "cycle and flow process" are phrases that characterize this paradigm shift. The goal defines the process – projects cannot be managed without clearly defined performance targets. For many SMEs performing construction, facade or interior finishing work, the way is still the goal. As they build; on some sites they may even know the estimate, they may have no idea how many target hours have been planned for fabricating the foundations, erecting the formwork, reinforcing steel, cast insitu concrete or erecting prefabricated concrete elements, and so on. When the hours are finally counted, they are surprised to find they are (say) 20% over budget, with some interior finishes firms often 20 to 40% over.

Of course this can be changed. This chapter will focus on work preparation planning (known as "top down") and work control on site (known as "bottom up") that use weekly and daily plans to successfully implement the desired targets through continuous improvement at foreman and team levels.

The construction workflow must be coordinated as a cycle and flow process between the respective trades at each phase of construction, or for each element, and for each individual job, together with the respective logistics. This cycle and flow process must be coordinated for both structural work and interior finishes, with their respective parallel and consecutive trades to ensure that no work time is lost.

This chapter will also address the issue of how to make construction sites more successful, and we will learn more about the concept of "Lean Construction" (used personally by the author for 18 years) on smaller and major construction sites. Examples for major construction site, where WPP has been successful, is the road overpass in Bangkok (Figure 7.1), a pumping station in Alexandria (Figure 7.2) and a high-rise building in Zurich (Figure 7.3).

The systematic application of such workflow processes on the construction sites of medium-sized and larger companies is crucial. The aim must be to ensure that the work adds value and does not generate losses – neither in terms of hours nor through inefficient work or inefficient cellphonlogistics because materials, tools, or equipment are missing. Cellphone logistics means intuitive decisions for logistics and using cellphone to solve missing supply for material or equipment which is not properly, instead using advance planed procurement.

Work preparation should build on a well-developed work execution estimation. Since the time for completing both tasks is limited, they require fast track performance alongside each other. The basic determinants for shaping construction production planning (Figure 7.4) are:

Figure 7.1   Road overpass – Bangkok.

Figure 7.2   Pumping station – Alexandria.

Figure 7.3   High-rise building – Zurich.

1. The external determinants:
   a. Contract
   b. Project concept or type of project
   c. Site and its utilization for setting up the production plants and the necessary auxiliary equipment, and housing the construction methods and workflows.
2. The internal determinants:
   a. Work calculation
   b. WPP (construction methods, resources, construction program)
   c. Availability of manpower, equipment and expertise
   d. Proprietary and subcontractor works

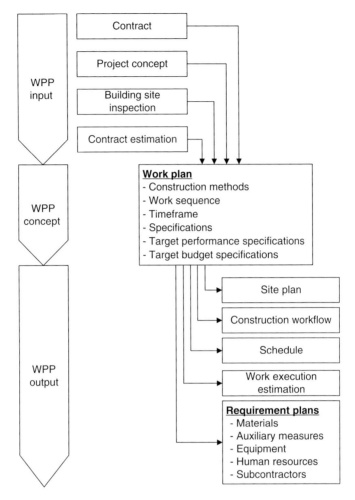

Figure 7.4   Work preparation planning.

Although most projects nowadays commence immediately after signing the contract to meet the owner's time-to-market requirements, construction enterprises should not try to speed up the project delivery time at the expense of execution preparation. Medium-sized construction enterprises often complain that they do not have any time for work preparation and have to commence improvised execution immediately. Unless the enterprise only builds one specific type of structure, this approach generally only produces a fleeting benefit that is not sustainable over the entire construction period. An example of this was the use of precast concrete for school buildings in Malaysia in the 1990s, for example, as shown in Figure 7.5.

The additional costs associated with sub-optimal choice of construction methods, or insufficient or superfluous resources, such as auxiliary materials, skilled labour, and so on, quickly negate the initial benefits. Efficient work planning often enables reduced execution costs and times, allowing a contractor to reduce its own costs, gain significant competitive advantages, and actually achieve or even surpass the targeted results.

Figure 7.5   Precast concrete school buildings in Malaysia.

Further success can be achieved by eliminating idle times and disruptions. Moreover, the workforce is more motivated when it can focus on its own work performance instead of having to permanently troubleshoot.

The factors for assuring the success of a best practice construction production process are (i) good execution preparation tailored to the specific project, with appropriate performance and cost target specifications, (ii) logistics for providing the right volume of resources in the right place and just in time, and (iii) all flanked by a continuous improvement process. It is not until you start planning construction production that you gain real insight into the project, allowing you to specify appropriate construction processes and methods and the required workforce, to order the right quantities of necessary machinery, auxiliary and construction materials for delivery at the right times, and to complete the workflow on schedule and budget. Controlling procedures are enabled by knowing the target hours. Designing the necessary workflows generates the following benefits:

- The construction task becomes more transparent for the construction managers, foremen, excavators, and so on, as they are familiar with the project and the weekly/monthly scope of work
- Arranging the provision of manpower and equipment, and commissioning subcontractors and suppliers can be assured in good time
- Partial and final invoices can be drawn up promptly, with all the economic benefits this entails.

Work preparation involves delegation of the following assignments:

- Workflow planning by the responsible construction manager/construction process manager
- Process control in accordance with the workflow chart by the people responsible on the building site, for example, construction manager, foreman, excavators.

A project structure plan with clear assignment of activities must be drawn up to estimate the work execution, design the controlling concept, and plan the schedule and resource requirements. To do this, the project must be broken down into its material and production-related components. Material components encompass the individual structures, building parts, and job/procurement specifications for the project. The material components are then broken down into workflows and auxiliary processes, based on a numerical system. Such an execution-oriented project structure plan should be linked to the owner's project structure plan to facilitate communication between the owner's and the executing enterprise's documents. The same should apply equally to the material components of the contract, such as structures, building parts and units items, wherever possible. The construction enterprise is free to design the production-related structure as it chooses.

Based on these preliminary steps, the following work planning phases can be performed:

- Construction workflow planning: Descriptive specification of the process activities with notes on execution
- Construction process planning: Choice of construction methods for elements and sections, with performance specifications, deadlines for provision of materials and auxiliary equipment, and deployment of workforce, machinery, and subcontractors
- Schedule: Chart tracking the passage of time with incorporation of described activities, complete with duration, derived from the material and production-related breakdown
- Requirements plan: Task-specific lists derived from the workflow plan and schedule to ensure reliable provision on time of the right quantities of materials, manpower, subcontractors, and machinery
- Special plans: Formwork, steel and timber structures, chutes and tracks, and so on, for on-time provision of efficient auxiliary materials
- Site installation plan: Plot plan with site infrastructure, cranes, containers, storage areas, routes for efficient insitu production workflows
- Preparation of the work execution estimation and controlling.

These planning tasks cannot be performed in isolation of each other, as each one influences the other. Planning a site infrastructure, for example, requires knowledge of the fabrication and construction methods that are going to be used as they dictate what machinery is needed on the construction site and the level of manpower to be expected. Then again, workflow planning depends on the site infrastructure. As such, construction process engineering is a key issue when planning the construction production/fabrication process.

In construction, the production process is dictated by the construction method with regard to the deployment and combination of the production factors (manpower, machinery, equipment, tools, fixtures) for machining and processing building materials. During the construction production process, input is transformed into output, a process that creates an asset for which the customer is paying an agreed price. Executing a construction job involves a series of sub-jobs which, in turn, are broken down into individual processes and single activities (elementary processes).

Construction production planning is performed at three stages of the construction process. In the case of major projects, a contractor must draw up a preliminary production plan for the project in question during the bidding phase. This bid-phase production plan enables determination of the specifications for estimating resource needs (equipment list, team size, total wage hours, equipment hours, standby times, etc.). Production planning is therefore a *top-down* process for designing the basic production plan as a target specification, and, at the same time, a *bottom-up* process organized as a cybernetic continuous improvement process to enable achievement of the target specifications.

During execution, production must be planned and implemented on the construction site by the local team in a *bottom-up* process in construction phases, construction stages, construction elements, work cycles, weekly/daily workflows, and in a cycle and flow process. The production plan must also be adjusted to reflect any disruptions, and uncertainties that arise. It involves organizing the work plan derived from weekly (detailed) plans and daily specifications based on the basic execution process and resource plan, which is updated monthly (*top down*).

The weekly plans are drawn up by the construction managers and foremen charged with performing the works, and are agreed with the site managers on a weekly basis (*bottom up*). They extend at least two to four weeks ahead, and plan both workflows and logistics, and the procurement and provision of equipment and materials. With the aid of the work prep target construction program and of implementation based on weekly target plans, self-caused deviations can be identified, but so can specific disruptions to the construction workflow, as well as indirect changes caused by the owner to the purchase order. This enables causes to be tracked in writing to justify subsequent claims.

The target basic execution process plan (BEPP) drawn up by WPP compared to the actual construction program each month to identify deviations and assess the impact of improvement measures, as well as to enable performance deficits and deviations from the target plan to be made good. Reference must always be made back to the target BEPP to highlight possible effects on contractually agreed milestones and the final completion date. Thus, contractors and the owner can take action early to put the schedule back on track. Any necessary acceleration measures must be initiated and paid for by the contract partner who caused the delay. The target and actual construction programs are planned in detail with regard to calling up equipment, teams, materials and subcontractors for a three-month forward period. The weekly and daily construction programs are then developed on the basis of these specifications.

One of the biggest mistakes in proper project management is the bad habit that many project managers have of drawing up a new target program each month. Ultimately, this results in nobody knowing the base target deadlines any more, or whether deviations have occurred. Project and construction managers often consciously adopt this approach in order to conceal the fact that the project has gone off track.

7.1.2   Construction production process – principles and sequence

Given the elements to be built and the various methods/processes for producing each element, construction production is a highly variable process. Even if the structure is already specified by the structure plan, and the main target defined as *minimal construction production costs*, it is still not easy to observe the economic minimum principle. Apart from which, generally only a limited number of construction methods are known. As such, even with the use of analytical simulating tools, the decision will always be based on limited rationale (Weber, 1968). This does not detract from the value of systematic, target-oriented procedures; it is simply a basic fact governing human actions in social and technical systems.

Figure 7.6   Systematic construction production planning.

Systematic, analytically generic construction production planning that is structured as illustrated in Figure 7.6 is necessary to ensure a rational decision-making
process when choosing the production method. This procedure is explained below.

7.1.3   Systematic basic production process planning – steps

When developing the basic construction production process and resource plan, the
analytical, generic *top-down* procedure for designing or planning construction processes must comprise the following analyses in sequence (Figure 7.6):

1. Break the building down into modules, element groups, and construction elements
2. Identify construction production processes for the elements
3. Coordinate the construction production processes for the individual elements at element group level (e.g., all vertical elements or all horizontal elements on one floor of the building)
4. Determine the generic fabrication sequence based on constructive, structural stability and technical requirements
5. Specify the timing milestones for trades groups to be fabricated in main processes due to the specified total production time (owner's/investor's master plan).
6. Break the main process of each trades group down into:
    a. Logical module processes, sorted by element, to determine the fabrication sequence based on constructive, structural stability and technical considerations
    b. Choice and allocation of the construction methods to the individual parts and sections
    c. Logically generic allocation of the elementary processes for the fabrication of the elements to the individual parts and sections, for each respective construction method, for example, in cycle: assemble formwork, calculate, apply concrete, hydrate
7. Define/specify hourly performance factors for equipment and the work performance factors for the teams, depending on the chosen construction method
8. Define the number of teams and equipment chains to achieve the milestones/interim deadlines and meet the final deadline based on the hour/work performance factors for the specific construction method
9. Define the resources and time needed for the elementary processes based on the defined groups and compositions of teams/equipment
10. Develop risk overviews and identify uncertainties; estimate the impacts of the chosen construction methods and assumed hour/work performance factors
11. Examine the main process duration, bearing in mind the fabrication sequence and the duration of the elementary processes, as well as the uncertainties (probabilistic buffer); adjust resources if the aggregate duration of all main processes exceeds the specified total production time, or key milestones are missed.
12. Iterative commercial optimization of construction production in line with the economic minimum principles:
    a. comparison of various production process variants in order to identify the robust, optimal construction production process
    b. optimized details of the selected, optimal construction production process
13. For the optimal construction production process for each main process, bearing in mind the interactions between the main processes on the target construction production process plan, the next steps involve developing:
    a. Schedule – specifying the timing of the elementary processes
    b. Resource plan – material specification of the resources (manpower, equipment, auxiliary aids, operating equipment) in terms of quantity and quality
    c. Logistics plan – timing and allocation of space for resources in terms of quantity and quality

    d. Logistics infrastructure – developing the site infrastructure with office and social facilities, as well as logistics infrastructure, defined storage areas, and site transport facilities.

7.1.4 Continuous construction process management

During construction production, the production process must be organized in detail using a *bottom-up* process based on weekly work plans that incorporates a cybernetic continuous improvement process (CIP). This is achieved as follows.

- Building on the basic construction production process plan, the resource plan and the target work specifications (*top-down* WPP), together with the monthly and weekly work plans must be developed *bottom up* by the people executing the work (operatives).
- The monthly work plans containing the elementary processes, resource allocations and target performance specifications serve as the basis for the weekly work plans for each team. The latter assigns manpower, equipment and auxiliary materials to the activities slated for that week on a daily basis.
- On a weekly ongoing basis with (for example) a two-week look-ahead, the target weekly work plans are populated with details of those activities that are specified in hours, based on the target performance specifications/actual performance. The weekly work plans are populated with details for each day. The weekly work plans capture the target fabrication of the elements and the cycle sequence, together with the key individual work steps. The work steps and necessary resources (manpower, materials, equipment, auxiliary materials) are listed on a daily basis, together with the target work hours derived from the work execution estimation. In addition, the teams assigned to the weekly work plans must coordinate their activities in order to eliminate activities that do not add value.
- The target weekly and monthly work plans are reviewed and continued at weekly/monthly intervals, based on the *bottom-up* process. If deviations occur, corrective or risk mitigation/prevention actions are taken to ensure adherence to the milestones, specified total duration, and target budget.
- The monthly work plans include a look-ahead, and form the basis for planning the provision and coordinating the call-up of materials, specialists, subcontractors, equipment, and auxiliary materials.
- The weekly work plans define detailed deadlines for collaborative teams working in sequence, and for the works themselves. In addition, teams working in parallel must coordinate the shared use of the general site infrastructure as well as their own activities and jobs in terms of timing and space requirements. Resource call-ups are timed precisely in the weekly work plans in a *bottom-up* process relative to the overarching management of purchase orders, subcontractors, equipment and manpower deployment, in order to eliminate loss times, for example, through having to wait.

The target weekly work plans must be drawn up and revised each week. The teams or section leaders and foremen should revise the plans towards the end of the working week. The revisions are then discussed at the construction managers'/foremen's meeting with the site manager at the end of the working week. The target specifications are then binding.

The construction production process is designed cybernetically, based on the *top-down* basic construction production plan, using:

- *Bottom-up* elementary processes and activities, as well as their degrees of achievement, together with suggestions for achieving the targets through CIP.
- *Top-down* examination of the effect of achieving the weekly targets on the specifications governing the overall construction process and of feeding back the *bottom-up* suggestions in terms of their effect on overall target achievement. The aim is to ensure the team feels responsible for achieving the targets.

7.2   Construction production process planning procedure

Construction production is planned on the basis of the planning process illustrated in Figure 7.6, which in turn builds on the axioms of construction production theory and planning. Planning the construction production process can be broken down into the following nine steps:

1. **System structure** – Break the entire system (building as the final product) into (Figure 7.7):

    a. Sub-systems: Floors, basement, roof, rooms, facade
    b. Modules/elements: Ceilings, walls, staircases, lift shafts, facade elements Flooring, windows, doors, plaster/paint
    c. Properties: Physical, technical, architectural

2. **Process structure** – Break the construction production process (the construction task) down into module and elementary processes and jobs/activities (Figure 7.7, Figure 7.8, and Table 7.1) for fabricating the individual elements.

    The construction task "building a bridge" serves as an example to demonstrate process classification. Building the bridge is viewed as the construction production process (Table 7.1). This construction production process can be broken down into the following main processes:

    a. Setting up temporary site infrastructure for insitu production facilities
    b. Fabrication of the permanent sub- and superstructures for the bridge

    Further module processes relating to the individual elements are necessary to realize the individual main processes, for example, fabricating the foundations, abutments and pillars, as well as the necessary storage areas, drainage.

    Breaking the module processes down into elementary processes (Table 7.2) is crucial for identifying suitable combinations of methods, comparing them to each other, and ultimately selecting the most suitable. A basic requirement of process engineering is to use the simplest and most robust means possible to achieve a practical solution that takes account of all relevant influencing variables. This requires knowing which options are available for decision, which impacts must

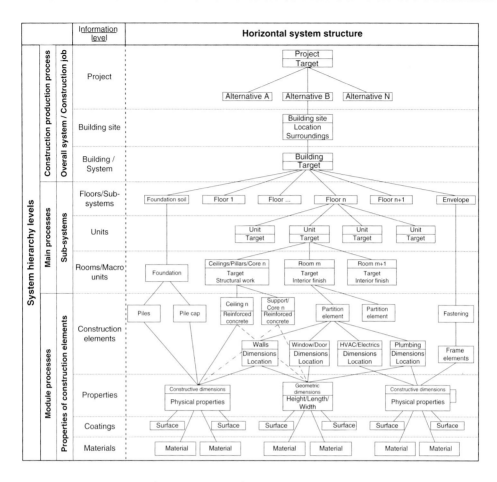

Figure 7.7   System structure of a construction task.

be expected from choosing one of these options, and which decision to make if specific criteria are given. Methodical decision-making procedures (operations research) can help here (Table 7.2).

3. **Fabrication sequence** – The fabrication sequence for the construction elements/parts is broken down into generic sequential levels determined by physical and locational aspects (Figure 7.7):

   a. Supporting structures from bottom to top
   b. Constructive sequence conditional upon stability
   c. Finishing work in layers, sequentially from the construction to the surface level
   d. Fastening elements prior to element assembly
   e. Partition walls – studs and one-sided cladding
   f. Slits must be cut and shafts semi-bricked in masonry to make room for ducts
   g. Lay plumbing, heating and ventilation pipes and ducts
   h. Close partition walls and clad
   i. Pour screed

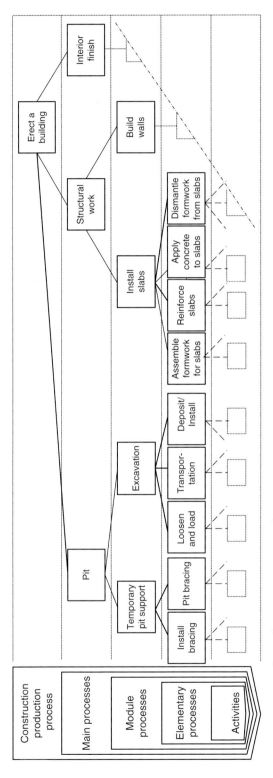

Figure 7.8   Process hierarchy in construction production.

Table 7.1 Process structure of a construction task

Construction job	Construction sub-jobs		Elementary processes	Activities
Construction production process	Main processes	Module processes		
		Site logistics	Set up cranes	Prepare
			Build construction roads	Transport
	Site infrastructure	Site social facilities	…	…
			Set up containers	…
		…		
			…	…
		…		
		Foundation	Surround the pit	Loosen
			Excavation	Load
		Abutments	…	Transport
			Backfill	…
		Columns	…	Interim store
	Substructure		Assemble formwork	…
Bridge construction		Storage	Reinforce	Re-assemble
			Apply concrete	…
		Drainage	…	…
			Install drainage	…
		…		
			…	
		…		
		Box girder	Assemble formwork	…
			Reinforce	Set up scaffolding
		Drainage	Apply concrete	…
			…	Prepare formwork
	Superstructure	Road surface	Install drainage	…
			…	Lay reinforcing steel
		E&I	…	…
				…
		…		Dismantle formwork
		…		

    j.   Paint walls

    k.   Install flooring

    l.   Install fittings (plumbing, heating and electrics)

4.   **Interdependency** – Module and elementary processes, as well as activities (Figure 7.8 and Table 7.2), are interlinked in a cycle and flow process due to following interdependencies:

    a.   Upstream – predecessor resp. superior dependency

    b.   Downstream – successor resp. subordinate dependency

    c.   Lateral – neighboring dependency at the same hierarchy level

Table 7.2 Fabrication variants

Elementary processes	Fabrication options			
	1	2	3	4
Reinforce	reinforce on site	install prefabricated reinforcing steel	some on site, some prefabricated	____
Assemble formwork	Board, plank, square timbers (conventional)	Large board formwork on movable working platforms	Climbing formwork	Sliding formwork
Apply concrete	with bucket	with concrete pump	____	____
Transport logistics	Tower crane on tracks	Climbing crane on elevator shaft	Mobile crane	____

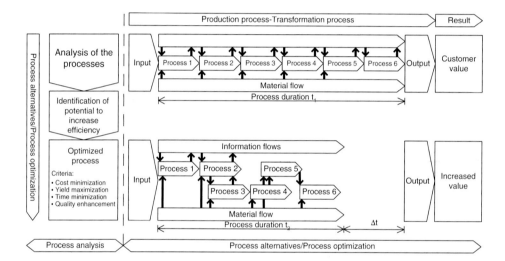

Figure 7.9 Production process analysis.

For constructive, structural stability and technical reasons relating to the fabrication sequence, these interdependencies must be identified and collated if the processes are to be optimized and executed in parallel (Figure 7.9).

In order to define the production process interdependencies at module and elementary process level, as well as at activity level, the generic dimensions relating to system structure, process structure and fabrication sequence must be set in relation to their temporal interdependencies (Figures 7.10 and 7.11).

5. **Generic axiomatic relationship** – In order to provide the information and workflows specifically needed for the individual processes in projects involving a tight schedule, and planning and execution processes that are being performed alongside each other (simultaneous engineering), then (Figure 7.10) user and system requirements, draft parameters, planning process, and inspection and approval process must be coordinated. To do this, the interaction among the trades must be identified for each part during the planning and execution process. The information, planning and work results of the upstream activities that are needed for this must be coordinated in good time in the process, and the

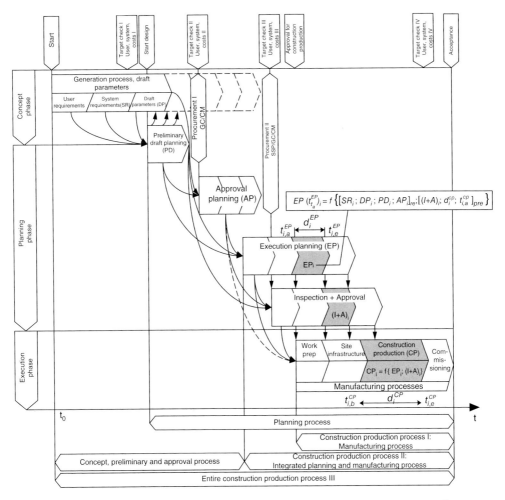

Figure 7.10   Generic axiomatic relationship of planning – to construction process in terms of timing, and to system requirements, draft parameters, preliminary and approval planning in terms of content and timing.

generic interdependencies highlighted. This is the only way to avoid non-value adding activities (waste) due to changes in planning and production.

6. **Estimate the main process duration** – This requires an iterative *bottom-up* approach to estimate the duration of the main processes based on floor areas or construction volumes, floors or volumes, together with the associated works. This procedure produces provisional milestones for the main processes that must be slotted into the specified total production duration (master plan).

7. **Iterate the main process duration** – In keeping with the generic timing interdependencies, planning the module processes within the main processes can commence. To do this, the elementary processes are generically structured for the module process relating to a part (Tables 7.1 and 7.2). The duration of all module processes comprising a main process must be achievable within the main process

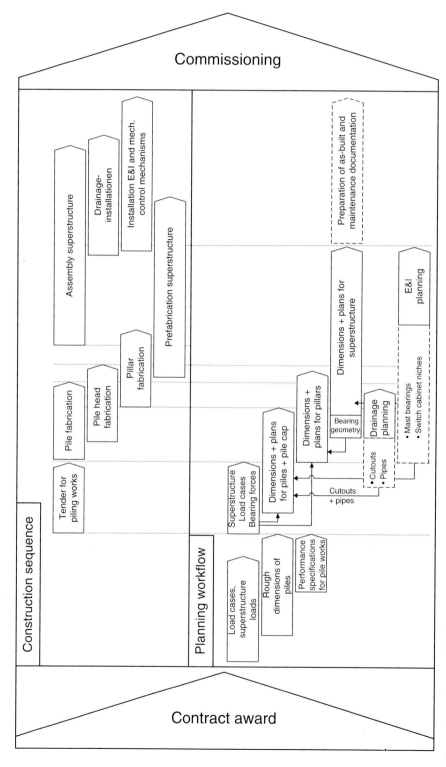

Figure 7.11 Fast-track bridge construction – optimized production process for insitu concrete pillars and prefabricated superstructure.

milestones, bearing in mind the sequential and parallel interdependencies. If this is not the case, the first cybernetic iteration process successively adjusts the resources and sequence of elementary processes and module processes – bearing in mind the generic interdependencies – until adherence to the timeframe dictated by the main process is assured. This procedure is repeated for each main process (Figure 7.9) and ultimately results in the basic production workflow (Figure 7.11).

8. **Leveling resources** – Next, the (even) utilization of resources within the main processes is examined. If team deployment is not even (slack time) the corresponding, interdependent elementary processes must be changed to level the resources (teams/equipment/auxiliary materials). In doing so a distinction must be made between two cases:

   a. The main process duration is shorter: no further action needed.
   b. The main process duration is longer:
      i. need to examine whether leveling other main processes enables time savings in order to comply with the contractually agreed total production time.
      ii. If the total production time is also exceeded, the option of increasing resources in one or several main processes while maintaining even utilization needs to be examined to ensure adherence to the total production time.

9. **Cost analysis** – As the outcome of the construction production process (the structure) is specified in the production phase by the tender documentation, contract, and approval and execution plans, the *economic minimum principle* must be applied. This can be performed in two steps:

   a. Selection of the lowest-cost construction method(s) from various alternatives.

      The various methods and processes that satisfy the project-specific conditions are subjected to cost analysis. In addition to identifying the construction production method and associated resources (manpower, equipment, etc.) and costs, the "sensitivity" of each construction production process must be examined relative to the risks of delays caused by the weather or potential disruptions to workflows. Figure 7.12 illustrates one such sensitivity analysis aimed at containing the cost risk. When using the minimum cost principle, comparison should encompass, not just a deterministic cost, but also the potential bandwidth.

   b. Selection of the lowest-cost construction method by varying the total duration or partial duration of the main processes as a result of optimizing resources with regard to reducing the fixed production costs by reducing the standby time for the site infrastructure and management, and the variable costs through use of high-performance equipment, and so on.

The outcome pinpoints the optimal process according to the minimum cost principle. When planning the construction production process, make-or-buy-or-cooperate decisions must also be made for each module and elementary process, namely, make it yourself or use third-party production resources.

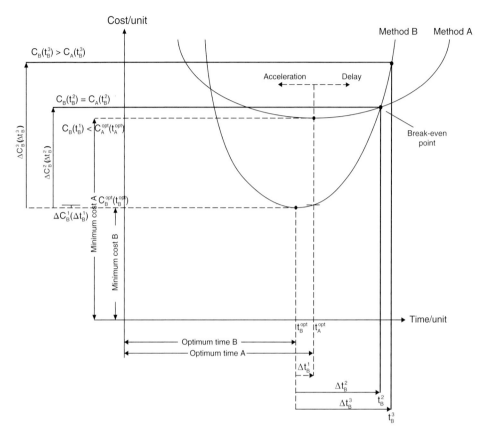

Figure 7.12   Sensitivity of construction methods in respect of time and cost.

The work prep tasks should be spread as follows:

- Site manager and WPP team: WPP work plan with specification of target performance, cost limits and resources
- Construction manager or foreman: Control of work by initiating and overseeing work activities on the construction site based on the work prep plan

If this division of labour is to be fruitful, the site and construction managers must study the project together, in advance, and make sure all specifications are feasible from a practical point of view. The interaction among planners and executors must be ensured if compliance with the specifications is to be assured.

7.3   Work process planning (WPP) – work execution estimation

Once the constructions methods have been selected on the basis of the minimum cost principle, the workflows, resources and target specifications are drawn up interactively in order to perform the work execution estimation and to prepare the schedule. The construction manager is therefore provided with a tool, in the shape of a work execution estimation, to enable them to manage and control the building site (Figure 7.13).

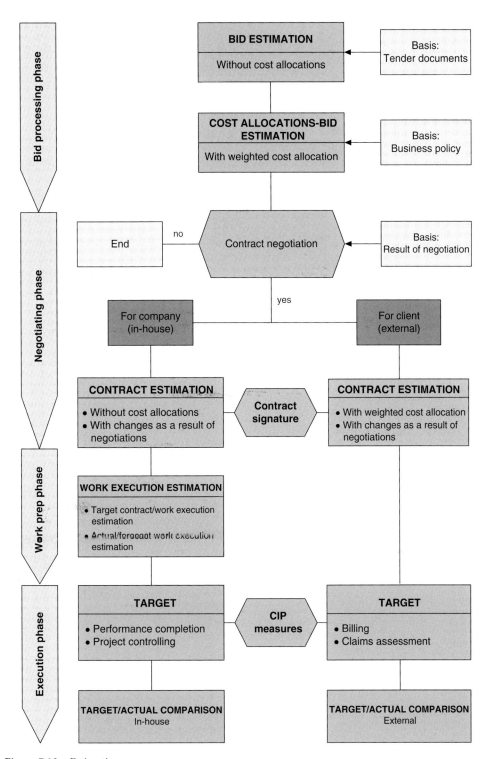

Figure 7.13   Estimating process.

The work execution estimation (Girmscheid & Motzko, 2013) specifies binding targets for the construction manager when managing and controlling the building site, and serves to estimate the cost of own and third-party works, as well as creating a control tool to analyze cost efficiency during execution. As such, it constitutes the basis for target-oriented forecasting, and a management tool for:

- the target performance specifications for individual workflows relating to the chosen construction methods
- the target budget specifications for each unit item and auxiliary activity in respect of hours and wages, material and auxiliary material costs, equipment hours and costs, and third-party costs
- the target budget and timeframe for the type and size of the site infrastructure
- the award limits for commissioning subcontractors for specific unit items

The current or modified bid estimation, and the bid estimation of cost allocations that are valid at the time of contract negotiation/award, form the basis for the contract estimation (Figures 7.13 and 7.14). The original estimations of the bid and cost

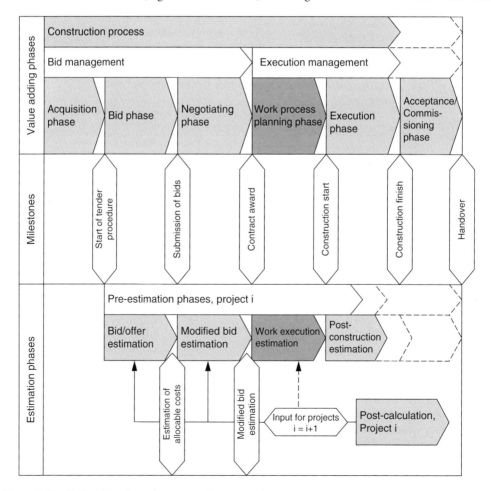

Figure 7.14   Estimation phases in construction enterprises.

allocations remain unchanged as long as the contract negotiations/bidding rounds do not produce any changes. Changes may become necessary, for example, if the contract negotiations result in:

- Price reductions
- Adjustments to payment terms
- Changes in volume
- Changes in the scope of work
- Flat rates for specific works
- Changes in quality standards
- Deadlines being moved

The bid estimation without a cost allocation, and the estimation of a cost allocation with assignment of the costs for the purpose of deriving unit prices, are transferred to the modified bid estimation by incorporating any estimation-relevant outcomes of the negotiations or bidding rounds. The modified bid estimation is for internal use by the enterprise, and contains budgeted work performance and cost factors for deriving un-weighted prices (Figure 7.13).

The contract estimation is used as the basis for settling with the owner during execution; it contains the weighted work performance and cost elements for potentially increasing the invoice total (external). Contract award procedures usually require submission of a sealed contract estimation to the owner/client. The contract estimation reflects all changes/corrections developed/agreed during the negotiating process.

In the contract estimation, the entrepreneur has weighted the site overheads, supervision and management costs in the hourly rates and cost types for each unit item, based on the bid/modified bid estimation. As such, the contract estimation only serves as an external information and billing tool in the event of claims for (i) the property owner, and (ii) the construction site. The contract estimation cannot be used to internally manage the construction site as it only contains weighted budget figures for external purposes, rather than any target specifications. The internal modified bid estimation, which is transferred to the work execution estimation, is used to manage the construction site during work preparation once the contract has been awarded (Figure 7.13).

When preparing the work execution estimation, any deviations or mistakes in the bid estimation must be rectified. The work execution estimation should form the basis to enable the construction execution management team to plan execution; during execution, it serves as the basis for controlling progress. The work execution estimation extracts the work performance and hourly performance factors from the individual unit items and assigns them to the fabrication workflows for the individual elements. The work execution estimation therefore serves as the basis for specifying the target hours for each element and their respective workflows in the cycle.

One key task of the work execution estimation is to specify targets for performance and cost limits as benchmarks for evaluating internal and external performance and the actual costs of a project. Focus must center on specifying cost and performance limit targets in order to enable effective controlling procedures on the construction site. The work execution estimation is not meant to be used for drafting, documenting or analyzing target/actual performance or for comparing costs; it provides guidance with regard to construction progress and the success of the construction project.

The following checks must be performed before developing the work execution estimation:

- Have deviations occurred/mistakes been made in the estimation?
- Check the contract performance specifications
- Which alternative items has the client confirmed?
- Conduct a final comparison of construction methods with work preparation to select the optimal construction method
- Which unit items are going to be performed by the enterprise, and which are going to be purchased in the market?

To this end, the work execution estimation must satisfy the following conditions:

- Throughout the entire construction phase, the work execution estimation must document – in detail and without being altered – the target performance and cost specifications for all internal and external work performance. This procedure ensures that the work performance and cost targets of the commissioned contractual works are documented, unchanged, for the controlling process. In addition, it must be formatted vis-à-vis the (modified) bid estimation such that the construction manager can immediately recognize the limit specifications for individual works and costs, the award limits for subcontractor works, and the total performance factors for the relevant works, which are then incorporated into the controlling process as target specifications.
- The work execution estimation must act as a basis and screen for defining the actual costs of comparable, differentiated works, and using the actual values to predict future progress to the end of the construction phase. Regular comparison with the target specifications is necessary as part of the controlling process.

In order to satisfy the sometimes contradictory requirements of work execution estimation, it must be broken down into:

- Target contract work execution estimation (TWE – Target work cost)
- Actual forecast work execution estimation (AWE – Actual work cost)

The TWE incorporates all budget specifications from the modified bid estimation for the practically differentiable works on the construction site in the form of fixed target specifications. The TWE contains the work performance and hourly factors for each element, together with the individual fabrication steps. These work performance and hourly specifications are extracted from the various unit items in the performance specifications of the modified bid estimation. They remain frozen throughout the construction period, and are only changed if subsequent works are commissioned. They contain the target performance and target earnings. The AWE is subject to constant adjustment from start to end of the project to reflect new lessons learned; it enables the ultimate costs, contribution margins, and results to be forecast to the end of the construction period.

The work execution estimation makes the unit items of the tender "transparent" by indicating all of the auxiliary measures that are not put out for tender separately but must be included in the billing procedure to enable the fabrication process. The individual unit items are broken down into elementary processes and their activities as part of the work execution estimation. The elementary process of "assembling formwork", for example, is broken down into the following steps:

- Prepare the formwork on the construction site (unload/assemble)
- Assemble the formwork, for example, on each floor
- Dismantle the formwork, for example, on each floor
- Rework the formwork (disassemble/clean/load)

together with the respective target hours.

The elementary processes (e.g., formwork, reinforcing steel, concreting) and associated activities can be estimated if quantification is sufficiently accurate and with the aid of target work and hourly performance factors (e.g., derived from actual figures or post-calculation of similar projects), with the aim of determining the target duration for completing the individual items and sections, and of defining as accurately as possible the target cost of each work. The individual billing items are broken by target work estimates including (i) ages, (ii) materials, (iii) auxiliary materials, (iv) equipment, and (v) subcontractor work. The target contract work execution estimation and the actual forecast work execution estimation both have the same format and both build on the modified bid estimation.

The target contract work execution estimation TWE(0) reflects the modified bid estimation as a frozen target specification of the hour and work performance factors, and the costs (0 = zero version). Zero means at begin of project and frozen till end of project. It should be structured to reflect the workload and cost involved in conducting all of the workflows needed to complete one unit item (including auxiliary tasks). Changes in construction methods, for example, *insitu* concrete instead of prefabricated parts, or subcontractor instead of own work performance, must be incorporated as target specifications. As such, all new lessons learned from contract work prep are already incorporated into the target contract work execution estimation (TWE(0)), for example, changes in work and cost estimates for changed construction methods, prices for subcontractor works and material inputs. If any grave estimation errors are discovered, they must be eliminated for the construction site. Such corrections impact the contribution margin for the enterprise and do not affect the profit or loss made by the construction site. Mistakes discovered in the modified bid estimation must be evaluated in respect of work performance and cost in the target modified bid estimation, and the impact on the enterprise's contribution margin analyzed. These lessons learned already allow an initial zero forecast of the expected operational profit/loss. The TWE serves as the work specification for the construction manager and the controlling department throughout the entire construction period up until final invoicing. It may only be altered to reflect approved claims based on the post-construction estimation. Enterprises approach this issue differently. The TWE should also be drawn up by the construction manager and the planner, at the latest, two or three weeks after signature of the contract. It must be presented to, and approved by, the management. Only then is it filed by management and construction management as an unalterable specifications document. The TWE forms the basis for the target weekly work plans on the construction site.

The format of the actual forecast work execution estimation is a mirror image of the target contract work execution estimation (TWE(0)). The actual forecast work execution estimation (AWE (i)) is updated regularly during construction in order to forecast the ultimate probable site result to the end of the construction period, and to perform target/actual comparisons based on the TWE(0). This enables early identification of deviations that may impact profit or loss, and the implementation of control measures. A continuous improvement process (CIP) can thus be initiated on the construction site to generate cost savings, which can be extensive in the case of repetitive works. This process requires motivation of the entire team.

The requirement in terms of hours that is thus documented can subsequently be used for target weekly work planning and as a possible tool for controlling individual execution steps (e.g., preparing the formwork, erecting/dismantling the formwork in (hour h/m^2), reworking the formwork; installing the reinforcing steel in (tonne/h); applying the concrete in (h/m^3)) on the construction site. The calculated costs are structured and processed in line with enterprise-specific specifications. Structuring them by trade or type of cost also enables their use as a budget-planning tool within the enterprise, and to evaluate award limits for subcontractor performance (Figure 7.15).

The control options presented by specifying the hours and costs can be converted into various tools for control on the construction site in order to compare target with actual hours and costs, and to compare the actual versus the scheduled construction workflow. The work performance and cost factors for controlling purpose are determined at least monthly.

A further output of the work execution estimation exercise is the specification of requirements in respect of materials, auxiliary measures and equipment (cf. (Girmscheid, 2010) for an explanation of the detailed work execution estimation procedure).

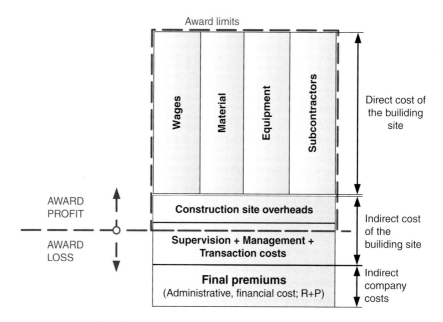

Figure 7.15   Cost limits for awards to subcontractors.

7.4   Work process planning (WPP) – planning the processes and construction methods

WPP forms the link between modified bid estimation/work execution estimation and construction execution (Girmscheid, 2012). The purpose of work preparation is to optimize the deployment of resources needed for construction operations by planning the work such that the lowest possible construction production cost is incurred. Furthermore, the target specifications from the work execution estimation can be used to prepare suitable tools for construction site control (Figure 7.16). Work preparation can be understood as a cybernetic control loop model consisting of planning, monitoring and control, which is subject to constant disruption due to the unique nature of each structure. The task of this cybernetic control loop is to achieve the forecast target as documented in the target contract work execution estimation with the required level of quality, and in line with the calculated budget and schedule.

The purpose of WPP is to streamline the construction production process by choosing the best possible construction methods, by using mechanized workflows and streamlining auxiliary systems, and prefabricating parts, to integrate the work activities in a cycle and flow process, and to design them such that the lowest possible construction production cost is incurred. These streamlining measures should be industrialized wherever feasible, with as many parallel flow processes as possible, and using high-performance, mechanical equipment. The extent to which industrialized prefabrication of parts might speed up the construction workflows should be examined. Prefabrication is suitable for supports, staircases, ceilings or filigree slabs, facade elements, bridge elements, and so on, which can be fabricated while the excavation and foundation phase is ongoing. In doing so, workflows can be decoupled and works performed in parallel. Such delivery methods are also known as fast track projects. Refer to the literature for an explanation of the structure and delivery of such projects (Girmscheid, 1997; Girmscheid & Hartmann, 1999).

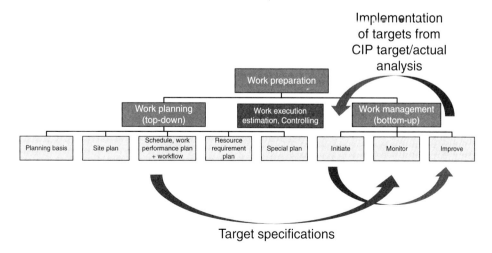

Figure 7.16   Cybernetic functions of work preparation.

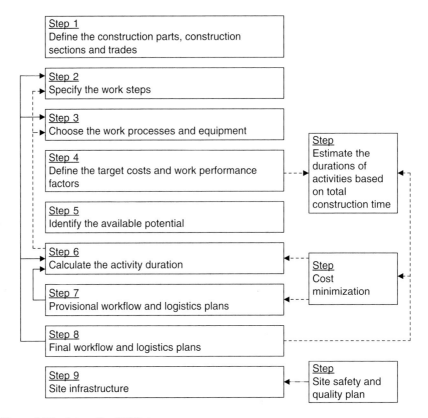

Figure 7.17   Interactive WPP steps.

The execution workflow must be broken down into cycles and organized as work steps in a flow process that are independent of each other in respect of both technology and timing. To do this, a construction workflow concept must be developed prior to detailed construction production planning, which breaks the execution task down into the following steps (Figure 7.17):

Step 1:   Break the construction task down into building parts, building sections, and trades.

Step 2:   Specify the execution sequence and main construction phases of the project.

Step 3:   Choose the potential construction methods and check their compatibility with the boundary conditions of the contract, building site, surrounding area, statutory provisions, and the fine tuning of the workflows.

Step 4:   Estimate the target work and hourly performance factors for the potential construction methods, based on the performance requirements for equipment/construction aids and processes (Girmscheid, 2003) and examine their impacts on the master schedule. Cost impacts must be examined iteratively based on the work execution estimation. The site installations for the potential construction methods must be included in the cost considerations.

Step 5:	Check the availability of equipment, manpower and expertise within the enterprise or on the market.
Step 6:	Examine the duration of the potential construction methods to determine the most probable minimum/maximum. These values are taken as fuzzy variables and should be examined with regard to the impact of possible disruptions to construction workflows (maximum) on the cost and schedule stability of the entire construction workflow. This process is repeated and optimized until it produces the most cost-efficient construction workflow, with the aim always being to minimize production cost.
Steps 7 and 8:	Develop the provision cycle and flow process with workflow and logistics plans. Once the construction methods and workflows have been aligned to the final work performance and capacity plans, development of the final workflow plans and logistics plans based on the phases and schedule and incorporating all commissioned trades can commence, together with the control tools. For major projects – and especially GC projects – a flow process should be developed, complete with logistics plan, that incorporates all trades involved in structural and interior finishing work. The flow process with logistics plan is used to coordinate the supply, storage and workflows for the various activities and trades that are being performed in parallel, to ensure they do not get in each other's way. The aim is to implement the economic minimum principle while complying with the performance and quality specifications.
Step 9:	Once the construction methods have been chosen, the flow process for the site infrastructure must be planned on the basis of the contractual plans and conditions, lessons learned from the site inspection, and the requirements relating to logistics and the construction methods. If the construction workflows are relatively complex, operations plans and brief descriptions should be developed to illustrate the breakdown of the cycles and work steps involved in the construction workflows. In addition to the detailed schedule and site infrastructure, work safety and quality plans must be developed to address the safety measures specific to each construction phase.

Work planning involves developing the site infrastructure, cycle, workflow, requirements, logistics, execution and special plans, together with the schedule. Work preparation is a fundamental prerequisite for cost-efficient design of interdisciplinary interdependent construction production and the smooth-running success of a construction site. It also serves as a tool for continuous control of work during execution. This requires the development of a systematic control system derived from the work execution estimation and target performance specifications (SBV, 2001), which implements and controls the agreed specifications, and monitors target achievement. As such, work preparation evolves into a cybernetic control loop during execution. This process is necessary given that unscheduled events and internal and external impacts may necessitate modification of the construction workflow, in spite of the work preparation. Since virtually every building structure is unique, not every event can be detected, and its impacts planned, no matter how good the work preparation is.

7.5   Planning the execution process

A construction job is composed of two different processing phases:

- Planning process/planning phase – Structural engineering
- Construction production process/execution phase – Structural production

The planning process incorporates the draft and engineering process to enable realization of the owner's idea; it incorporates the latter's ideas with regard to functionality and aesthetics, and takes account of local and local conditions.

The construction production process (fabrication process) brings the plan to reality (materialization) using construction production aids, with the aim of completing the required construction works with the required quality, within the specified time, and at minimum cost. This must be achieved through production planning and the execution of the fabrication process.

Depending on the structure in question, the planning and construction production processes are dependent on each other to differing extents. In building construction, focus is generally on the design, whereas the structures in heavy or bridge construction are strongly influenced by the fabrication options. Since construction generally involves the production of unique parts rather than mass or serial production, planning construction production based on the specific contract and property is absolutely crucial to ensure cost efficient execution. It defines:

- How the property is to be executed.
- Which production aids and resources are to be used for its purpose.
- The sequence the construction workflow.
- How the sequence of works is to be incorporated into a flow process.
- Which target performance is required, and how many target hours are available.
- How high the cost limits are.
- How the schedule for the main and module processes is to be defined.

Production planning for construction execution, that is, the construction production process, involves planning fabrication with the construction methods, provision planning, workflow planning and site infrastructure planning, together with target performance and cost limit plans (Figure 7.18). These four planning tasks cannot be performed in isolation of each other, as each one influences the other. Planning site infrastructure, for example, requires knowledge of the fabrication/construction methods that are going to be used as they dictate what machinery is needed on the construction site and the level of manpower to be expected. The site infrastructure, in turn, dictates the workflow plan (Girmscheid, 2012).

When performing the WPP for the fabrication process, the optimal choice of construction methods is therefore at the forefront. For any project, planning execution requires answers to the fundamental *WWW* questions:

- **WHAT** is being built?
- **WHERE** is it located?
- **WHICH** boundary conditions need to be taken into consideration?

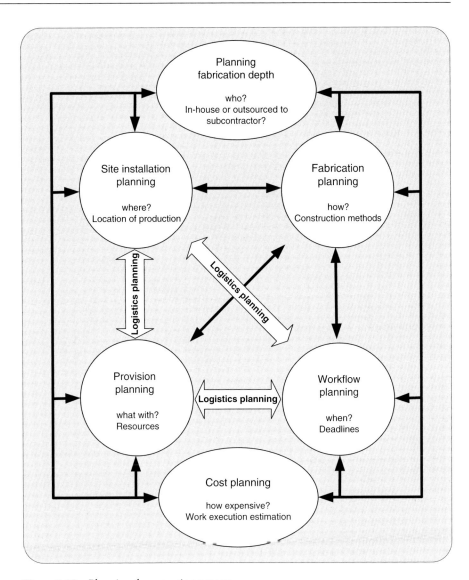

Figure 7.18 Planning the execution process.

Using a specific project the following section shows which conditions have to be taken into consideration in respect of the fundamental questions of "what, where, which" when planning construction work (WPP) and conducting the fabrication process.

Brief description of the project (WHAT)
The project description describes the project and its function and engineering elements.

Location of the construction project (WHERE)
The project is located in a residential area in the city of Zurich. Access to the construction site is provided by public roads (Figure 7.19)

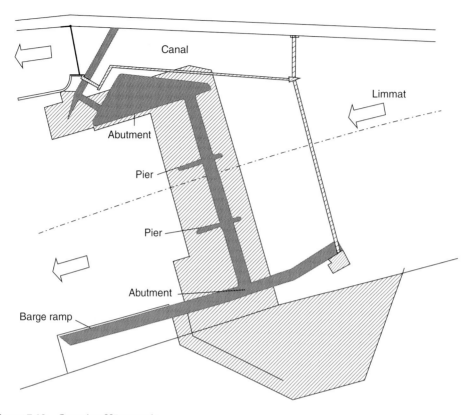

Figure 7.19   Overview Höngg weir.

Boundary conditions for production (WHICH)
The type of construction project and its location produce numerous natural, technical and environmentally-related boundary conditions that must, under all circumstances, be adhered to, and which influence both the construction workflows and methods (Figure 7.20).

Natural boundary conditions:

- Geology
- Hydrology

Technical and environmentally related boundary conditions:

- Site infrastructure
- Access routes
- Planted greenery
- Public pedestrian connections
- Works lines
- Weir bridges
- Electrical installations
- Water pollution
- Noise control
- Pile driving

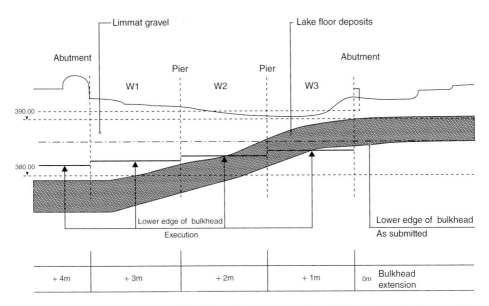

Figure 7.20   Overview of geology and hydrology (cross-section through the Limmat in the area of the weir).

How the building materials are processed during the execution process is determined by selecting the construction methods, including the deployment and combination of production factors (manpower, machinery, equipment, tools, fixtures).

Bearing these project boundary conditions in mind, the construction methods and workflows are then planned, together with the site installations for:

- Setting up the local construction site (installations, logistics)
- Methods for digging the pit
- Soil and excavation methods, including transport concept
- Fabrication processes, such as concrete, reinforced concrete, and prestressed concrete-bearing structures
- Target contract/actual forecast work execution estimation for controlling purposes

In construction, the production process is dictated by the construction method with regard to the deployment and combination of the production factors (manpower, machinery, equipment, tools, fixtures) for machining and processing building materials. During the construction production process, input is transformed into output, a process that creates an asset for which the customer is paying an agreed price. Executing a construction job involves a series of sub-jobs which, in turn, are broken down into individual processes and single activities (Table 7.1). Wherever possible, these are then incorporated into a cycle and flow process in order to minimize non-value adding activities.

Breaking the processes down like this into main, module and elementary is a prerequisite for ensuring suitable combinations of construction methods, comparing them with each other, selecting the most suitable method, and incorporating it into a cycle and flow process. Building a bridge is viewed as the construction production process (Table 7.1). This construction production process can be broken down into the following main processes:

- Setting up temporary site infrastructure for insitu production facilities
- Fabrication of the permanent sub- and superstructures for the bridge

Further module processes are necessary to realize the individual main processes, for example, fabricating the foundations, abutments and piers, as well as the necessary storage areas, drainage.

A basic requirement of construction process engineering is to use the simplest and most robust means possible to achieve a practical solution that takes account of all relevant influencing variables, and which incurs minimum production cost as part of a cycle and flow process.

7.6   Procedure for selecting construction methods and processes

A practitioner will generally choose the respective construction method, equipment concept or formwork system intuitively, based on experience, and without performing any particular comparative analysis. This often produces very good results, since the team and its leaders are very familiar with working with the equipment, building and auxiliary materials, and with the workflows. Experience does, however, frequently restrict vision, and prompts a tendency to reject anything new. Today's construction market is strongly driven by price competition and intentionally, albeit often unconsciously, striving for cost leadership as the "strategy for survival". Given that inventors always attract imitators, striving for cost leadership inevitably necessitates an ongoing and continuous improvement process (CIP) to sustain a leading position in the market. If new methods are being used, systematic analysis of both own strengths and weaknesses, and competitors, is necessary, as is an assessment of future market potential. If several construction methods are available for achieving the ultimate targets, a simple, but systematic, analysis should be conducted to identify a project-specific, cost-efficient solution in line with the economic minimum principle (Figure 7.6).

7.6.1   Objectives when comparing construction methods

Comparing construction methods requires a decision theory approach. With the aid of the construction method comparison and using the target function of "minimum production cost" bearing in mind the project-specific and environmentally relevant determinants, the most effective construction method in terms of both engineering and cost can be selected by integrating the interactive process interdependencies. The methodical approach to comparing construction methods requires the following cybernetic considerations and steps (Figure 7.21):

- Target definition process
- Construction method comparison process

Two variants are available as shown in Figure 7.22. They differ in that the simpler variant (individual target system) only focuses on one target with selection criterion (e.g., cost). Various target criteria are employed, weighted and assessed in a multiple

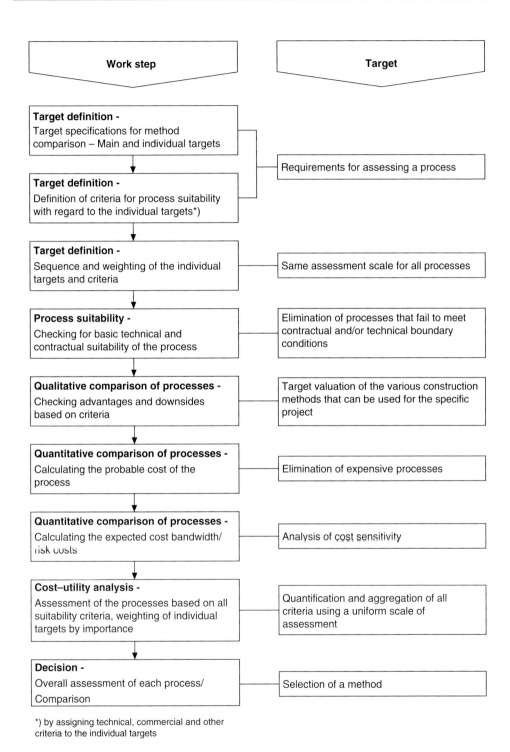

Figure 7.21 Methodical approach to comparing construction methods.

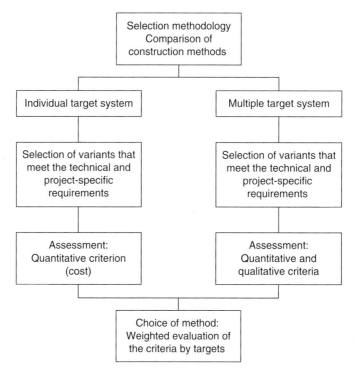

Figure 7.22   Procedure and variants for selecting construction methods.

target system. A fundamental condition of both variants is that only construction methods are compared that satisfy all project-specific, engineering and physical requirements. The target definition process specifies (Figure 7.21):

- the main target
- the individual targets
- the order of targets
- the assessment criteria
- the weighting of the criteria

bearing in mind the project-specific and environmentally relevant determinants (contract specifications, natural and artificial conditions, etc.) as well as the interactions dictated by the production process. In addition, the benchmark for evaluating the targets and criteria is set as shown in Figure 7.21.

7.6.2   Methodological approach to comparing construction methods

The systematic sequence for comparing construction production methods is shown in Figure 7.23. In a comparison of construction methods aimed at producing the most cost-efficient construction production process, the main target is defined as the "economic minimum principle". Ultimately, we are looking to find the most cost-efficient construction production method or process for a specifically defined project.

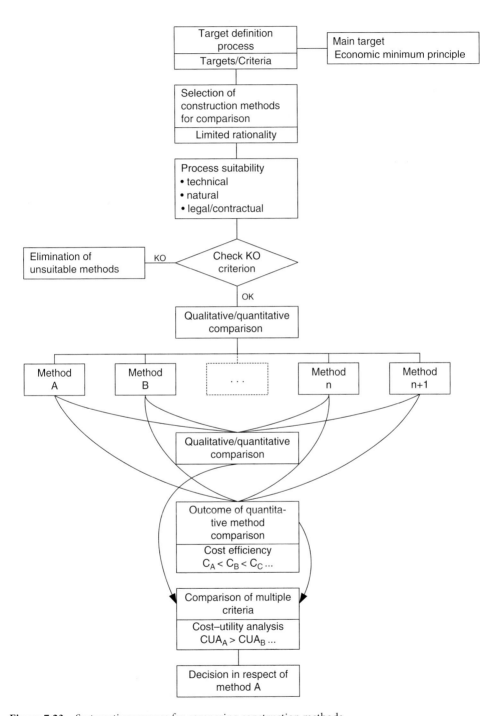

Figure 7.23   Systematic sequence for comparing construction methods.

Table 7.3 Construction method selection matrix for fabricating a rectangular silo

Elementary processes	Fabrication options			
	1	2	3	4
Reinforce	reinforce on site	install prefabricated reinforcing steel	some on site, some prefabricated	____
Assemble formwork	Board, plank, square timbers (conventional)	Large board formwork on movable working platforms	Climbing formwork	Sliding formwork
Apply concrete	with bucket	with concrete pump	____	____
Transport logistics	Tower crane on tracks	Climbing crane on elevator shaft	Mobile crane	____

Possible construction methods for the job in question are compiled systematically, or using creative methods such as brainstorming. The construction processes can then be broken down, first into sub-processes and subsequently into different variants in terms of their production/process engineering (Table 7.3). Given, however, that there are a multitude of conditions and requirements in terms of the construction job, laws, the contract, the environment, accident prevention, scheduling and quality requirements, a comparison of construction methods must follow specific steps (Figure 7.24):

- Process suitability
- Qualitative comparison of processes
- Quantitative comparison of processes
- Choice of method.

7.6.2.1   Process suitability

First, the suitability of the process in terms of production engineering must be checked, based on project-specific conditions and requirements, for the following KO criteria (Table 7.4):

- Technical – Is the method technically capable of performing the construction job (right method)?
- Geometric – Can the method be adopted, and exploited to the full, given the site and space conditions?
- Natural – Is the method suitable for the hydrological, topographical and meteorological conditions?
- Legal/contractual – Does the method comply with official provisions, and owner's, environmental and accident prevention requirements?

The additional measures to support satisfaction of the conditions can also be defined. This preliminary decision-making process should define those construction methods that are, in principle, suitable, given the specificity or all-round fitness for purpose without substantial additional measures.

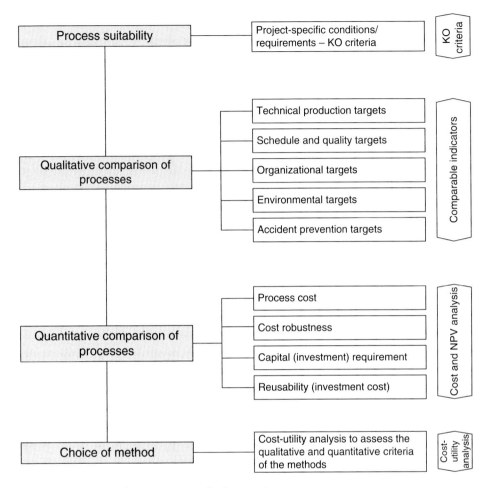

Figure 7.24   Overview of construction method comparison.

7.6.2.2   *Qualitative comparison of processes*

The extracted construction method options for the specific project are then compared in terms of both quality and quantity. They are first subjected to qualitative assessment (Table 7.5). The qualitative assessment dives deeper into the evaluation of the methods with respect to targets and criteria that are cost effective, and not directly cost effective. This involves in-depth examination of the technical production requirements that must be satisfied, and are satisfied, by the construction method. Taking account of the scheduling and quality requirements, this examination produces the requirements with regard to organization of the project in terms of manpower quality and quantity. Moreover, it produces the targets or degree of target achievement related to each respective construction method. Qualitative assessment criteria also include environmental, accident prevention and other targets.

Table 7.4   Construction process suitability – KO selection process

Targets	Criteria	Type	KO	yes	no	Additional measures
Project-specific requirement	**Technical suitability for:**					
	• Construction job	tech-nical				
	• Geometry/type of structure					
	• Available space					
	• Geology/Hydrology	natural				
	• Topography					
	• Meteorology					
	• Official requirements	legal/contractual				
	• Environmental requirements					
	→ Legal requirements					
	- Water pollution					
	- Air pollution					
	- Noise pollution					
	→ Neighborhood disturbance					
	- Risks to existing buildings					
	→ Risks to traffic					
	• Property owner's requirements					
	• Accident prevention requirements					
	→ Risks to the workforce					
	→ Risks to the project					
	→ Risks to the environment (traffic/neighborhood)					
	→ Scope of protective measures					
	Overall assessment					

Table 7.5   Targets and criteria for qualitative method comparison

Qualitative comparison of processes				
Targets	Criteria	Yes	No	Remarks
Technical production requirements/targets	**Construction process**			
	• Performance capabilities of the process			
	• Transport logistics requirements			
	• Coordination of process/work sequences			
	• Robustness vis-à-vis in-house/external disruptions			
	Space requirement			
	• Local geometric conditions			
	• Is space available for preparation?			
	• Is space available for interim storage?			
	• Does space have to be made available for all support units?			
	• Construction site accessibility			
	Process robustness/stability			
	• Start-up problems with new processes			
	• Operational robustness/resistance to disruptions (equipment failure, etc.)			
	• Climate sensitivity (cold, heat, rain, etc.)			
	• Maintenance requirements			
Schedule and quality requirements/targets	**Timing requirements**			
	• Can legal deadlines be met?			
	• Can timing benefits be generated (premiums, avoiding bad weather)?			
	Quality requirements			
	• Does the process assure the required quality (surfaces, texture, dimensional stability, etc.)?			

(continued)

Table 7.5 (*continued*)

Qualitative comparison of processes				
Targets	Criteria	Yes	No	Remarks
Organizational requirements/targets	**Manpower requirements**			
	• Is the workforce available in the company?			
	• Are qualified workers available on the labor market?			
	• Skilled or unskilled labor?			
	WPP and site supervision			
	• Can WPP prepare or plan the process?			
	• How big is the workload for WPP?			
	Susceptibility to disruptions if delays occur			
	• Can the workforce be deployed on another building site?			
	• Is the service equipment (cranes, blenders) sufficient for uninterrupted fabrication?			
	• How great is dependence on the infrastructure (traffic, water, electricity, disposal)?			
Environmental requirements/targets	**Scope of environmental protection measures**			
	• Water pollution			
	• Air pollution control			
	• Noise pollution limits			
	• Neighborhood disturbance			
	• Risks to traffic			
Accident prevention requirements/targets	**Scope of accident prevention measures**			
	• Risks to the workforce			
	• Risks to the neighborhood/third parties			
Other requirements/targets	**Inage**			
	Market power			

7.6.2.3   Quantitative comparison of methods

Building on the qualitative considerations, cost efficiency can then be compared (Table 7.6), based on cost comparisons. Cost robustness in the event of disruptions or risk occurrence is also examined. If investments in construction production processes are planned, the capital requirement also needs to be considered, together with future utilization, or the extent to which the investment fits into the enterprise's future development concept.

7.7   Conclusions to Chapter 7

Thus the TWE delivers the target work hours and expected target costs for the individual item/elements, together with their fabrication steps. In the course of

Table 7.6   Quantitative comparison of processes

Quantitative comparison of processes					
Targets	Criteria	Amount [€]	Yes	No	Remarks
Cost efficiency	Process cost				
	• Learning cost				
	• Amortization and interest				
	• Rental cost				
	• Wage costs				
	Cost robustness				
	• Is the process the cheapest?				
	• Is the process cost robust if disruptions (causing delays) occur?				
	• Is the process cost robust, e.g. in the event of geotechnical variability, and if so, within what limits?				
	Capital requirement for investments				
	• Is capital available to purchase the equipment?				
	• Increase in working capital				
	Reusability of investments				
	• Does the investment fit in with the strategic orientation?				
	• Can the equipment be used subsequently?				
	• Is the same degree of utilization likely in the future?				
	• Can the equipment be written off over the specified service life?				

optimizing the processes and construction methods, these data serve as the specification for ultimately selecting the most cost efficient method that is, moreover, resistant to disruption. The expected target work hours and target costs are then adjusted again in detail in the work execution estimation as the specification for this construction method. The lead construction production processes may now be identified in order to draw up a schedule and a resource plan for the flow process of fabrication while optimizing the processes and methods, and this will be the focus of the planning process given in Chapter 8.

References

Girmscheid, G. (1997). Fast Track Projects – Anforderungen an das moderne Projektmanagement (Requirements for Modern Project Management), Bautechnik 73, H. **8**, S. p471 - 484.

Girmscheid, G. & Hartmann, A. (1999). Fast Track Projects im Brückenbau – Anwendung und Bauprozess der Segmentbauweise mit externer Vorspannung. (Fast Track Projects in Bridge Construction– Application and Construction Process of the segmental Construction Method with external Pre-stressing.) Bauingenieur 74, H. 7/8, S. p332–344.

Girmscheid G. (2003). *Leistungsermittlung für Baumaschinen und Bauprozesse (Assessment of Construction Machines and Processes)*, 3rd edition. Springer-Verlag, Berlin.

Girmscheid, G. (2010). Arbeitskalkulation. Vorlesungsskript. *Institut für Bauplanung und Baubetrieb, Work Cost Estimation (Lecture notes. Institute of Construction Planning and Operation)*, ETH Zürich, Switzerland.

Girmscheid, G. (2012). Bauproduktionsprozesse des Tief- und Hochbaus Construction Processes of Civil Engineering Underground and Surface), Vorlesungsskript. *Eigenverlag des Instituts für Bau- und Infrastrukturmanagement (ETH Zürich)*, Zürich, Switzerland.

Girmscheid, G. & Motzko, C. (2013). *Kalkulation und Preisbildung in Bauunternehmen (Calculation and Pricing in Construction Companies)*, 2nd ed., Springer-Verlag, Berlin, Germany

SBV (Hrsg.) (2001). *Standard-Analysen SBV ST-WIN Hochbau 2001 (Betrachtungsprogramm mit Datenbank)*. Zurich, Switzerland.

Weber, M. (1968). *Gesammelte Aufsätze zur Wissenschaftslehre (Collected Essays on Epistemology)*, J.B.C. Mohr, Tübingen, Germany.

Chapter 8

Lean Construction – Industrialisation of On-site Production Processes

Part 2. Planning and Execution of Construction Processes

Gerhard Girmscheid

ETH, Swiss Federal Institute of Technology, Zurich, Switzerland

Based on the *top-down* workflow program in Chapter 7, this chapter presents the planning and execution of the construction process. The top-down work planning process includes the site infrastructure planning to enable effective industrialized execution of the construction works on site, based on the demands of cycle and flow work processes, and the sequence of the same, together with the resource and logistics plans and time scheduling. This top-down work planning process is derived from, and based on, the work estimation which provides the site execution team with the key performance factors in terms of timing and cost targets for executing the different construction elements and modules. The cycle and flow process of the construction program and work preparation resource provision planning process must be translated *bottom up* by the on-site construction team into weekly and daily work plans for the structural and interior finish phases of works. To do this, working *top down* from the work estimation planning, the key performance figures must be given for the execution of the individual work components and construction workflows. Site management is responsible for the execution of the *bottom-up* work and flow processes on a weekly and daily basis. Based on the top-down work plan, they have to detail the daily work plan with all required resources. This also requires a forward plan to sufficiently consider the lead times for material, equipment and auxiliary installations, to make sure they are on site in time to enable workflows to continue uninterrupted. In an effort to steadily improve the cycle processes on site, a continuous improvement process must be established by the leadership of the site management in cooperation with the foremen for the different trades. This chapter describes the systematic procedure for ensuring coordinated, value adding collaboration on a construction site to complete the construction work in an efficient industrialized cycle and flow production process.

Modernisation, Mechanisation and Industrialisation of Concrete Structures, First Edition.
Edited by Kim S. Elliott and Zuhairi Abd. Hamid.
© 2017 John Wiley & Sons Ltd. Published 2017 by John Wiley & Sons Ltd.

Finally, the finishing works in building construction, in particular, require highly coordinated organization and planning so that the various teams can work without the usual interference, which frequently causes inefficient and idle time for the teams. This phase in the daily and weekly cycle and flow processes, in particular, where the teams are working consecutively, is most critical and important to ensure a highly value added construction process.

8.1 Introduction – top-down / bottom-up work planning scheduling and resource planning

Chapter 7 described in detail how the work planning process has to be structured, and which work-planning processes have to be involved. The top-down work-planning process includes the site infrastructure planning to enable execution of the construction works on site based on cycle and flow work process and the sequence of the same, together with the resource and logistics plans, and time scheduling. This work planning process is derived from the work estimations.

Work estimation incorporates the target resources which have been allocated for the different types and phases of works. The top-down work planning process provides the fixed base for economical and efficient processing of the interdisciplinary construction production works, to ensure site success without friction. Further, the top-down work plan serves as an instrument for continuously steering and controlling the execution of the weekly and daily work plans for the different work teams on site. To ensure efficient and target-oriented control, every cycle and flow process for the different types of works must be enhanced with the target hours for the workers, as well as for equipment deployment in reference to the volume of works planned, as derived from the work estimations. These targets for team work hours and machine hours in reference to the volume of works planed, also help to ensure detailed controlling of the target and actually achieved hours on site, which in turn supports continuous improvement of the site steering process.

Based on the top-down work plans and work estimations for the cycle and flow processes, site management then has to organize the daily and weekly works for their teams bottom-up. Drawing on the top-down work plans for the cycle and flow process, and incorporating the target performance, resource plans and the targets derived from the work estimation for the production of the individual construction elements and construction sequences, site management has to adopt a bottom-up approach to preparing the weekly and daily work plans for the construction works as well as for the finishing works on site for the different, interacting work teams. This chapter explains how a lean construction approach by site management can make sure that the work flows continuously, with no losses in terms of time or materials. The chapter further explains how the site manager and site supervisors are responsible each week for advance planning the next week's work on a daily basis and a projection for the next three to four weeks coming, with clear assignment of resources, materials, equipment, and auxiliary materials.

Based on these daily work plans with clear target hours for both the work teams and equipment deployment for the planned volume of works, it is very easy to control actual work achievement versus targets. Drawing on these controlling results, site management can discuss and decide measures each week to improve the processes on site with a view to assuring that the targets defined in the top-down plans are, at least, met. This chapter describes the systematic procedure for ensuring coordinated and value adding collaboration on construction sites to complete both the structural and finishing works in an efficient cycle and flow production process.

8.2   Scheduling and resource planning

Developing a top-down work plan requires the design of a clear cycle and flow process for the different construction elements that considers the sequence in which construction elements have to be erected in light of the spatial conditions and from a structural strength perspective, and the sequence for applying the different layers. However, it is very important that we identify the lead process in the construction activities. These lead processes are particularly important activities that are characterized by:

- Their position in the schedule on the critical path – any change in the timing of the lead process in terms of duration, start or finish will influence the downstream activities.
- The – usually considerable – portion of the total cost needed to perform the activities. If lead processes get behind, downstream work performance must be increased by raising capacities – space permitting – in order to rectify the situation and stay on track with the project targets; otherwise the "quality" has to change. These corrections are generally not achievable without additional cost or impairment.
- Once construction projects reach a certain size and complexity, the duration of the activities (Girmscheid, 2003) that comprise the lead processes should be described in the network plan as a stochastic variable with three event parameters – shortest, most probable, and longest duration – together with the associated estimated probabilities. The lead process durations are calculated or specified as a weighted statistical mean value.

The overall project duration can then be simulated using the critical path, meta potential or PERT method (Schmidli, 2001). This produces the overall duration with stochastic distribution, which allows detection of the sensitivity to disruptions of the examined/selected construction method with regard to the construction program and the final deadline. The results are usually plotted in Gantt charts as the most common form of scheduling. The construction workflow is plotted on a time axis (X) and work steps axis (Y). The duration of an activity is indicated by a bar (Figure 8.1) and its timing specified by the start or end date. Network planning techniques can help to highlight interdependencies within this Gantt chart (Loschert, 1999).

The resource plan serves as the basis for the logistics concept for the construction site, i.e. for ensuring the provision of the right quantity and required quality of resources and information on time and in the right location, and without getting in each other's way. The resource planning process also includes organization of the material deliveries, and manpower and equipment deployment, together with provision of the approved execution and work plans, and the provided materials or building areas as contractually agreed, especially permits affecting execution, financial

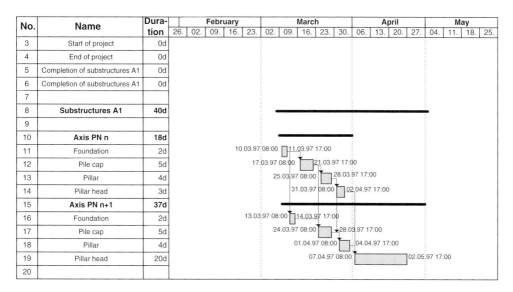

No.	Name	Dura-tion	26.	02.	09.	16.	23.	02.	09.	16.	23.	30.	06.	13.	20.	27.	04.	11.	18.	25.
3	Start of project	0d																		
4	End of project	0d																		
5	Completion of substructures A1	0d																		
6	Completion of substructures A1	0d																		
7																				
8	**Substructures A1**	**40d**																		
9																				
10	**Axis PN n**	**18d**																		
11	Foundation	2d																		
12	Pile cap	5d																		
13	Pillar	4d																		
14	Pillar head	3d																		
15	**Axis PN n+1**	**37d**																		
16	Foundation	2d																		
17	Pile cap	5d																		
18	Pillar	4d																		
19	Pillar head	20d																		
20																				

February: 26. 02. 09. 16. 23. | March: 02. 09. 16. 23. 30. | April: 06. 13. 20. 27. | May: 04. 11. 18. 25.

Task dates shown on chart:
- 11 Foundation: 10.03.97 08:00 – 11.03.97 17:00
- 12 Pile cap: 17.03.97 08:00 – 21.03.97 17:00
- 13 Pillar: 25.03.97 08:00 – 28.03.97 17:00
- 14 Pillar head: 31.03.97 08:00 – 02.04.97 17:00
- 16 Foundation: 13.03.97 08:00 – 14.03.97 17:00
- 17 Pile cap: 24.03.97 08:00 – 28.03.97 17:00
- 18 Pillar: 01.04.97 08:00 – 04.04.97 17:00
- 19 Pillar head: 07.04.97 08:00 – 02.05.97 17:00

Figure 8.1   Project Gantt chart.

planning, etc. Network planning programs make manpower and equipment deployment relatively easy to plan by weighting the activities.

Manpower deployment must be planned such that waiting times or absences are avoided. In addition, buffer times should enable handling of extremely short notice manpower and equipment requirement peaks. The work cycles of teams should be optimally matched to each other in terms of capacities, or it should be possible to resort to other construction sites close by. Moreover, the workflow and team deployment should be planned such that they can perform their workflows optimally in the available space, without getting in each other's way.

Where possible, **equipment** waiting or idle times should be eliminated, albeit that is not entirely possible in the case of many types of equipment (Girmscheid, 2003). The equipment must be reported back and returned from the construction site as soon as it has completed its tasks in order to reduce the rental cost, enable deployment of the equipment on another construction site, if appropriate, and to free up further storage space or work areas on the building site.

When planning the material deliveries, a distinction must be made between the lead times for commissioning and delivery, and the availability for deployment on the construction site. In the case of certain materials or special equipment, such as tunnel driving machinery, the following lead times between ordering and availability on the construction site must be taken into account:

- Contract processing time
- Planning and approval time for prefabricated parts and special construction machinery or installation units
- Production preparation and production time
- Transport time
- Assembly or installation time on the construction site in the case of large units and special machinery that cannot be delivered assembled

On confined buildings sites, just-in-time delivery is essential to avoid interim storage and possibly costly relocation of the materials. This form of delivery, necessitates very detailed execution preparation and logistic planning, with no mistakes whatsoever, and should be adopted for material deliveries where such planning depth can be guaranteed.

Planning the delivery of the execution/work plans is also part of the scheduling process (Table 8.1) and is based on the part production planning process. The construction production planning schedules are adopted as milestones for the delivery dates in the execution and work plans (Girmscheid, 1997). The following workflows should be specified between the planner and the construction site when planning execution:

1. Planning and approval process sequence, with times
2. Change management process, broken down:
    a. without renewed examination in the case of minor changes that do not affect utilization or the approved (tested) structural stability concept,
    b. with renewed examination in the case of changes that deviate substantially from the tested concept. Advance agreement of the specifications with the inspections engineer based on defined cases may be useful, especially if the reinforcing steel on the construction site is set to undergo engineering modifications.
3. Planning schedule tracking concept
4. Plan circulation (to whom, how many copies)
5. Plan status identification for information and inspected/approved for construction
6. Plan references for which plans belong to a part and which plans belong to the adjacent structure?
7. Internal inspection procedures for structural stability: formal inspection, comparative calculation and building plans: Cross-validation/dimensional test
8. Tracking changes – index change box

Organizing the execution planning for Total Contractor (TC) projects poses special challenges for planners and the coordinator responsible on the construction enterprise side. These challenges are addressed in more detail in (Girmscheid, 1997). Special permits that are needed for execution also need to be taken into consideration. These include, for example:

- Traffic blocks
- Special transports
- Erecting particularly high cranes, for example, in flight paths
- Connections to supply and disposal systems.

A further resource-planning task is to calculate the approximate **funding requirement** for the construction site. The funding requirements are derived from the project cash flow, by individually weighting the activities on the Gantt chart – bearing in mind the weighted mean durations – with the cost types (wages, materials, equipment, subcontractor work) and their distribution over time. Monthly totals are then aggregated by adding the costs for all activities over the respective month, and plotted in a

Table 8.1 Control of workflows and schedule planning

No.	Axis	Activity	Construction date	Plan to PM		Plan to person i/c checking		Plan from person i/c checking		Plan to building site	
			Completion	Target	Actual	Target	Actual	Target	Actual	Target	Actual
PN n		**Foundation**	10.03.97								
		Structural stability of foundation		16.12.96		23.12.96		21.01.97			
		Depth of pile foundations		30.12.96		06.01.97		03.02.97		03.03.97	
Pn n		Overview of pile caps		30.12.96		06.01.97		03.02.97		03.03.97	
		Pile cap formwork plan		30.12.96		06.01.97		03.02.97		03.03.97	
		Survey coordinates		30.12.96		06.01.97		03.02.97		03.03.97	
		Pile cap	17.03.97	23.12.96		30.12.96		27.01.97			
		Structural stability of pillars + foundation		06.01.97		13.01.97		10.02.97		10.03.97	
		Pit lining		06.01.97		13.01.97		10.02.97		10.03.97	
		Pile cap armoring		06.01.97		13.01.97		10.02.97		10.03.97	
		Pillar formwork plan		06.01.97		13.01.97		10.02.97		10.03.97	
PN n		Pillar armoring		06.01.97		13.01.97		10.02.97		10.03.97	
		Pillar drainage		06.01.97		13.01.97		10.02.97		10.03.97	
		E&I ducts									
		Pillar	25.03.97	14.01.97		21.01.97		18.02.97		18.03.97	
		Pillar head formwork		14.01.97		21.01.97		18.02.97		18.03.97	
		Pillar head armoring									

bar chart. Aggregating the cost bar chart then produces the cumulative cost for the project. The cost bar chart and the cumulative cost also serve as work performance and cost budgets for the project for controlling purposes. Inflows are then plotted on this cash flow chart as a line of steps. This line of inflows is derived from the monthly bills invoicing work performance and taking any delays into account resulting from:

- time spent on drafting detailed monthly invoices with dimensional documentation,
- time spent by the owner on checking that the invoice is correct,
- time taken for payment to be made once the owner has accepted the invoice.

This usually produces a time lag of about one and a half months between cost-effective work performance and receipt of payment. The total funding requirement is the aggregation of the gap between cumulative cost and incoming payments curve. The monthly funding requirement constitutes the difference between the costs and payments bar charts.

8.3   Site Logistics

8.3.1   Logistics planning

For a construction enterprise, logistics planning comprises construction work, equipment yard, procurement, and disposal (Figure 8.2). Logistics are particularly important when planning construction workflows for a project as a cycle and flow process. As such, logistics planning must be a cross-divisional and interactive process conducted during work preparation. In addition to providing the right production equipment and site infrastructure for proper construction production, logistics planning as part of WPP also focuses on supply, disposal, transportation, storage, and turnaround activities on the construction site (Figure 8.3). Logistics links and supplies the stakeholders on a construction site with a smooth flow of information, materials, energy and construction work. The aim of logistics is to save resources and optimize costs.

Individual surveys (Blecken, Boenert, & Blömeke, 2001) confirm the intuitive impression created by many construction sites that productivity relative to the billable unit item is often very low. The problems associated with this weak performance of billable items usually arise as a result of unproductive activities, such as

- Clearing materials up and out of the way as they are hindering work
- Searching for materials
- Transporting materials long distances between the warehouse and the location of installation
- Disruptions and interruptions caused by missing materials and tools

Controlling procedures usually cannot/do not identify these unproductive times; they can only be detected on site by performing time studies. Systematic logistics aligned to activities that are subject to change in terms of timing and location can

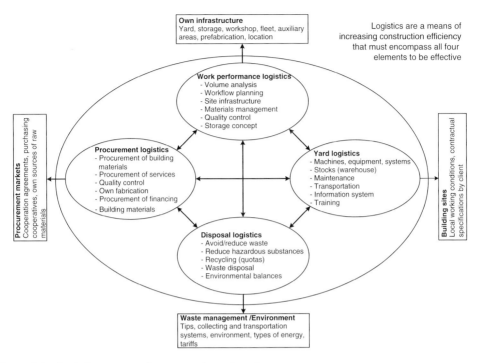

Figure 8.2   Logistics concept of a construction enterprise.

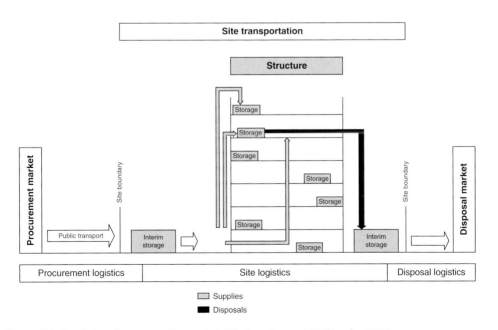

Figure 8.3   Logistics of a construction project (Blecken, Boenert & Blömeke, 2001).

activate enormous potential for enhanced performance. Planning logistics processes for stationary operations with defined stationary processing equipment and storage areas is not a trivial task, and poses a real challenge for a mobile industry such as the construction sector, where the building is fabricated on the site of utilization.

Fabrication workflows on a construction site are subject to constant change in terms of timing and location. Changes during fabrication in a stationary industry usually only relate to timing. The product is moved between stationary fabrication plants in the production hall, all of which work under clearly defined conditions. As such, logistically planning the material and intangible flows on a construction site will never achieve the same improvements in efficiency as a stationary industry.

Logistics problems relating to subcontractors are particularly noticeable among Total Contractor (TC)/General Contractor (GC) forms of project delivery. This is often due to the absence of a clearly planned cycle and flow process during the interior finishes phase with clear timing and logistics specifications for the interior finishes firms. Problems particularly arise with the flow production and logistics of the interior finishes firms, who often store their materials in and around the building, regardless of whether they are in another trade's way. This increases the workload involved in moving stuff around, and the waiting times, and thus the costs.

8.3.2   Transport logistics

The success of a construction site is due in no small part to central logistics management as part of the work preparation/site management, which bundles all project-relevant information in digital form, calculates the quantities of required materials, submits orders that take account of lead times, calls them up from the dealer/manufacturer just in time, coordinates the transport activities, and hands the materials over to the responsible person on the construction site, or delivers them to a predefined storage area. This enables material, transport and circulation costs to be reduced, and deliveries to be coordinated just in time.

Transport efficiency on a construction site is dependent on the following criteria (Blecken, Boenert, & Blömeke, 2001):

- Location of the material sources and the construction site
- Existing infrastructure
- Local conditions on the construction site; owner's specifications
- Quantities and properties of materials in transit (gas, fluids, parcels, bulk)
- Size, location and state of the lay-down areas for possible interim storage
- Timing specifications derived from the project workflow

Building on this, work preparation for a cycle and flow process must develop the following basic specifications for a logistics concept:

- Delivery areas and handover points must be easy to reach
- Transportation means and equipment on the construction site must match
- Route planning and area utilization outside and inside the building
- Traffic safety and site access rules

This all produces a huge need for coordination and communication to reduce space and timing constraints when ordering materials. Transport reports with quantities and tours, site access rules, and storage area signage all have to be developed and coordinated. Even before leaving, or while still en route, vehicles should radio through to the coordinator in order to avoid unnecessary waiting times when unloading. The results of implementing these concepts are (Loschert, 1999):

- The construction site is supplied with the necessary materials as required during the construction process
- Bottlenecks in material deliveries can be identified and avoided
- Early coordination of material transportation and storage across all trades
- Overview of all material storage locations and quantities through barcodes with central registration of material flows.

8.3.3   Delivery, storage and turnaround logistics

Construction site logistics can be broken down into planning the storage areas, and planning site transports. When planning the material flow on a construction site, the material units must be specified, the means of transport selected, transport chains developed, and material storage areas defined at the location of use.

One fundamental principle should always be observed: Wherever possible, store materials right where they are going to be installed to avoid additional wage costs for moving them. Since this is, however, not always possible, given the space constraints and the changeability of the project timing/spacing, transportation means and chains must be planned, together with the associated equipment and storage areas. Logistics planning as part of WPP is a process that flanks construction and which should be planned and adjusted on the construction site, given the prevailing conditions there.

Open or fully encased pallets, crates and containers are suitable as means of transport. For reasons of work safety and work efficiency, materials must be stored on pallets and encased in foil to prevent them falling during transportation.

According to Blecken, Boenert, and Blömeke (2001), delivering the required material to the specified storage or work area is crucial for guaranteeing proper material supply. Barcodes attached to the loading units are ideal for this. They store the data that matter in material flows, such as content, quantity and execution sections, where the material is ultimately to be delivered. The use of hand-held registration devices enables identification of the loading units at any time on the construction site, as well as the exact purpose/destination, and registration of the consumption and relocation to the installation site, with ultimate daily transfer back to the central logistics computer to determine the overall balance.

Planning the transport chain includes material turnover on the construction site, vertical transportation of the materials to the individual floors, and horizontal distribution of the materials throughout the floors of the structure.

The following principles must be observed:

- Transport materials on the shortest route
- Avoid manual transport; use technical means of transport
- Try to set up small material storage areas close by the production site

- Select reasonable transportation units
- Ensure the storage areas are kept in good order and set up cross-trade logistics on each floor.

"Suitable unloading equipment must be available to ensure smooth unloading of materials being turned over on the construction site. Call-up, transport registration and early distribution of the materials in suitable units ensure that whoever is responsible is ready to accept deliveries at any time. If trucks are not fitted with their own unloading equipment, the materials can be unloaded using the forklifts, wheel loaders, truck or tower cranes provided by the logistics coordinator. Depending on the available space, and provided suitable openings and mobile floor loading platforms are provided, materials should be delivered directly to the floors, if suitable cranes are available on the site or the truck's own unloading equipment has sufficient reach. As such, WPP and logistics planning should ensure sufficiently large openings in the facade for the respective periods. Wheel loaders and forklifts can move the materials to the vertical transport equipment. Lists of available equipment and transport units simplify transport chain planning and equipment selection" (Blecken, Boenert, & Blömeke, 2001).

A hoist or crane with floor platforms can be used to vertically transport materials. The material must be delivered to a location on the site installation plan from where, for example, the crane can move it to the storage areas. Depending on local conditions, the material may have to be interim stored, or is directly lifted up to the floors for processing. The same applies when delivering raw materials; storage close by the processing equipment and vertical transportation via hoists, elevators or cranes.

The best way to horizontally distribute material on the floors is by forklift, lift truck, and so on, depending on the materials properties and distance of travel. The units of materials should be taken on unpacked by the vertical transport equipment, and moved to the storage areas. The traffic routes and storage areas must be sufficiently large.

8.3.4   Planning storage areas – storage space management

Material stores form the buffer between irregular material deliveries and fluctuating consumption during production. This poses the following requirements in terms of storage space management (Loschert, 1999):

- Coordination meetings and agreement with all trades
- Provision of required building materials in the requisite quantities in the window needed for the relevant construction phase by matching supply to requirements through just in time delivery
- Dividing storage areas up into spaces for different materials
- Assigning storage areas on the floors of the building in the various time windows as construction progresses without the actors getting in each other's way
- Entering the delivery location and storage areas, together with the construction roads and location of the equipment for the individual construction phases on the site installation plan
- Shortening travel routes

- Storage areas that are decentralized, and close to the workplace on the floors and/or locations of installation, for example, strive for vertical distribution immediately upon delivery of the materials at the handover points.

This is all guaranteed by developing floor logistics concepts for the respective construction phase, which should incorporate the following principles (Blecken, Boenert, & Blömeke, 2001):

- Plan transport routes and lay-down areas for each construction phase
- Avoid moving materials from one storage place to another
- Make sure the interior finishes trades working in parallel do not get in each other's way
- Safeguard materials and completed construction works against damage
- Keep the building site clean
- Enhance health and safety
- Reduce storage space requirements on the construction site
- Keep the routes clear (supply and disposal)
- Reduce lay-down space requirements by coordinating storage.

The interior finishes schedule forms the basis for planning floor logistics. The area must be divided up into work areas and storage based on the workflow sequence. The storage space requirement is determined on the basis of the work to be performed by each trade during the relevant construction phase, the space needed to execute the work, and the necessary transport routes. Depending on the construction phase, the storage and work areas, and the transport routes needed for horizontal material distributions are marked on the storage plan for each floor.

8.3.5   Disposal logistics

Disposal logistics is responsible for ensuring proper disposal of all materials left over after completion of construction production. The cost efficiency of cleverly designed disposal logistics concepts is secured by separating waste properly for recycling or final disposal.

8.4   Weekly work plans

8.4.1   Lean construction – weekly work program

The cycle and flow production of the structural and interior finishing work should be planned regularly and in detail for a defined, immediately consecutive period. One form of such planning is the weekly work program, which basically reflects the schedule and capacity plans, but is more detailed. Weekly work programs for the appropriate unit items can, however, only be planned shortly before execution of the respective works, thus enabling influencing factors to be taken into account, such as, for example, the weather, shutdowns, status of interdependencies with other activities. By the same token, such detailed work preparation and logistics planning guarantees that the materials, equipment and skilled labour are available at the right time. Material shortages can be ordered on time, and capacities assured.

The weekly work program plans the daily activities of the teams for their respective sections and elements by specifying the work in detail, for example, disassembly and installation of m² of formwork, installation of tons of construction steel, application of m³ of concrete. In addition to specifying the teams' workloads, the specifications for ordering concrete (volume, timing) and the necessary equipment, such as pump, poker vibrator, truck cranes, and so on, are also identified. As such, the site manager or WPP logistics coordinator can coordinate the shared resources needed by the various execution teams, who each prepare their own weekly work programs. Equally, the storage, transportation and installation of materials can be coordinated between the trades, including the floor-by-floor provision of work and storage space. Logistics planning is performed as part of the weekly WPP revision that flanks the construction process. Coordination takes place during the weekly construction managers' meetings.

Given the levels of intensity and interaction, and interdependencies, a distinction must be made between structural work and interior finishes when organizing work in a cycle and flow process (Figure 8.4).

Overall management in the shape of the owner's construction manager, or the TC or GC, is responsible for work organization on the construction site. Overall management is responsible for ensuring smooth, performance-oriented, efficient and interactive collaboration among the teams in the various disciplines. The aim of this work organization must be to increase the value added by the involved enterprises, to eliminate waste time, and to align everyone to the success of the project.

In order to implement such coordinated, value adding collaboration on a construction site, the construction manager, or TC/GC, must take the following steps:

1. WPP – draw up a construction program and work execution estimation (target specifications) (*top down*)
2. Workshop 1 – Present the construction program and define basic rules of collaboration to inspire shared team responsibility for the project among everyone involved
3. WPP – Prepare the installation teams on the basis of the construction enterprise's program with volume and work performance definitions, and definition of installation times and team sizes for each section and element.
4. Workshop 2 – Agreement and coordination of the work teams, and preparation of the first weekly plan with three- to four-week look ahead
5. Weekly meetings:
   a. Agreement of weekly and daily plans on Friday afternoons with preparation of the installation teams before the meeting in respect of types/quantities of materials, work performance, time requirement and team size
   b. Agreement of the interactive jobs with window on each work day
   c. Serve to prevent obstacles for the teams, to raise efficiency, and to eliminate non-value adding activities
   d. Target/actual analysis with continuous improvement process (CIP) should be used systematically as a value adding activity
   e. Systematic advance planning of the materials in respect of quantities, types and grades, and of the utilization of the logistics infrastructure (transport and storage)

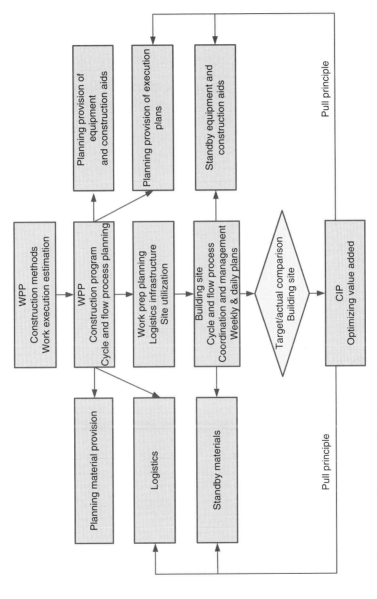

Figure 8.4   Cycle and flow process planning and implementation in construction production.

8.4.1.1   Structural work planning

8.4.1.1.1   WPP – Top-down production planning and detailed construction program
During the structural work phase, the work that the construction enterprise must
perform to erect the structure forms the lead process. The HVAC, plumbing and electrical works are subordinate to this lead process. The cycle and flow process developed
in WPP and on the basis of the work execution estimation for erecting the structure, with sections, elements, and jobs or sub-jobs (formwork, reinforcing steel, insitu
and precast concrete, brickwork) determine the timing of the installation works for
electric ducts, ventilation ducts, passive energy storage, and so on, for the purposes
of coordinating the construction enterprise, electrician, HVAC and plumbing teams.
The construction enterprise manager needs to draft a top-down, detailed construction program (Figures 8.6 and 8.9) as part of the work preparation. This program
must include details of the fabrication sections and elements. To do this, the individual, interrelated fabrication of foundation plate, wall and ceiling sections must be
described as activities with start and end points (Figures 8.6 and 8.9).

Sub-job lists are prepared as part of the WPP, and to secure the content and scope
of work of each activity or sub-job in the construction program (Table 8.2). These

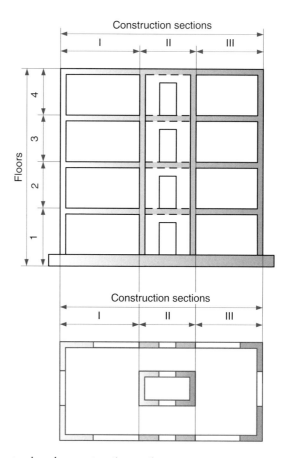

Figure 8.5   Structural work – construction sections.

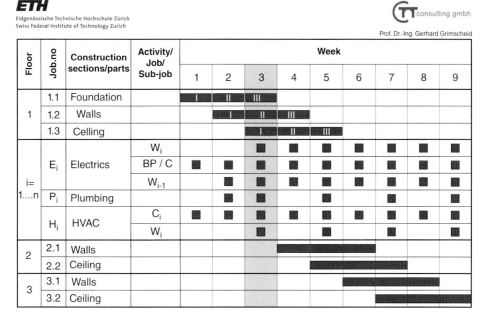

Floor	Job.no	Construction sections/parts	Activity/ Job/ Sub-job	Week								
				1	2	3	4	5	6	7	8	9
1	1.1	Foundation			▰ I ▰ II ▰ III								
	1.2	Walls				▰ I ▰ II ▰ III						
	1.3	Ceiling				▰ I ▰ II ▰ III						
i= 1....n	E_i	Electrics	W_i			▪	▪	▪	▪	▪	▪	▪
			BP / C	▪	▪	▪	▪	▪	▪	▪	▪	▪
			W_{i-1}		▪	▪	▪	▪	▪	▪	▪	▪
	P_i	Plumbing			▪	▪		▪		▪		▪
	H_i	HVAC	C_i	▪	▪	▪	▪	▪	▪	▪	▪	▪
			W_i			▪		▪		▪		▪
2	2.1	Walls					▬▬▬▬▬					
	2.2	Ceiling						▬▬▬▬▬				
3	3.1	Walls							▬▬▬▬			
	3.2	Ceiling								▬▬▬▬		

Figure 8.6   WPP top-down schedule and construction program for cycle and flow production.

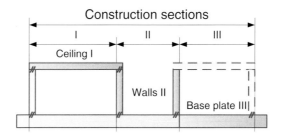

Figure 8.7   Weekly plan – week 3 – work sections.

sub-job lists include the following details for each activity or partial activity affecting, for example, work sections or elements:

1. Structural work phase
2. Sub-jobs for the section, broken down by discipline
3. List of sub-jobs with indication of:
    a. Fabrication equipment
    b. Construction aids for fabrication
    c. Materials, with quantities, dimensions, grades

This forms the basis for planning and controlling production on the construction site, on which the weekly and daily plans can build, complete with all materials, equipment and construction aids. Equally, the procurement, provision and call-up logistics can build on this basis and be controlled by targets.

Sub-job no.	Activity/Job/Sub-job		Week									
			Mon		Tue		Wed		Thu		Fri		Sat/Sun
			am	pm	am	pm	am	pm	am	pm	am	pm	
1.1.III	Foundation plate III	Assemble formwork	▓										
		Reinforce 1		▓▓▓▓								
		Reinforce 2						▓▓▓				
		Apply concrete								▓		
E:BP1	Electrics						▓					
HVAC:BP1	HVAC						▓					
1.2.II	Walls II	Assemble formwork 1	▓										
		Assemble formwork 2								▓		
		Reinforce 1		▓▓▓								
		Reinforce 2					▓▓					
		Apply concrete								▓		
E:W1-II	Electrics					▓						
P:W1-II	Plumbing					▓						
HVAC: W1-II	HVAC					▓						
1.3-I	Ceiling I	Assemble formwork		▓▓▓▓								
		Reinforce 1			▓▓							
		Reinforce 2						▓▓				
		Apply concrete								▓		
E:C1-I	Electrics					▓						
HVAC: C1-I	HVAC					▓						

Figure 8.8   Bottom-up weekly and daily plan of cycle and flow production of the sub-jobs in the work sections.

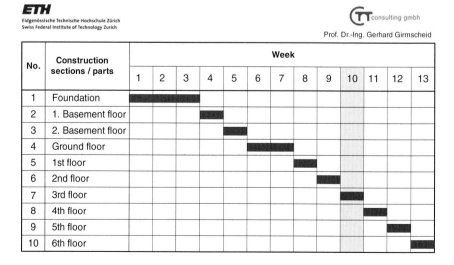

Figure 8.9   High-rise structural work – WPP cycle and flow process.

Table 8.2 WPP sub-job list, structural work – Work steps, equipment, construction aids, materials and quantities for fabricating an activity (in this case: wall A1) in the construction programme

ETH
Eidgenössische Technische Hochschule Zürich
Swiss Federal Institute of Technology Zurich

⊤⊤ consulting gmbh

Prof. Dr.-Ing. G. Girmscheid

No.	Sub-job Description of activity	Quantity	Materials/equipment/aids		Responsibility	Remarks
			ordered	on standby		
Job: Concrete works						
1	Assemble formwork Frame formwork Anchorage 2 x concreting platforms	H=2.75m, L=10.00m 22 pcs. L=10.00m				
2	Reinforce Vertical Horizontal Brackets	Ø 22, l=3.25m Ø 12, l=10.50m, N=11 Ø 10, N=60				
3	Apply concrete C20/25	d=25cm V=6.875m^3				
Job: Electrics						
4	Lay ducts	Ø 25, l=20m				
5	Install sockets	4 pcs.				
Job: HVAC						
6	Build ventilation shaft	20/100, l=3m				
7	Riser ducts	Ø 25, l=3m				
Job: Plumbing						
8	Wastewater downpipe Lay on each floor	Ø 25, l=3m				
9	Water supply pipes Lay on each floor	Ø 25, l=3m				

8.4.1.1.2 *Work organization on the construction site – Bottom up*

The sub-jobs in the weekly construction site plans, which are drafted by the construction enterprise's management team and build on the WPP construction program, can then be broken down into work steps (Figure 8.8 and 8.10). Based on the construction enterprise manager's weekly plan (Figure 8.8), the installation teams must then plan their quantities, work performance and manpower deployment. After all, once the first layer of reinforcing steel is in, for example, in ceilings, the installation works are generally next. In the case of walls, installation is performed on the side of the respective room that will subsequently be used.

The weekly work plans of the interacting site teams must be lead managed by the owner's construction manager, or the TC/GC. The construction enterprise manager is responsible for coordination with the installation firms. If disputes arise, the construction manager must suggest solutions, and make the final decision.

Based on the construction enterprise's weekly plan (Figure 8.10), the weekly meetings are then used to coordinate the installation firm's electrics, plumbing and HVAC in respect of exact days, windows, and work areas. The installation firms must define their types and quantities of materials, work performance and team sizes for the respective windows for their installation works. Changes in the duration of the windows may need to be agreed to eliminate unnecessary waste. Added to which, work must be planned for three to four weeks in advance in order to secure material availability and the provision of special equipment, and so on (Figure 8.11). This is why work implementation and organization on a construction site are so important, even as part of lean management. After all, the WPP construction and work program (Figures 8.6 and 8.9) must be broken down into weekly and daily targets (Figure 8.8) on the construction site.

This work preparation and work organization process on the construction site is described for the building in Figure 8.5. The WPP schedule and construction program for this project is shown in Figure 8.6. This WPP program contains the sections and parts with their respective jobs. The foundation, for example, is fabricated in three consecutive steps. The total duration of this job is comprised of the following sub-jobs: formwork, reinforcing steel and insitu and precast concrete per section.

The cycle and flow processes for these sub-jobs are not planned in detail in the WPP construction program, and are the responsibility of the construction site as part of its work organization. Having said that, the durations of the jobs, complete with sub-jobs, are derived from the target work hour specifications for the sub-jobs from the work execution estimation and the team size selected during WPP. The times needed to perform the respective installation works such as electrical ducts, plumbing lines and HVAC pipes and ducts are also taken into consideration (Figure 8.6).

Translation into the target weekly and daily work plans on the construction site (e.g., for week 3) is shown in Figures 8.7 and 8.8. Work organization based on the target specifications derived from WPP and the work execution estimation is planned by the site team, consisting of the site manager, section leader(s), foreman for construction, and project managers for the electrics, plumbing and HVAC.

The construction enterprise manager forms the lead team for the structural work phase, together with the section construction managers and foremen. These target weekly and daily plans (Figure 8.8) are detailed by sub-job. The sub-jobs of the construction enterprises constitute the lead cycle and flow process as it is here that the core

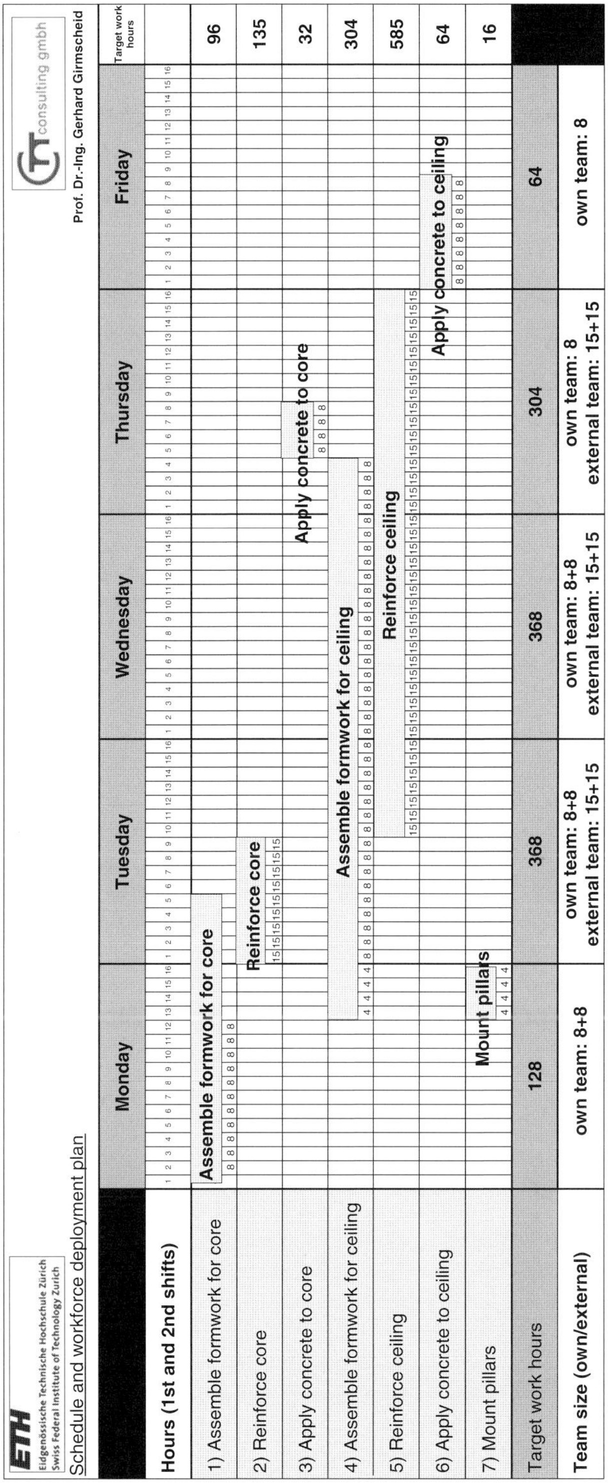

ETH Eidgenössische Technische Hochschule Zürich
Swiss Federal Institute of Technology Zurich

consulting gmbh
Prof. Dr.-Ing. Gerhard Girmscheid

Schedule and workforce deployment plan

Hours (1st and 2nd shifts)	Monday	Tuesday	Wednesday	Thursday	Friday	Target work hours
1) Assemble formwork for core	Assemble formwork for core (8 8 8 8 8 8 8 8 8 8)					96
2) Reinforce core		Reinforce core (15 15 15 15 15 15 15 15)				135
3) Apply concrete to core				Apply concrete to core (8 8 8)		32
4) Assemble formwork for ceiling		Assemble formwork for ceiling (8 8 8 8 …)				304
5) Reinforce ceiling		Reinforce ceiling (15 15 15 …)				585
6) Apply concrete to ceiling					Apply concrete to ceiling (8 8 8 8 8 8 8 8)	64
7) Mount pillars	Mount pillars (4 4 4 4 / 4 4 4 4)					16
Target work hours	128	368	368	304	64	
Team size (own/external)	own team: 8+8	own team: 8+8 external team: 15+15	own team: 8+8 external team: 15+15	own team: 8 external team: 15+15	own team: 8	

Figure 8.10   Lean construction – third floor execution organization with weekly construction program on the building site.

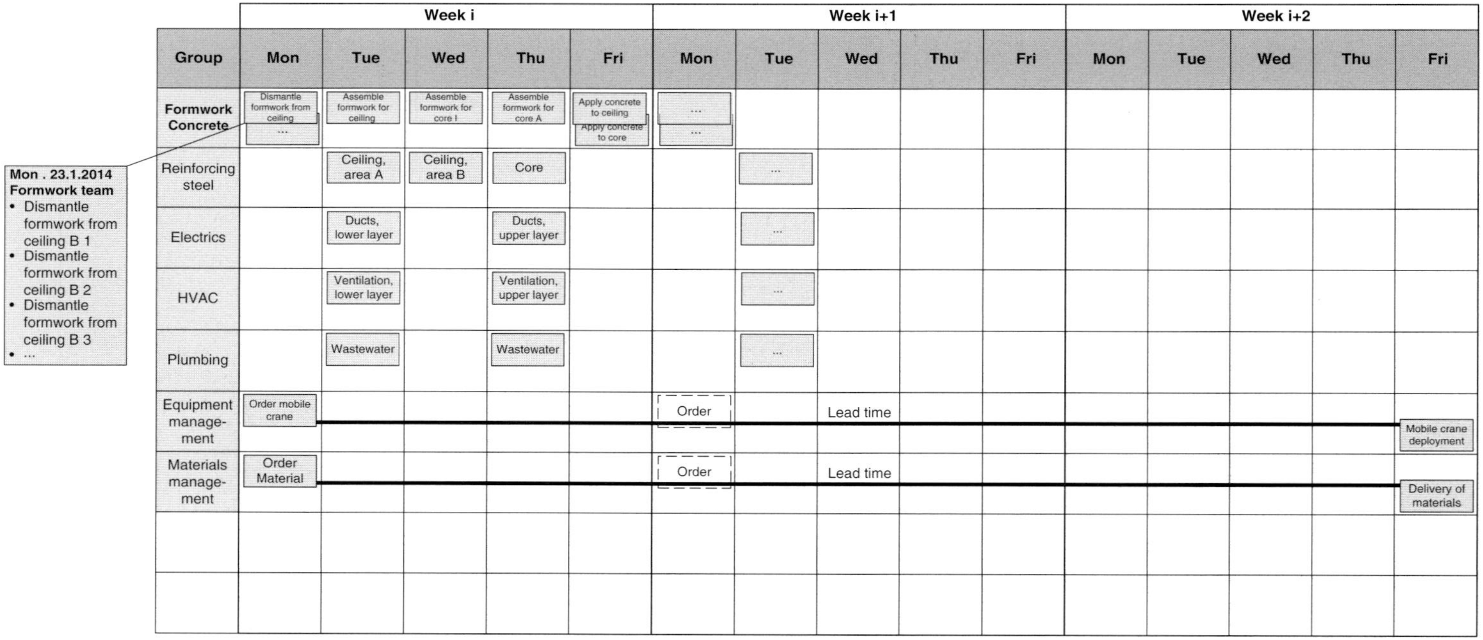

Group	Week i					Week i+1					Week i+2				
	Mon	Tue	Wed	Thu	Fri	Mon	Tue	Wed	Thu	Fri	Mon	Tue	Wed	Thu	Fri
Formwork Concrete	Dismantle formwork from ceiling ...	Assemble formwork for ceiling	Assemble formwork for core i	Assemble formwork for core A	Apply concrete to ceiling / Apply concrete to core	... ...									
Reinforcing steel		Ceiling, area A	Ceiling, area B	Core			...								
Electrics		Ducts, lower layer		Ducts, upper layer			...								
HVAC		Ventilation, lower layer		Ventilation, upper layer			...								
Plumbing		Wastewater		Wastewater			...								
Equipment management	Order mobile crane					Order		Lead time							Mobile crane deployment
Materials management	Order Material					Order		Lead time							Delivery of materials

Mon . 23.1.2014
Formwork team
- Dismantle formwork from ceiling B 1
- Dismantle formwork from ceiling B 2
- Dismantle formwork from ceiling B 3
- ...

Figure 8.11 Lean construction – third floor execution organization with weekly and daily program on the building site (week 10), as well as material and equipment provision.

of the building is being erected by the construction enterprise. The interior finishes firms – electrics, plumbing and HVAC – must fit their installation works into the target cycle and flow process of the construction enterprise. The individual sub-jobs and their durations are listed in the target weekly and daily work plans, and are broken down by individual sections and their related sub-jobs: formwork, reinforcing steel, and concrete, and the electrical, plumbing and HVAC installation works, based on the target work hours specified in the work execution estimation (Figure 8.8).

The installation works concentrate on the following elements:

- Foundations – electrics and heating
- Walls – electrics, plumbing and HVAC
- Ceiling – electrics and HVAC

The plumbing installations are located in Section II – Core. The HVAC and electric risers are also in the core. The electrical lines and HVAC are channeled on each floor from the core to each room via a sub-distributor. The electrical duct, plumbing and HVAC installation works are generally performed following completion of the first reinforcing steel layer. The second layer is then applied once the installation works are finished. The installations must be fixed firmly and securely to the reinforcing steel with spacers. The target weekly fabrication cycle and flow process is shown in Figure 8.8.

A work execution estimation is crucial for this process to specify clear target work hours for the fabrication of the elements and their respective work steps, for example, formwork, reinforcing steel, concrete, or electrical ducts per concreting section. The individual sections and parts/elements are then broken down into weekly performance specifications, based on the construction program. The weekly plan should specify the target weekly performance and thus the target work steps per day, which are underlaid with target hours, based on the work execution estimation. On the one hand, this determines the team size per work step, and the target duration. The foreman then determines the individual work steps, complete with team size and target times per work day, for each weekly plan.

The work program on the construction site for a target weekly cycle in the cycle and flow process for a building floor (Figure 8.10) is developed in a further example, based on the WPP construction program (Figure 8.9). This target workflow was planned by the site team of the construction enterprise executing the work, based on the WPP. The target top-down construction program and the work hours specified in the work execution estimation were taken into account. The electrical, HVAC and plumbing installation works were planned on this basis.

Work charts with target performance are often drawn up for the individual work teams for each part as part of lean construction or lean management (Figure 8.11). Like Figure 8.8, these work charts define the windows for executing the individual sub-jobs. In addition, each work chart features the key works, materials, aids and equipment.

This is an excellent way to prepare work on a daily basis as it ensures that (Figure 8.11):

- Lead time – materials are ordered on time
- Lead time – equipment and construction aids are organized in good time

- Preparation – each subsequent work day is prepared the evening before in terms of materials management, equipment, hand tools, and assuring access to the work area

where the day's tasks are listed on a work chart (see Figure 8.11) for each and all teams involved.

In the case of structural work for a building project, the construction manager's team is the lead team, since the electrics, ventilation and plumbing teams must install their ducts and pipes, as construction progresses, on the lower layer of ceiling reinforcing steel before the upper layer can be installed. The presence of the technical trades (electrics, ventilation, plumbing) is coordinated to the exact day or half-day to ensure that non-value adding waiting times are reduced for all enterprises involved. Further coordination then takes place during the Friday coordination and weekly planning meetings. The electrical contractor might suggest laying all its ducts on the Tuesday to optimize its presence on the construction site. This suggestion would, however, need to fit seamlessly into the overall context of both the processes and the other stakeholders. This ensures that all dependent stakeholders think and act interactively. Added to which, all stakeholders must roughly sketch their weekly plan for the next two to three weeks to ensure the equipment, auxiliary aids and materials are ordered on time and provided just in time (Figure 8.11).

Any disruptions occurring while implementing the daily plan are remedied immediately, if at all possible, for example, overtime. If delays occur, nevertheless, that are obviously going to affect the next day, the downstream contractors (e.g., electrics) must be notified, for example, to send their team in the afternoon rather than the morning.

A further advantage of detailed weekly work preparation and logistics planning is the ongoing control of specified deadlines and, as such, the means to adjust the schedule. Delays are detected early on, and steps can be taken, if necessary, to catch up (by adjusting the resources: additional teams, equipment, hours), and thus restore adherence to the overall master schedule. The performance capabilities of workflows and construction methods are shown transparently, enabling them to be revised and improved. This procedure even allows interdependent, parallel construction works between the teams and subcontractors to be properly coordinated, and mostly without conflict.

8.4.1.2   Interior finishes work planning

Organizing the interior finishing work of the various interacting teams in the same building sections, floors, rooms and elements such that they do not get in each other's way is particularly challenging. It is during this phase that most construction sites/ventures experience the most serious problems with efficiency, and the most non-value adding work hours are accumulated. These disruptions are triggered by two influencing factors:

- Getting in each other's way when working in the same rooms and through each contractor's material stores
- Inadequate performance by the upstream contractors results in interruptions to individual jobs.

It is during the interior finishes phase, in particular, that many construction sites are felt to be chaotic. Searching for materials, moving materials around, and hindering each other with interrupted workflows due to incomplete upstream works are daily occurrences. Most enterprises see their hours taking off. Construction sites that are exposed to fierce competition are particularly frequent loss makers. Most have a simple excuse to hand: "It's the competition's fault." That is generally not true. In the case of such construction sites, usually either the property owner has commissioned individual contractors, or the TC/GC has planned poorly. The SMEs then try to organize themselves to improvise and make the best of the situation.

The solution lies in planning the workflows of the trades/disciplines and teams as a cycle and flow process. The cycle and flow process and associated logistics must be planned in terms of space – by floor and by room, and in terms of workflows – to incorporate the teams and their job packs and sub-jobs.

It is essential that the jobs are coordinated interactively and integratively with regard to both content and location to ensure that:

- the individual workstations are separated in respect of both space and timing to prevent them getting in each other's way
- the job packs are structured in layers to enable the next contractor to install in parallel or in the next layer without interfering
- the job packs are compiled in steps, without interference, and with optimal content to ensure each discipline can complete its tasks without interruption
- the job packs are optimally coordinated with the next team/discipline so that the downstream contractor can perform the subsequent work without structural interference and completely, where possible.

In order to coordinate the job packs in terms of content, space and timing, the following WPP activities must be performed interactively and integratively with the affected disciplines in a workshop:

- Definition of the job packs and sub-jobs per floor and per room
- Definition of the order of work of the disciplines with their job sub-packs and sub-jobs
- Definition of the duration of the respective job sub-packs and sub-jobs

Disciplines involved in the interior finishing work, and the job packs or sub-packs in logical sequence include:

1. Fabrication of masonry partition walls with installation slits and door frames
2. Fabrication of installation shafts
3. Installation of partition wall frames and cladding on the first side, and installation of the soundproofing, heat insulation, and door frames
4. Installation of the plumbing lines
5. Installation of the electrical ducts
6. Installation of the HVAC pipes
7. Partition walls: Closing the second side of the partition wall
8. Brickwork: Closing the shafts and slits
9. Installation of partition ceiling
10. Application of sound insulation to the floor, followed by screed

11. Soundproof sealing of the plumbing lines, electrical ducts and, possibly, ceiling and wall openings for electrical installations
12. Plastering and painting the walls
13. Laying the flooring
14. Installation of the doors
15. Installation of the fittings:
    a. Plumbing – washbasins, toilets, and so on
    b. HVAC – outlets, radiators, and so on
    c. Electrics – switches, sockets, and so on

Once the interior finishes sub-jobs and their sequence have been defined, the individual teams or disciplines must draft the sub-jobs list as the next step. This list of interior finishes sub-jobs (Table 8.3) includes a systematic compilation of the required quantities and auxiliary materials for each discipline and for each work step. The list of sub-jobs forms the basis for each team to plan its performance, with duration and team size, and to manage the logistics requirements, material orders and provision together with the necessary storage logistics, and the reservation of the required logistics infrastructure.

The next step defines the geometric-spatial order of job packs for each discipline. Job area plans (Figure 8.12) are good for this, as the individual job packs and sub-packs can be assigned timed spaces within these areas. These spaces serve as temporary work and storage areas for the disciplines and are structured on a cycle basis. The teams may use these areas for work and storage during the times indicated for the individual sub-jobs in the cycle plan. Unless otherwise indicated, the material belonging to each team must be removed once the sub-job is completed. Added to which, each team must clear and clean up the work areas they have been occupying.

Assigning temporary storage space to individual teams is particularly important to keep the distances short when working. Allowing for example plasterboard for cladding, the second side to be stored in rooms that first have to undergo installation of the electrical ducts, heating pipes or plumbing lines is not permissible. In such cases, the downstream electrics, plumbing and heating teams would have to access the rooms several times over to perform the same types of work because of severe interference caused by the upstream contractor storing materials for downstream works.

The capacities needed for, and durations of, the individual sub-jobs are then specified in the next step of the cycle planning process. These WPP tasks are prepared by the individual disciplines following the initial workshops, based on the sub-job lists. The cycle and flow plan, and job area assignments are then discussed in a second workshop, and coordinated in terms of timing and space.

Each individual team must internally specify the individual workflows and resources needed to adhere to the relevant cycle times, based on the sub-job lists. To do this, the workflows must be specified in weekly and daily plans. An electrician commissioned to install ducts on a hotel floor, for example, must specify what works are needed, and in which order:

• Route the cable support system from floor distributor down the corridor
• Lift ducts in bundles into the cable support system
• Lay the ducts from the floor distributor to the individual room distributors
• Lay the ducts from the room distributors to the individual rooms

Table 8.3   Sub-job list for electrical interior finishing work – work steps, equipment, construction aids, materials and quantities for each discipline

ETH
Eidgenössische Technische Hochschule Zürich
Swiss Federal Institute of Technology Zurich

TT consulting gmbh

Prof. Dr.-Ing. G. Girmscheid

No.	Sub-job Description of activity	Quantity	Materials/equip- ments/aids		Respon- sibility	Re- marks
			ordered	on standby		
Job: Electrics						
1	Lay ducts from distribution points to the rooms	Ø 25, l=25m				
2	noise insulations around duct penetrationsin slabs	d=25mm, b=30cm, l=3m				
3	Cable pulling Install room distributors Lay wiring Install sockets in walls	 1 pc. 25m 5 pcs.				
4	Install electrical fittings Switches Sockets Controller	 4 pcs. 5 pcs. 1 pc.				

This requires:

- thinking through the work steps
- specifying the time needed, based on the work execution estimation
- specifying the team size per work step
- specifying works that can be performed in parallel by several teams
- determining quantities
- determining work aids, such as platforms, equipment, and hand tools

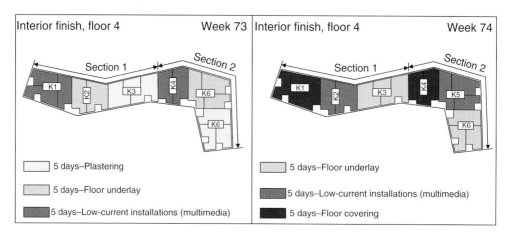

Figure 8.12   Job area plans – extract from the plot plan of a building site.

This produces an initial cycle and resource plan, which generally requires optimization to comply with the economic minimum principle. The target work hours are derived from this target plan. They must be consistent with the specifications from the work execution estimation, or else the cycle plan must be revised. These work performance/hour targets, which are interactively connected and are not changed in respect of either performance or hours, form the basis for the daily and weekly target/actual comparisons. The planned cycle and flow process is implemented as per the jointly agreed work prep and weekly work plans. The basic planning of the cycle and flow process, and the coordination and weekly control of the disciplines and their teams is the core task of the property owner's overall manager, or TC/GC.

The planning and coordination meeting run by the construction manager must take place at the end of each working week, attended by all interacting interior finishes firms. Items on the agenda include:

- Target/actual comparison
- Reasons for delays and interferences
- Action plans to reach targets, with three- to four-week look-ahead
- Status of material orders and deliveries; materials in store
- Cleanliness and safety on the construction site
- Storage utilization; logistics infrastructure utilization plans

The individual contractors must prepare these planning meetings themselves for their disciplines and sub-jobs about one day before the end of the working week. The purpose of the meeting is to identify target/actual deviations and plan actions to ensure target achievement, as well as to work on suggestions for improving the overall cycle and flow process. By adopting such a systematic procedure, the target defines the path to success. The improvisations witnessed on building sites today – especially during the interior finishing work – waste resources, and especially profits, which no sensibly operating enterprise or corporate management can afford. The same applies to all heads of department, regional office managers or project managers. Any project manager/site manager is also acting as an entrepreneur with responsibility for the relevant construction production of a project. Project, site and construction managers must all think and act as entrepreneurs. The tools they need are clear:

- A top-down work preparation process with cycle and flow process planning, logistics and logistics infrastructure planning, based on the work execution estimation, and matched to the phases and requirements of the relevant construction site.
- A work execution estimation with clear target specifications for each sub-job derived from the individual unit items on the bid estimation.
- A link-up between target work performance and target work hours for each building section, assembly, and element, as well as the associated sub-jobs (works).
- A bottom-up production cycle and flow organization on the construction site for each contractor and their discipline, as well as interactive, integrative coordination of the interacting processes.
- Weekly and daily work plans for the individual sub-jobs, and a three- to four-week work look-ahead to enable equipment and materials to be ordered.
- Weekly target/actual controlling procedure with analysis of deviations, including plans of action if deviations occur, and CIP measures.
- Weekly coordination among the various contractors with regard to cycle workflows, logistics infrastructure, storage areas, safety, and tidiness.

These principles ensure the success of a project for both the property owner and construction manager, as well as for the contractors involved.

8.4.1.3   Example: weekly and daily flow plan – total installation of the interior finishes job packs

The following analysis focuses on the coordination of the various interacting teams in terms of space and workflow, to ensure efficient performance of the interior finishing work without reciprocal interference, using a hotel floor as an example. This involves planning the cycle and flow process and associated logistics on a floor-by-floor basis, and developing a weekly plan for the interior finishes job packs for overall installation of the partition walls, plumbing, HVAC and electrical installations, screed and architectural finishes.

The layout of the floor in question is shown in Figure 8.13. The solid walls are shaded gray. Drywall construction will be used to build the interior walls. The supply and disposal media will be fed in as follows:

1. Service room – this is the location of the central riser zone and floor distributors for:
   a. electrics, with floor distributor
   b. heating and ventilation, with floor junctions and shutoff fixture
   c. water, with floor junctions and shutoff fixture
2. Corridor walls – the wastewater disposal downpipes are located here.

All media, water, heating with feed and return, and electrics with power, TV and radio, internet, and sprinkler alarms are fed to the room distributors. The room distributors with fuse and shutoff fixtures are located in the top or bottom cupboard in the wardrobe in each room.

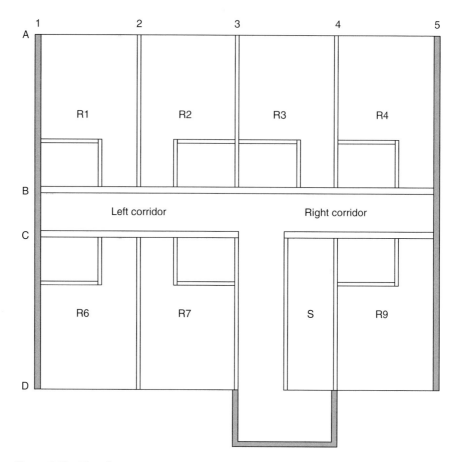

Figure 8.13   Floor layout.

The explanations below use the following definitions:
Job pack: Total task of a discipline, for example:

- Construction enterprise: Build the structure
- Electrics: Install all ducts, lay all cables and distributors, install all switch and consumer fittings
- HVAC: Lay all heating pipes from floor distributor to room distributor, and into the individual rooms, lay all ventilation ducts, install all fittings
- Plumbing: Lay all water and wastewater lines and install all plumbing fittings

Sub-job: Sub-job pack derived from a job pack, for example:

- Construction enterprise: Build ceiling on fifth floor, comprising sub-jobs: formwork, reinforcing steel, concrete
- Electrics: Lay ducts in ceiling, install wiring
- HVAC: Lay heating pipes
- Plumbing: Lay wastewater and water pipes

Work steps: Within a sub-job, for example, electrics:

- Lay the ducts in the corridor
- Assemble the line
- Lay the ducts in room 1-n, and so on

The weekly plan documents the various sub-jobs, composed of individual work steps, together with the selected team sizes and relevant durations, vertically in chronological order. The sub-jobs are shown horizontally in the sequential and parallel order of their cycle and flow production. An extract of a weekly plan containing sub-jobs 1 to 3 relating to the installation of all partition walls in a hotel is shown in Table 8.4. The sub-jobs are shaded blue in each case.

8.4.1.3.1 Sub-job 1: partition wall installation
The installation of partition walls by the drywall team comprises the work steps shown in Table 8.4. With the selected team size of eight people for the drywall construction team, frames can be built for four rooms in one day. The frames are built for R1, R2, R6 and R7 on day 1, and for R3, R4, S and R9 on day 2. As shown in Figure 8.14, the frames for each pair of rooms facing each other are stored in the corridor. The door frames for 15 doors are installed the next day, and one-sided cladding of the frames commences at the same time in rooms R1, R2, R6 and R7 and on the next day in rooms R3, R4, S and R9. This step is completed in two days for all eight rooms. The soundproofing and heat insulation is scheduled for the next two days.

8.4.1.3.2 Sub-job 2: installation of plumbing pipes – wastewater
The next step is the installation of the wastewater pipes (Figure 8.15). While sound-proofing and heat insulation are being installed in R3, R4, S and R9, the plumbing team can start installing the wastewater downpipe for the floor, together with the wastewater lines for the toilets, washbasins and baths/showers in R1, R2, R6 and R7. The plumbing pipes are stored in the rooms as shown in Figure 8.15.

The routing of the wastewater pipes to the wetrooms is shown in Figure 8.16. The wastewater from the washbasin and shower is channeled into a single downpipe. A separate downpipe is installed for the toilet. Sub-job 2 is completed by a team of eight in two days in all eight rooms.

8.4.1.3.3 Sub-job 3: installation of plumbing pipes – water
While sub-job 2 is being performed in rooms R1, R2, R6 and R7, sub-job 3 can commence with laying the water pipes in the corridor. In the corridor, the pipes are laid between the ceiling and the suspended ceiling. This work step involves routing the water supply from the service room in the entrance hall off to the left and right down the corridors to the rooms (Figure 8.17). Next day, the water supply lines for the toilets, washbasins and baths/showers (Figure 8.16) can be installed in rooms R1, R2, R6 and R7 while the wastewater plumbing works continue in rooms R3, R4, S and R9. Sub-job 3 is completed in three days with a team of eight.

8.4.1.3.4 Sub-job 4: Installation of the heating pipes
While the water pipes are being laid in the wetrooms on Tuesday and Wednesday of week i + 1, the heating lines are also being routed from the service room through the entrance hall and left and right on down the corridors to the rooms. Installation starts in the service room, where the riser and floor distributor from the riser zone are

Table 8.4 Extract from weekly plan – Sub-jobs 1-3

ETH
Eidgenössische Technische Hochschule Zürich
Swiss Federal Institute of Technology Zurich

CT consulting gmbh
Prof. Dr.-Ing. G. Girmscheid

Phase: Interior finish

Project: Hotel

Floor: k

Interior finish job packages		Week i					Week i+1				
Sub-jobs	Workgroup	Mon	Tue	Wed	Thu	Fri	Mon	Tue	Wed	Thu	Fri
1 Drywall installation	Drywall construction										
1.1 Build frames	8 people	R1, R2, R6, R7	R3, R4, S, R9								
1.2 Install door frames	8 people			15 door frames							
1.3 Clad side 1	8 people			R1, R2, R6, R7	R3, R4, S, R9						
1.4 Install sound and heat insulation	8 people					R1, R2, R6, R7	R3, R4, S, R9				
2 Lay plumbing lines	Plumbing, wastewater										
2.1 Wastewater downpipe for floor	8 people						R1, R2, R6, R7	R3, R4, S, R9			
2.2 Wastewater downpipe for toilet	8 people						R1, R2, R6, R7	R3, R4, S, R9			
2.3 Wastewater downpipe for washbasin	8 people						R1, R2, R6, R7	R3, R4, S, R9			
2.4 Wastewater downpipe for bath/shower	8 people						R1, R2, R6, R7	R3, R4, S, R9			
3 Lay plumbing lines	Plumbing, water										
3.1 Lay lines left and right	8 people						Lines L+R				
3.2 Water supply to toilet	8 people						R1, R2, R6, R7	R3, R4, S, R9			
3.3 Water supply to wash basin	8 people						R1, R2, R6, R7	R3, R4, S, R9			
3.4 Water supply to bath/shower	8 people						R1, R2, R6, R7	R3, R4, S, R9			

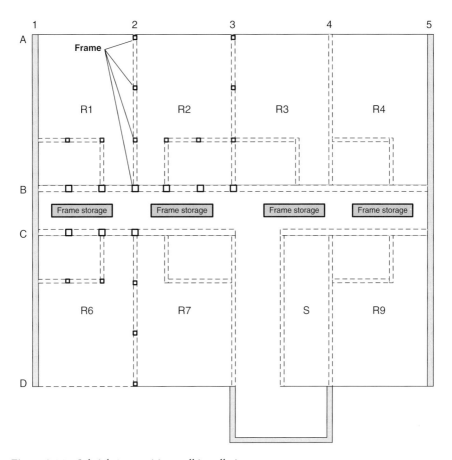

Figure 8.14 Sub-job 1 – partition wall installation.

located. The riser is extended and the connections and shutoff valves for supplying the floor are installed. At the same time, the feed and return piping system is hung and fastened to the ceiling. These feed and return lines are laid from the floor distributor to the room distributors. In the rooms, installation proceeds from the room distributors to laying the floor loops (Figure 8.18). The last work step comprises installation of the ventilation ducts, complete with ventilators, in the evacuation shafts on the floor. As shown in Table 8.5, this sub-job takes five days with a team of four.

8.4.1.3.5   Sub-job 5: Installation of the electric ducts
In the case of the electric ducts, installation also starts with the line from floor distributor in the service room to the entrance hall, and then left and right on down the corridors up to the room distributors in the rooms, before finally installing the ducts for electricity, TV and radio, internet and telephone from the room distributor to the individual consumer points in the room (Figure 8.19). Four men take four days to complete this work (Table 8.5).

To ensure efficient installation of the ducts, make sure the electric ducts do not cross the water and heating piping systems (Figures 8.20 and 8.21).

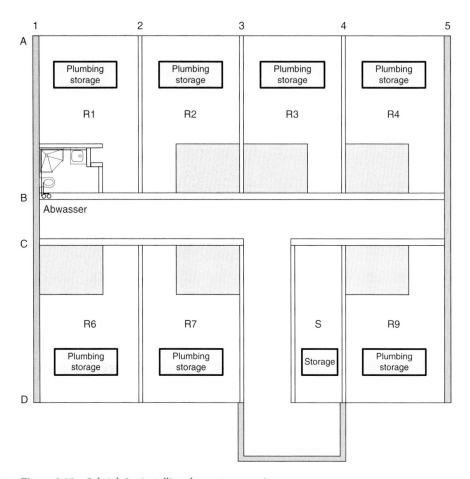

Figure 8.15   Sub-job 2 – installing the wastewater pipes.

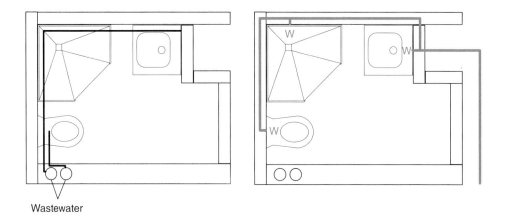

Figure 8.16   Wet room with wastewater and water pipes.

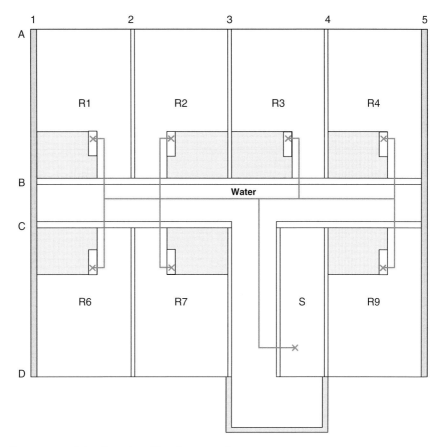

Figure 8.17 Sub-job 3 – installing the water pipes.

8.4.1.3.6 Sub-job 6: close the partition wall
Once all the ducts have been laid, the second side of the partition wall must be clad to close the wall.

8.4.1.3.7 Sub-job 7: installation of suspended partition ceiling
As soon as the partition wall is closed, the suspended partition ceiling can be installed. The suspended ceiling is installed in the entrance hall and corridors leading to the rooms, as well as in the entrance areas to the rooms (halls, opposite the bathrooms).

8.4.1.3.8 Sub-job 8: fabricate floor screeds for room and corridor
The first work step involves applying sound insulation to the floor in the rooms before pouring the floating screed. Sound insulation is then applied in the corridor, before being covered with screed. As shown in Table 8.6, this sub-job takes five days with a team of four. The screed can then set at the end of the working week.

8.4.1.3.9 Sub-job 9: soundproof seal
Sub-job 9 comprises fitting all of the (plumbing, HVAC and electric) lines in the ceiling and wall openings with structure-borne sound insulation on the Monday and Tuesday of week i + 4. This work can be completed in two days by teams of four from each trade (Table 8.7).

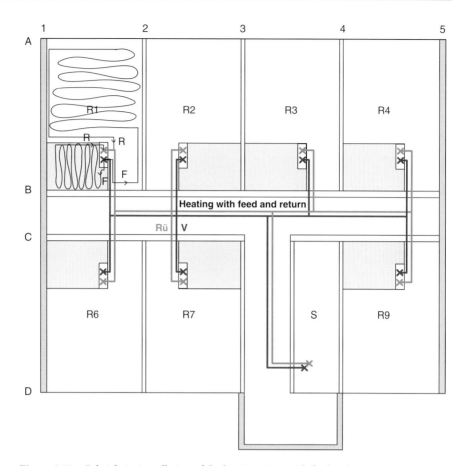

Figure 8.18   Sub-job 4 – installation of the heating pipes with feed and return.

8.4.1.3.10   Sub-job 10: lay wiring
While the soundproofing seals are being applied, the wiring can also be laid in sub-job 10. The first step involves installing the room distributors, before the wiring is laid and the sockets installed in the walls. This sub-job takes four men two days, as shown in Table 8.7.

8.4.1.3.11   Sub-job 11: plaster and paint
The walls are plastered and painted in the next sub-job. Four men are occupied with this task for four days (Table 8.7).

8.4.1.3.12   Sub-job 12: lay the flooring
The next step needs four men to lay the flooring in all rooms, followed by the corridor, which all takes three days (Table 8.7).

8.4.1.3.13   Sub-job 13: door installation
At the same time, two men can hang and align the 15 doors all in one day (Table 8.7).

8.4.1.3.14   Sub-job 14: install the fittings
The final sub-job involves installing the fittings. This includes plumbing fittings such as wash basin taps, towel holders, and so on, HVAC fittings, such as ventilation outlets,

Table 8.5   Extract from weekly plan – Sub-jobs 4-7

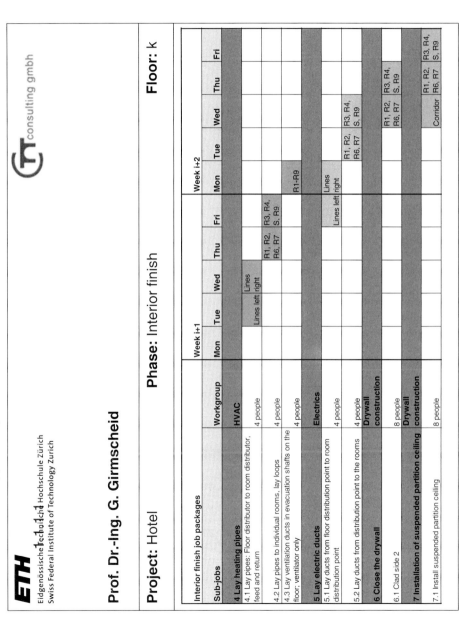

ETH
Eidgenössische Technische Hochschule Zürich
Swiss Federal Institute of Technology Zurich

Ⓣ consulting gmbh

Prof. Dr.-Ing. G. Girmscheid

Project: Hotel   **Phase:** Interior finish   **Floor:** k

| Interior finish job packages | | Week i+1 | | | | | Week i+2 | | | | |
Sub-jobs	Workgroup	Mon	Tue	Wed	Thu	Fri	Mon	Tue	Wed	Thu	Fri
4 Lay heating pipes	HVAC										
4.1 Lay pipes: Floor distributor to room distributor, feed and return	4 people		Lines left right	Lines							
4.2 Lay pipes to individual rooms, lay loops	4 people				R1, R2, R6, R7	R3, R4, S, R9					
4.3 Lay ventilation ducts in evacuation shafts on the floor; ventilator only	4 people						R1-R9				
5 Lay electric ducts	Electrics										
5.1 Lay ducts from floor distribution point to room distribution point	4 people					Lines left right	Lines				
5.2 Lay ducts from distribution point to the rooms	4 people							R1, R2, R6, R7	R3, R4, S, R9			
6 Close the drywall	Drywall construction										
6.1 Clad side 2	8 people								R1, R2, R6, R7	R3, R4, S, R9		
7 Installation of suspended partition ceiling	Drywall construction										
7.1 Install suspended partition ceiling	8 people								Corridor	R1, R2, R6, R7	R3, R4, S, R9

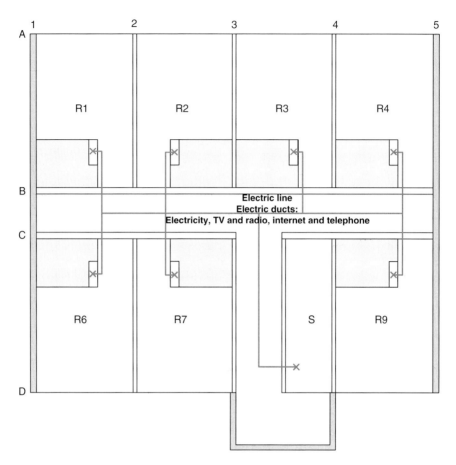

Figure 8.19   Sub-job 5 – installation of the electric ducts.

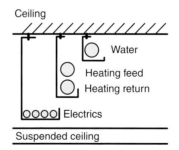

Figure 8.20   Cross-section through the ceiling with arrangement of the ducts.

valves, radiators, and so on, and the electrical fittings, such as switches, sockets and other controllers. This sub-job is completed in two days by teams of four for each trade (Table 8.7).

The individual teams prepare the daily work charts for the work targets, based on the sub-job list in Table 8.8.

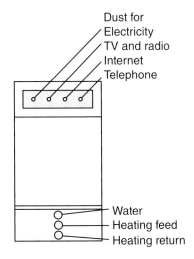

Figure 8.21 Arrangement of the electric ducts, water and heating pipes in the top and bottom cupboards in the wardrobe.

Table 8.6 Extract from weekly plan – Sub-job 8

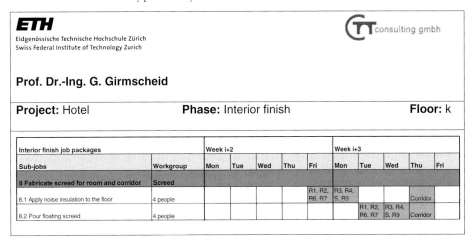

Table 8.9 shows such a daily work chart for a drywall team including work targets and steps, materials and quantities, aids and special tools. For purposes of clear presentation, the work charts are incorporated into a board with a look-ahead of about three weeks (Table 8.10). The charts are removed once the works have been executed. If work is not completed on a specific day, the chart is not removed in the evening, but is given a red dot. A note is also added to the chart detailing the problem that prevented completion of the task. In addition, the team meets at the board every evening to discuss the next day's work. This is also an opportunity to notify the downstream teams of any disruptions. Every evening, project management knows whether the

work is progressing on schedule, or whether problems have occurred. As such, immediate intervention is possible to discuss measures with the teams, and to instruct and
implement the actions. Each team is thus disciplined and motivated to perform its
work in line with the plan. Such detailed work plans are the only means of coordinating and managing a cycle and flow process.

8.4.1.4   Conclusion

By systematically planning the workflows of the trades and teams in a cycle and flow
process as shown in this example, the individual job packs and work steps can be
coordinated interactively and integratively in respect of content and space, and thus
the efficiency of the interior finishes process maximized.

8.4.2   Equipment and material call-up

Work prep and logistics planning also involves managing the deployment of the teams
and equipment, the subcontractors, and the material deliveries. Every schedule and
construction workflow plan is based on performance assumptions used to calculate
performance and plan the provision of resources, and duration of activities. During
the production phase of a structure, the scheduling and workflow planning processes
are exposed to the widest range of noise variables, such as poor weather, or learning
curves that cause deviations from the planned specifications.

Once the contract has been signed, WPP notifies the Human resources (HR)
and machinery departments, and the subcontractors of the anticipated approximate
dates for deploying/delivering the resources. Heeding the agreed lead times, the
construction site manager calls up the resources for the building site as construction progresses. In doing so, the construction manager must bear the following
in mind:

- When are the equipment/teams/materials needed?
- How long are the minimum call-up times for provision of the resources on the
  construction site?
- Are sufficient storage/lay-down areas and social facilities available?

When running a building site, a just in time delivery strategy should be attempted,
that is, materials are delivered straight to the location of installation – the construction
site – rather than being interim stored at the yard. This avoids the duplicate handling
and transportation of the material, and reduces the required capacity for interim storage at the yard, albeit lead times are key when it comes to calling up equipment and
materials. Logistics management during construction is essential to avoid unproductive work hours caused by having to wait for equipment or materials, in particular, or
by having to repeatedly move materials around because they are getting in the way of
the workflow. Construction site logistics are especially important in confined down-
town spaces and when several contractors are performing interior finishing work in
parallel on TC/GC contracts.

Table 8.7   Extract from weekly plan – Sub-jobs 10-14

ETH
Eidgenössische Technische Hochschule Zürich
Swiss Federal Institute of Technology Zurich

(T) consulting gmbh

Prof. Dr.-Ing. G. Girmscheid

Project: Hotel   **Phase:** Interior finish   **Floor: k**

Interior finish job packages		Week i+4					Week i+5				
Sub-jobs	Workgroup	Mon	Tue	Wed	Thu	Fri	Mon	Tue	Wed	Thu	Fri
9 Soundproof seal	**Plumbing, HVAC, electrics**										
9.1 Noise insulations around plumbing pipes penetrations in slabs	4 people	R1, R2, R6, R7	R3, R4, S, R9								
9.2 Noise insulations around HVAC pipes penetrations in slabs	4 people	R1, R2, R6, R7	R3, R4, S, R9								
9.3 Noise insulations around electric cables penetrations in slabs	4 people	R1, R2, R6, R7	R3, R4, S, R9								
10 Cable Pulling	**Electrics**										
10.1 Install room distributors	4 people	R1-R9									
10.2 Lay wiring	4 people		R1-R9								
10.3 Install sockets in walls	4 people			R1-R9							

Activity	Trade	People							
11 Plaster and paint	**Painter**								
11.1 Plaster the walls		4 people	R1, R2, R6, R7	R3, R4, S, R9					
11.2 Paint the walls		4 people		R1, R2, R6, R7	R3, R4, S, R9				
12 Lay the flooring	**Flooring**								
12.1 Lay the flooring		4 people			R1, R2, R6, R7	R3, R4, S, R9	Corridor		
13 Door installation	**Doors**								
13.1 Hang and align the doors		2 people					15 doors		
14 Install the fittings	**Plumbing, HVAC, electrics**								
14.1 Install sanitary fittings (wash basin taps, towel holders, etc.)		4 people						R1, R2, R6, R7	R3, R4, S, R9
14.2 Install HVAC fittings (ventilation outlets, valves, radiators, etc.)		4 people						R1, R2, R6, R7	R3, R4, S, R9
14.3 Install electrical fittings (switches, sockets, controllers, etc.)		4 people						R1, R2, R6, R7	R3, R4, S, R9

Table 8.8   Sub-job list for electrical interior finishing work – work steps, equipment, construction aids, materials and quantities for each discipline

	ETH					
	Eidgenössische Technische Hochschule Zürich Swiss Federal Institute of Technology Zurich			TT consulting gmbh		
Prof. Dr.-Ing. G. Girmscheid						
No.	**Sub-job** **Description of activity**	**Quantity**	**Materials/equip-** **ments/aids**		**Respon-** **sibility**	**Re-** **marks**
			ordered	on standby		
Job: Electrics						
1	Lay ducts from distribution points to the rooms	Ø 25, l=25m				
2	Noise insulations around duct penetrations in slabs	d=25mm, b=30cm, l=3m				
3	Cable Pulling Install room distributors Lay wiring Install sockets in walls	 1 pc. 25m 5 pcs.				
4	Install electrical fittings Switches Sockets Controller	 4 pcs. 5 pcs. 1 pc.				

Logistics management during construction, which is adjusted every week as part of the WPP revision and review process, is crucial for transforming time wasting activities often undetected by conventional controlling processes, into productive work performance and thus sustainably improving the monetary success potential of the contractors.

If a construction site is short on storage capacity, the delivery and storage of subcontractors' materials must be clarified when agreeing the contract. If subcontractors

Table 8.9 Daily work chart – Drywall construction

ETH
Eidgenössische Technische Hochschule Zürich
Swiss Federal Institute of Technology Zurich

(T consulting gmbh
Prof. Dr.-Ing. G. Girmscheid

Daily work chart: Drywall construction

Floor: k

Day: Monday Week i

Team: 2 people

Room R1 –Set up frames

Works:

1. Measure location points
2. Install U profiles in floor and ceiling
3. Measure location of props
4. Install props and fasten with L brackets

Aids:

1. Theodolite/measuring tapes
2. 2 step ladders
3. 2x Hilti drills
4. 2x Hilti screwdrivers

Material:

	Type	Quantity	Individual length	Total
1. U profile	U 5x10	5	3.00 m	15.0 m
2. Props	□ 10x10	8	2.73 m	21.9 m
3. Fastening bracket	L 10x10	16		

Problems:

are commissioned during construction execution, the possible delivery dates for sub-contractor materials must be taken into consideration. A subcontractor will only place firm orders for materials once it has received order confirmation; this lead time must be taken into consideration when determining the final commissioning date, and must be clarified in the contract.

Production and delivery times must be heeded when calling up materials. Production times are dictated by the product requirements, performance capabilities of the contractors, and the quantities. They are joined by delivery times, which can take several weeks, depending on the product or its origin. Care must be taken to ensure the availability of sufficient storage space, unloading equipment and, if necessary, additional labour, on the day of delivery.

When calling up major pieces of equipment, the time needed to set up and install the kit must be incorporated into the schedule. For example, setting up particularly

Table 8.10 Work chart board for interior finishing work on floor K – Execution organization with weekly and daily plans

ETH
Eidgenössische Technische Hochschule Zürich
Swiss Federal Institute of Technology Zurich

CT consulting gmbh
Prof. Dr.-Ing. G. Girmscheid

Workgroup	Week i					Week i+1					Week i+2				
	Mon	Tue	Wed	Thu	Fri	Mon	Tue	Wed	Thu	Fri	Mon	Tue	Wed	Thu	Fri
Drywall construction Team 1	Frames R1, R2	Frames R3, R4	Doors R6, R7 / Cladding R6, R7	Cladding R3, R4	Insulation R1, R2	Insulation R3, R4							Cladding R1, R2, R6, R7	Cladding R3, R4, S, R9	
Drywall construction Team 2	Frames R6, R7	Frames S, R9	Cladding R1, R2	Cladding S, R9	Insulation R6, R7	Insulation S, R9							Suspended ceiling, corridor	Susp. ceiling R1, R2, R6, R7	Susp. ceiling R3, R4, S, R9
Plumbing Team 1						WW down-pipe R1, R2, R6, R7	WW down-pipe R3, R4, S, R9								
Plumbing Team 2						WW pipe R7	WW pipe R3, R4, S, R9								
Plumbing Team n						Water piping L + R	W pipe R1, R2, R6, R7	W pipe R3, R4, S, R9							
HVAC Team 1							Heating piping L1	Heating piping R1	Loops R1, R2	Loops R3, R4	Ventilation R1, R2, R6, R7				
HVAC Team 2							Heating piping L2	Heating piping R2	Loops R6, R7	Loops S, R9	Ventilation R3, R4, S, R9				
Electrics Team 1										Electric lines L1	Electric lines R1	E ducts R1, R2	E ducts R3, R4		
Electrics Team 2										Electric lines L2	Electric lines R2	E ducts R6, R7	E ducts S, R9		
Screed Team															
Painter Team															
Flooring Team															

large cranes can easily take a week. Auxiliary production equipment can be even more difficult, for example, a concrete mixing plant or even a plant for prefabricating concrete parts. Care must be taken to ensure the availability of sufficient and clear installation space on the day of delivery. This holds especially true for special formwork used in building construction, or preparatory installations for self-climbing core formwork.

8.4.3   Organizing the construction workflow, construction methods, and health and safety

The primary task facing execution management is to optimize the selected construction methods and to adapt the construction workflows as flexibly and beneficially to performance as possible to the dynamic process, which is often exposed to internal and external disruptions.

Work performed on a construction site is assessed in terms of both timing and cost as part of the work execution estimation exercise. The organization of the construction workflow is aligned to these approaches; construction workflows and construction methods must be regularly reviewed. Care must be taken to ensure that the planned and estimated performance specifications for the construction methods can be implemented, and the concepts realized, as planned in WPP and the work execution estimation.

Since each project is unique, a certain learning period must be taken into account for each construction method. Even if the construction methods are familiar to the site team, new team members or new conditions on site necessitate the incorporation of learning phases. Start-up difficulties must be expected if new construction methods or systems (e.g., formwork) are being used. Site management is responsible for minimizing the learning phase. Regularly weekly work meetings can be used to address and clarify technical problems. As far as the subcontractors are concerned, these meetings are particularly important to integrate them more closely into the construction workflow (e.g., in a cycle process). Once the learning phase is over, the construction production works must be subjected to a continuous improvement process (CIP). This requires the analysis of routinely recurring works to identify potential for improvement. If the construction workflow is particularly stressful for the workforce, personal rewards, for example, bonuses, or other performance-enhancing measures should be introduced.

Implementing a health and safety plan is a major responsibility of site management. Protective clothing, safeguarded areas, warnings and compliance with safety instructions must be checked at regular intervals. The site manager is responsible for his workforce and must not succumb to the belief that they are sufficiently safety conscious themselves. The health and safety concept should be based on preventive and impact-reducing measures. The aim of this step-by-step procedure is to avoid the occurrence of incidents with the highest possible level of probability. If an incident occurs regardless, counter-measures must be available to keep the impact on people, structures and the environment as low as possible. In addition to complying with technical health and safety at the workplace and on tools, the workforce must also be trained to deal with various hazardous situations. Rescue plans should be prepared for major incidents (Girmscheid, 2012).

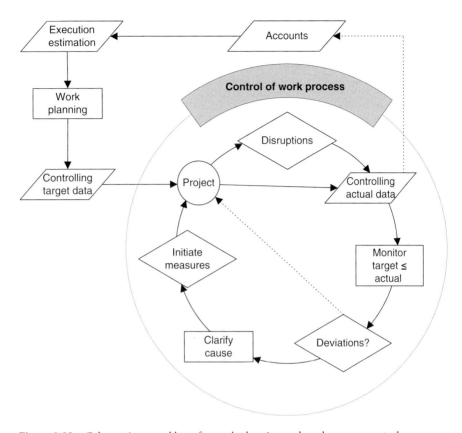

Figure 8.22 Cybernetic control loop for work planning and work process control.

8.5 Construction site controlling process

A continuous construction site controlling process is the only means of managing the optimally planned site workflows during the production processes in line with the targets, and thus taking regular controlling responsibility seriously. Efficient controlling requires the transfer of the specifications from the work execution estimation to a controlling sheet to enable early identification of deviations between target and actual work performance and initiation of counter-measures to improve results.

Regularly comparing the relevant target specification from the work execution estimation with the actual figures from the construction site at both weekly and monthly intervals is absolutely crucial as the only means of efficiently managing the construction site and implementing continuous improvements. While the target work hour specifications possess clearly defined control functions, the forecast figures are derived from the hours actually worked on the construction site and extrapolated to the fixed end of the construction period (Figure 8.22).

8.5.1 *Performance specifications*

Figure 8.23 shows the target work hour specifications derived from the work execution estimation for site controlling purposes (Girmscheid & Motzko, 2013). To do this, the

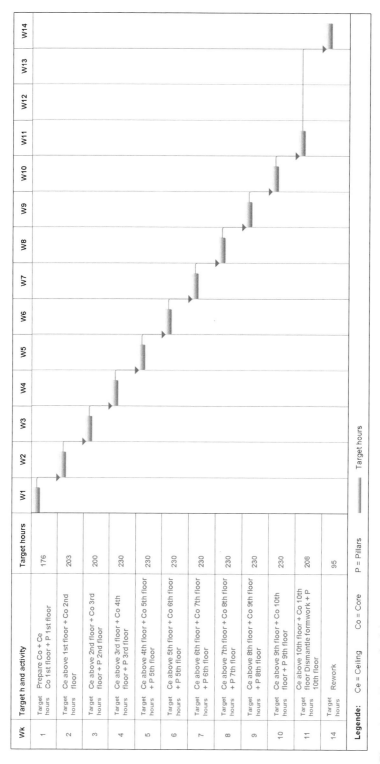

The following table data is shown in the figure:

Wk	Target h and activity		Target hours	W1	W2	W3	W4	W5	W6	W7	W8	W9	W10	W11	W12	W13	W14
1	Target hours	Prepare Co + Ce Co 1st floor + P 1st floor	176														
2	Target hours	Ce above 1st floor + Co 2nd floor	203														
3	Target hours	Ce above 2nd floor + Co 3rd floor + P 2nd floor	200														
4	Target hours	Ce above 3rd floor + Co 4th floor + P 3rd floor	230														
5	Target hours	Ce above 4th floor + Co 5th floor + P 5th floor	230														
6	Target hours	Ce above 5th floor + Co 6th floor + P 5th floor	230														
7	Target hours	Ce above 6th floor + Co 7th floor + P 6th floor	230														
8	Target hours	Ce above 7th floor + Co 8th floor + P 7th floor	230														
9	Target hours	Ce above 8th floor + Co 9th floor + P 8th floor	230														
10	Target hours	Ce above 9th floor + Co 10th floor + P 9th floor	230														
11	Target hours	Ce above 10th floor + Co 10th floor Dismantle formwork + P 10th floor	208														
14	Target hours	Rework	95														

Legende: Ce = Ceiling   Co = Core   P = Pillars   ──── Target hours

Figure 8.23 Target schedule.

specifications from the work execution estimation that constitute the target hours per week and floor, are transferred to a table. They can also be entered on a target schedule (Figure 8.23). The construction manager is responsible for comparing the actual hours worked on the construction site with the target specifications and for identifying any discrepancies between target and actual.

8.5.2   Controlling weekly work performance

Weekly work performance controlling procedures are used for bottom-up work organization and as specifications for the construction site. They can be used to manage the construction site at foreman level, and permit implementation of a weekly continuous improvement process, together with the foremen. These specifications also form an ideal base for planning the work of the teams.

Based on the work specifications (Table 8.11), a further step involves weekly performance controls of ceiling and core construction progress. Table 8.11 is used as

Table 8.11   Construction site controlling – target specifications in hours/week

Institut für Bau-und Infrastrukturmanagement Professur für Bauprozess-und Bauunternehmensmanagement Prof. Dr.-Ing. Gerhard Girmscheid		Wk 1	Wk 2	Wk 3	Wk 4	Wk 5	Wk 6	Wk 7	Wk 8	Wk 9	Wk 10	Wk 11	Wk 12	Wk 13	Wk 14	Target total
	Structural work, floors 1-10															
Prepa-ration	**Formwork preparation**															
	Core	27														27
	Ceiling	65														65
	1st floor															
	Core 1st floor + pillars 1st floor	84														84
	Ceiling above 1st floor		110													110
	2nd floor															
	Core 2nd floor		93													93
	Ceiling above 2nd floor			110												110
	3rd floor															
	Core 3rd floor + pillars 2nd floor			90												90
	Ceiling above 3rd floor				140											140
	4th floor															
	Core 4th floor + pillars 3rd floor				90											90
	Ceiling above 4th floor					140										140
	5th floor															
Con-struc-tion produc-tion	Core 5th floor + pillars 4th floor					90										90
	Ceiling above 5th floor						140									140
	6th floor															
	Core 6th floor + pillars 5th floor						90									90
	Ceiling above 6th floor							140								140
	7th floor															
	Core 7th floor + pillars 6th floor							90								90
	Ceiling above 7th floor								140							140
	8th floor															
	Core 8th floor + pillars 7th floor								90							90
	Ceiling above 8th floor									140						140
	9th floor															
	Core 9th floor + pillars 8th floor									90						90
	Ceiling above 9th floor										140					140
	10th floor															
	Core 9th floor + pillars 10th floor										90					90
	Ceiling above 10th floor											140				140
	Dismantle formwork from core + pillars 10th floor											47				47
	Rework															
Rework	Core											21				21
	Ceiling - 9th floor															
	Ceiling - 10th floor													95		95
	Auxiliary supports 8th floor															
	Total	176	203	200	230	230	230	230	230	230	230	208	0	0	95	2492
	Actual hours for the week															
	Target/actual difference															
	Aggregated difference															

a template for efficient weekly performance control. It systematically sorts the relevant work hour specifications for building the ceiling and core into jobs with the target work hour specifications and actual hours worked on building the respective elements, together with the sum total for the reporting month to date, and the forecast total to the end of construction. The forecast simulates the expected total hours worked up to completion, based on the actual result to the end of the reporting month and the circumstances surrounding construction, and taking account of improvements that have been put in place.

The job performance column aggregates the hours for the elements scheduled for production in the week in question in line $\sum$ (total) (Table 8.11). On the one hand, they serve as a target to limit hours for the work specification; on the other hand, actual hours worked are compared with the target/limit hours. In doing so, the target performance specified in the work execution estimation must be differentiated from actual performance in terms of actual hours worked. The associated degree of completion (in %) relates to work completed in the week under review.

Figure 8.24 charts the deviations in actual from target performance caused by the delayed provision of plans by the client (cause) and the resulting disruption to production and associated delay for the contractor (effect) who was unable to continue working. Because the plans were delivered late, and work was interrupted on the construction site as a result, the ceiling and core cannot be completed on time. The resulting performance shortfalls therefore have to be made good one week later (three additional working days + 2 idle days).

Despite the lack of special plans for user-specific ceiling openings and anchor brackets in the walls of the core, the ceiling and core teams started positioning the formwork and installing the standard reinforcing steel in week m. However, the late delivery of the plans, in respect of the additional reinforcement at the cutouts and brackets, and for the exact size and positioning of the cutouts, prevented the team from continuing their work. The special formwork with cutouts and additional reinforcing steel were not delivered until the Monday of week (m + 1); three days were lost performing the corrections and concreting the ceiling above floor (n − 1), as well as the core on floor n. As such, the team had two idle days at the end of week (m + 1) while they waited for the concrete to set. They could only start the normal cycle workflow on the Monday of week (m + 2) with the ceiling above floor n and core on floor (n + 1) (Figure 8.24).

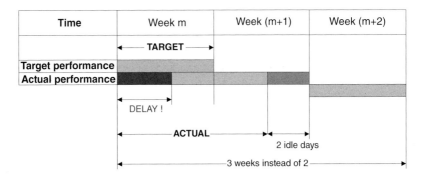

Figure 8.24   Target/actual deviation caused by delay on the construction site.

Because of the late delivery of plans, the team's work completion in week 8 only amounted to 56 target work hours for ceiling construction and 36 target work hours for building the core. Together, the total of 92 target work hours equates to a target degree of completion of 40%. Because of the impediment, however, only 175 actual hours were used as the team was able to reduce its normal working hours to 7 per day, thus minimizing the additional cost caused by the impediment.

As such, the degree of completion at the end of week 8 is as follows:

- All of the ceilings up to and including floor 6, and all cores up to and including floor 7 are 100% completed,
- Ceiling above floor 7 – 40% completed,
- Core on floor 8 – 40% completed.

In the following week 9, the outstanding target work of 60% (ceiling and core) must be completed with a value adding target quota of hours of:

- Ceiling: $h_{Ce, \text{ week } 9}^{\text{target}} = 140 - 56 = 84\ h$

- Core: $h_{Co, \text{ week } 9}^{\text{target}} = 90 - 36 = 54\ h$

- Total: $h_{\text{week 9, partially completed}}^{\text{target}} = 138\ h$

Actual hours again amounted to about 175 in week 9 since the team's production flow was hindered by idle time during the week while they waited for the concrete to set (last two days of the working week), instead of using the weekend for this process, as originally planned.

This relative estimate of weekly work performance is based on approximate allocations of the target hours to the individual ceiling and core sections. Such target work differentiation for unscheduled interim stages can be allocated arithmetically very precisely – for example, based on tons of installed steel, or volume of completed formwork – relative to the total scope, or estimated relatively accurately.

These interferences and associated non-value adding hours (h) are calculated as follows:

$$h^{\text{subsequent claim}} = (140 - 56) + (140 - 84) = 140\ h$$

The contractor must notify the property owner of the hourly figure when the impediment occurs, and again when it is eliminated (e.g., at the end of week 2). The contractor can subsequently claim the additional costs incurred by this additional workload.

The hours needed to complete the structure after each deadline must be listed in the forecast column (Figure 8.23), and the work hours to the end of construction must be anticipated. This reveals that the completion of work moves to one week later, and the total hours increase as a result of the additional work and idle times. As such, the construction manager is responsible for initiating counter-measures in good time to avoid further deviations from the target specifications, and to deliver the project successfully within the contractually agreed construction period.

In addition to presenting the target specifications and actual performance, together with deviations, in a table, they can also be plotted. In construction, the schedule is normally used to plot the target/actual status of performance. It can also be populated with target versus actual hours (Figure 8.25).

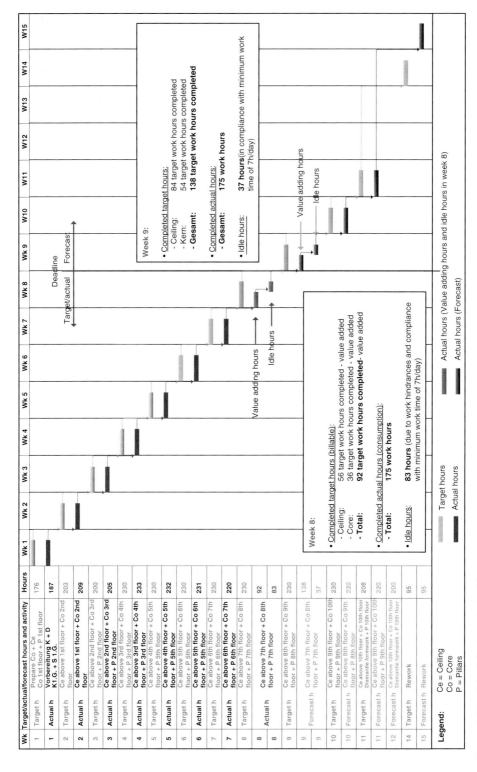

Figure 8.25   Schedule: target/actual hours.

Table 8.12 Controlling monthly performance

Month	Floor	Total degree of completion [%]	Week degree of completion [%]	Job performance Target [h]	Actual [h]	Aggregate performance Target [h]	Actual [h]	Final outcome Forecast [h]	Remarks
	Preparation	4	100	92	97	92	97		
	1	7	100	84	90	176	187		
1	2	15	100	203	209	379	396		
	3	23	100	200	205	579	601		
	4	27	40	92	93	671	694		
	Σ 1st month			671	694				
	4	32	60	138	140	809	834		
	5	42	100	230	232	1039	1066		
2	6	51	100	230	231	1269	1297		
	7	60	100	230	220	1499	1517		
	8	64	40	92	175	1591	1692	1692	DELAY
	Σ 2nd month			920	998				DEADLINE
	8	69	60	138		1729		1867	
	9	79		230		1959		2087	
3	10	88		230		2189		2307	
	11	96		208		2397		2507	
	Σ 3rd month			806					
4	Rework	100		95		2492		2602	
	Σ 4th month			95		2492		2602	

8.5.2.1 Controlling monthly performance

Monthly performance control is part of an enterprise's internal reporting procedures. It serves to review the status of progress and trend on the construction site relative to the performance specifications issued by the management in the form of the work execution estimation for the construction site.

Based on the weekly controls of the construction site, monthly performance is normally also controlled in order to present an overview of deviations of actual hours worked from the management's work targets. Table 8.12 shows an example of a monthly performance control report that visualizes the target/actual hours on the construction site, similar to the weekly performance control procedure.

Performing monthly controls is highly recommended, especially for structures with a construction period extending over one or several years. In addition to work performance, the costs incurred are frequently also controlled to enable counter-measures if any deviations occur. These controls are necessary to enable assessment of the financial situation of the construction site by the accounts departments.

In addition to monthly performance controls, the monthly reports also contain a brief systematic site report, which is generally structured as follows:

Performance:

- Brief description of work performed
- Reasons for work changes internally and externally
- Description of deviations from target specifications
- Actions taken to enhance performance

Property owner/planner:

- Project meetings – main agenda items
- Instructions issued by the owner

Impediments/claims:

- Notification of impediments – reasons
- Claims (usually in a separate list) include those in preparation, those already submitted and outstanding claims

Invoicing/payment receipt (separate list):

- Invoices submitted
- Payments received
- Late payments
- Outstanding receivables
- Reasons why the owner is not paying outstanding receivables
- Measures

Events worth mentioning:

- Accidents
- Visitors, and so on

In conjunction with the resource plan and schedule, construction site controlling procedures based on systematic, part-related work execution estimation enable the construction site to be managed systematically down to the level of foreman. The sequence and logistical interaction of the relevant work processes are coupled with the commissioned unit items for the elements under construction.

8.6   CIP – the continuous improvement process

The continuous improvement process must be embedded in the Quality Management Manual for the project and aligned to the enterprise's project targets. Improvement is only possible if target achievement can be measured ("If you can't measure, you cannot manage"). Improvements within a project are aligned to the respective project-related specifications, derived for example from:

1. Improving the performance of direct production workflows by simplifying activity sequences, changing processes, and so on
2. Improving the performance of logistics workflows in terms of
   a. Material ordering
   b. Material storage
   c. Equipment deployment
   d. Changes in workflows
   e. Coordinated use of construction site transport and other equipment between the various teams
3. Improved safety in the individual construction phases to reduce the risk of accidents and improve performance through enhanced safety
4. Improved quality to avoid the need for rework
5. Minimization of environmental impacts (noise, dust, etc.) to minimize complaints and interference on the construction site.

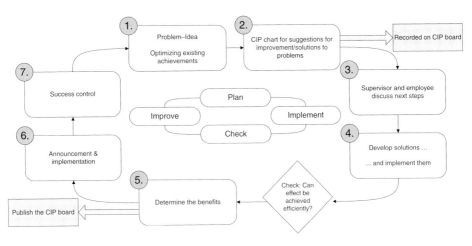

1. Problem/idea, suggestion for improvement
2. Employee with the original idea hands a completed CIP chart to their supervisor It is then recorded in IQ-Soft
3. Supervisor and employee discuss next steps
4. Develop solutions with the aid of facilitation or one of the known facilitation techniques, and implement
5. Determine the benefits
6. Announce & implement the measure with the necessary resources
7. Control success and, if further improvements are possible, start again at step 1

Figure 8.26 Procedure for implementing a CIP suggestion on a construction site.

A continuous improvement process (CIP) is usually structured as shown in Figure 8.26.

The problem solving process should be conducted as follows (Figure 8.27):

- Problem that causes inefficiency, and must be detected, for example, safety problem, waiting for materials, moving material around, more actual hours than target hours, rework due to flaws
- Brainstorm possible solutions in a team meeting and note them down on cards
- Evaluate the possible solutions in respect of effectiveness, costs, risks
- Select the solution
- Draw up an action plan for approval and implementation

Target versus actual deviations and problems that have arisen on the construction site are discussed in the weekly meetings between the site manager and foreman. The suggested solutions are then discussed and evaluated. Depending on the scope and effect of the CIP measures, they can be:

- resolved directly, for example, to immediately avert danger or directly improve the workflows or logistics flows without fundamentally changing the respective elementary processes,
- forwarded to work prep or the estimation department for further analysis, and to check the interactive effect on parallel or downstream workflows, as well as performance and cost effectiveness.

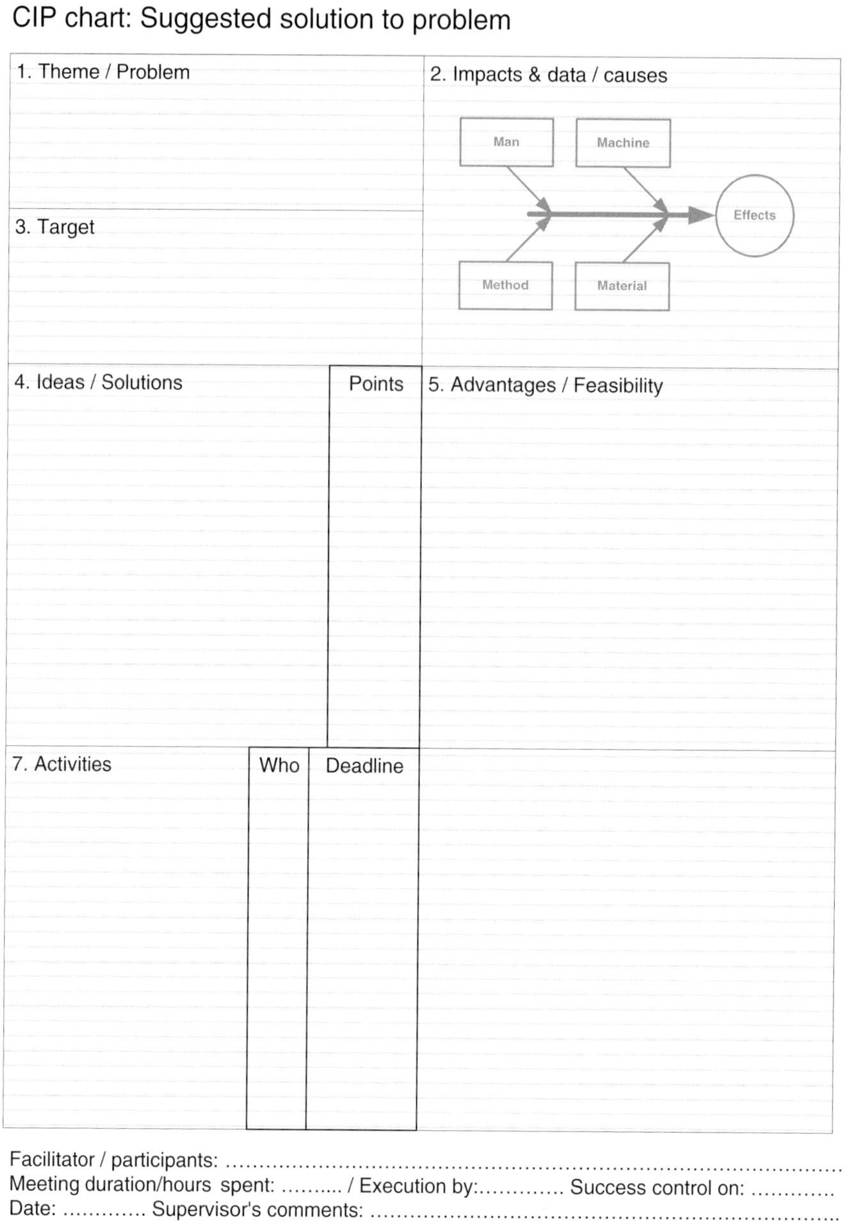
CIP chart: Suggested solution to problem
1. Theme / Problem
2. Impacts & data / causes
Man
Machine
Effects
Method
Material
3. Target
4. Ideas / Solutions
Points
5. Advantages / Feasibility
7. Activities
Who
Deadline
Facilitator / participants: ....
Meeting duration/hours spent: .... / Execution by:............. Success control on: ....
Date: .... Supervisor's comments: ....

Figure 8.27 CIP chart for construction sites: suggested solution to problem.

Every suggested solution to a problem is evaluated and assessed. The successful suggestions for solutions are announced by the site manager. Depending on resource requirements, they are:

- Announced formally and verbally in the weekly site meeting and the results recorded in the minutes

- Announced in writing together with instructions governing the provision of equipment, changes in workflows, and so on, as well as new target specifications and changes to the plans, and initiated.

As soon as the workforce has been instructed and implementation has commenced, the success of the measure in respect of complying with the target specifications is controlled, respectively the extent to which the CIP measures are proving to be effective and the anticipated improvement has occurred. Both the team and site management need to critically challenge whether the CIP measure has produced improvements.

A key factor for CIP is centering success on respectful treatment of the workforce. Then a foreman will be willing to suggest ideas for detailed workflows, and the construction manager can lead the overall concept. CIP is successful when this symbiosis actually works.

8.7   Conclusions

Nowadays, lean management and lean construction production for in situ construction are the modern way to combine the application of industrialized flow processes with continuous improvement in order to minimize construction industry losses and the waste of time, costs and material. Lean site management is a must for production processes that are economically, ecologically and socially sustainable. Lean management improves not only the workflow but also the logistics off site and the logistics infrastructure, while at the same time – last but not least – positively influencing work safety and improving the working conditions for the workforce. Lean management must be structured in a top-down process to plan the workflows for all of the works in each construction project, and in the bottom-up implementation, execution and coordination of the workflows on site to ensure the process is managed, controlled and continuously improved.

The top-down production planning process for construction sites requires systematic incorporation of the environmental conditions in the work preparation and logistics planning processes. Particular attention must be paid to What is being built, Where it is being built, and Which boundary and environmental conditions need to be considered. These three "Ws" influence the choice of construction method and logistics on a construction site.

Based on these preconditions, the construction task itself must be systematically broken down into main, module, and elementary processes. Once the processes have been broken down for the individual construction phases, components and manufacturing steps, the construction methods can be selected. In doing so, a systematic selection procedure for identifying the best possible construction methods under the given specific boundary conditions of the projects is crucially important. The options must be subjected to methodical procedures for assessing the best methods that are fit for purpose, and for performing both qualitative and quantitative multi-criteria comparisons. The selection of construction methods and the manufacturing workflows must meet or undercut the target specifications of the work execution estimation. Once the methods have been selected, the schedule and resources for executing the construction program must be planned – taking account of the temporal and spatial interdependencies of the various production processes – to produce the final work execution estimate. The key performance parameters – mainly defined in terms of man hours and machine hours – must be derived from this work execution estimation for the construction program with the selected construction methods as the target specifications for the site management. As execution process planning progresses, logistics planning plays a key interface role, interactively linking the other planning areas. For value creation in cycle and flow production, the success of the site is dependent on the construction methods and logistics planning being tailored to the requirements of the different workflows on site.

Based on the cycle and flow process program planned during the work preparation phase, the work must then be organized on site, based on weekly and daily workflow plans. This weekly and daily workflow planning is controlled by the key performance indicators for each type of work – mainly man and machinery hours. For both the structural work and interior finishing phases, site management must define specific time and performance targets for the work teams in respect of manufacturing the individual components and executing the construction during the workflow. The work has to be planned clearly with the different work teams, bearing in mind the key performance figures specified by the top-down work planning program and the key work estimation figures. In doing so, site management can control performance and fulfillment or non-fulfillment of the key performance indicators on a weekly basis, and can assess the extent of deviation. Site management must discuss with the work teams how to improve the construction workflow on site to minimize losses in terms of hours, materials and costs.

Based on these interactions, a continuous improvement process is established on site to improve the performance and quality on site, and to ensure the target figures are reached. A detailed weekly work program for the different construction teams must be developed and tracked on a daily basis to ensure cycle and flow production in the structural and interior finishing phases, together with a three- to four-weekly forecast to ensure compliance with the lead times for material deliveries. Further, the weekly workflow program promotes continuous workflows for the different works because the site infrastructure needed by the different work teams each forthcoming week can be properly planned so that no work team suffers interruptions in their workflow. Since the key performance figures for the teams are derived from the work execution estimate, the targets and actually achieved figures can be controlled on a weekly basis, and the teams can take steps in a continuous improvement process to reach the set target figures.

Further, the detailed weekly and daily bottom-up workflow programs enable implementation of a clear logistics program for the teams to ensure that material stores are set up in clearly assigned areas so that the teams are not obstructed during their work. For this reason, a clear storage program has to be developed to ensure that the workflows of the different work teams working in a cycle process do not obstruct each other. A logistics coordinator is therefore required on site to manage the assignment of the logistics areas and the use of the logistics infrastructure.

Successful construction execution in a lean construction process must adhere to all specified schedule, budget and quality targets. For this reason, a controlling procedure that compares the target and actual key performance figures of the actual work performed on a construction site is absolutely essential. This controlling procedure requires that the key figures of the work execution estimation and the construction time targets are monitored on site, which must involve a regular process of target and actual comparison on either a weekly basis or at least at monthly intervals to enable early detection of deviations. In the interests of continuous improvement based on such target/actual key figure comparisons, the implementation of appropriate countermeasures is then absolutely crucial, and is the responsibility of site management.

Implementing such a lean management process based on a top-down work planning process and bottom-up work execution organization with clearly specified key performance indicators that can be easily controlled promotes very efficient and successful site management, and is the way forward in modern industrialized site construction.

References

Blecken, U. Boenert, L. & Blömeke, M. (2001). *Studie zur Akzeptanz einer Dienstleistung Logistik in der Bauindustrie.* Universität Dortmund, Lehrstuhl Baubetrieb. (Study concerning the Acceptance of a Service Logistics in the Construction Industry. University of Dortmund, Chair of Construction Operation.)

Girmscheid, G. (1997). Fast Track Projects – Anforderungen an das moderne Projektmanagement (Requirements for Modern Project Management), Bautechnik 73, H. 8, S. p471-484.

Girmscheid G. (2003). *Leistungsermittlung für Baumaschinen und Bauprozesse (Assessment of Construction Machines and Processes),* 3rd edition. Springer-Verlag, Berlin.

Girmscheid, G. (2012). Bauproduktionsprozesse des Tief- und Hochbaus Construction Processes of Civil Engineering Underground and Surface), Vorlesungsskript. *Eigenverlag des Instituts für Bau- und Infrastrukturmanagement (ETH Zürich),* Zürich, Switzerland.

Girmscheid, G. & Motzko, C. (2013). *Kalkulation und Preisbildung in Bauunternehmen (Calculation and Pricing in Construction Companies),* 2nd ed., Springer-Verlag, Berlin, Germany

Loschert, P. (1999). *Terminmanagement im schlüsselfertigen Hochbau (Scheduling and Task Management in Turnkey Building Construction),* Dissertation, Technical University of Darmstadt, Germany.

Schmidli, A. & Schnüriger, W. (2001). *Projektmanagement Führung, Planung (Standard-Analysis SBV ST-WIN Building Construction 2001),* Kontrolle. Helbig & Lichtenhahn, Basel, Switzerland.

Chapter 9

New Cooperative Business Model – Industrialization of Off-Site Production

An Interdisciplinary Cooperation Network for the Optimization of Sustainable Life Cycle Buildings

Julia Selberherr

ETH, Swiss Federal Institute of Technology, Zurich, Switzerland

A new cooperative business model for the optimization of life-cycle buildings using modern industrialization technologies is developed. A business model is developed as a composition model with intended target-means-relations for shaping the socio-technical world. The model's scientific quality is ensured with triangulation through viability, validity and reliability. In the modelling process the so-called *outside view*, which describes the interaction with the environment, and the *inside view*, the cooperative value creation, are analysed separately. On the level of the *inside view* three sub-models are developed. The *initiation process model* describes the setup of the cooperation network as pool of potential system suppliers. The *project process model* aims at a project-specific, integral optimization using customized industrial production technologies. In this, each system supplier pursues the optimization of his subsystem whilst the system integrator is responsible for the identification and procurement of cross-system synergies. The *cooperation steering model* enables efficient collaboration while excluding opportunistic behaviour to a large extent. The new business model offers guidelines especially to small and medium-sized companies, including how to pursue a differentiation strategy by jointly creating sustainably optimized life cycle buildings with guarantees for the operation phase supported by modern industrialization technologies.

Modernisation, Mechanisation and Industrialisation of Concrete Structures, First Edition.
Edited by Kim S. Elliott and Zuhairi Abd. Hamid.
© 2017 John Wiley & Sons Ltd. Published 2017 by John Wiley & Sons Ltd.

9.1   Introduction

The building industry bears an enormous potential for sustainability optimization, which has been recognised but not seized yet. The existing building stock globally causes 30% of total CO_2 emissions and 40% of total energy consumption (United Nations Environment Programme, 2009). Furthermore, the building industry consumes 3000 Mt/a, about 50% of the total material consumption, more resources than any other economic activity (Pacheco Torgal and Jalali, 2011). Today's project execution is characterized by sequential, highly fragmented processes with many interfaces and the potential of modern industrialization technologies for off-site production is by no means seized to an appropriate extent (Rinas, 2011). Timber construction, for example, draws heavily on computerized methods of design and production. Steel construction as well routinely uses methods of industrialized prefabrication. In concrete construction prefabrication is nowadays only used for the production of elements but no system services are available on the market. Decisions are dominated by short-term yield expectations and are mainly focused on initial investment cost. No sustainable building optimization across the life cycle seizing potential synergies between the interdependent subsystems of a building is taking place (Selberherr and Girmscheid, 2012). In large healthcare projects, for example, the coordination of mechanical, electrical and plumbing systems is often a huge challenge (Khanzode, Fischer and Reed, 2008). Especially in concrete construction a lot of time-consuming re-work is done on site instead of systematically identifying all the requirements from the very beginning in order to integrate them in terms of "front-loading" (Selberherr, 2015).

The complexity of sustainably optimized buildings requires the interlinking of the relevant key capabilities, technically in terms of different trades and chronologically in terms of different phases of the building's life cycle, like design, execution, and operation (Selberherr and Girmscheid, 2014). The efficiency of the required coordination process between the diverse project participants (i.e. client, architects, structural and mechanical consultants, contractors) can be increased substantially using modern industrialization technologies of off-site production. In order to exploit this considerable and currently untapped potential, strategic as well as operational measures in the enterprises are necessary (Girmscheid, 2010). These measures involve the design of the service provision processes with design, off-site production, on-site assembly as well as management and support processes (Selberherr, 2014).

The construction management research community agrees on the fact, that the implementation of cooperative business models poses significant challenges for the enterprises often not delivering the expected positive outcomes (Gadde and Dubois, 2010). As enterprises require guidelines for the implementation of cooperative approaches to project delivery, a literature review (Selberherr, 2014) was conducted with the objective of identifying construction-specific process models, which describe the setup and/or realization of cooperative approaches. Several models dealing with cooperative approaches of value creation in general and specifically in the field of construction have been found.

Abudayyeh (1994) develops a process model for the setup of a partnering relation between one contractor and the client for the delivery of a single project. Crane *et al.* (1997) describe the steps necessary for the setup of a dyadic partnership with the main contractor from the client's point of view. Cheng and Li's (2004) model focuses on the

Figure 9.1   Erecting long-span prestressed concrete floor slabs (by Samsung Precast, Korea).

cooperation management processes for setting-up and realizing a strategic coopera-tion network. Dreyer (2008) developed a process model for the initiation and realiza-tion of a public-private partnership for communal street maintenance in Switzerland. Cho *et al.* (2010) created a process model combining the fast-track-approach with the partnering concept for public-sector design-build-projects in Korea. Figure 9.1 shows an example by Samsung Precast in Korea for the rapid construction of multi-storey car parks using precast concrete columns, beams and prestressed hollow core floor units, erected at the rate of about 1000 m^2 per week. Rinas (2011) developed a busi-ness model for cooperative and innovative sales concepts in the customizable precast concrete products industry.

None of the discovered models deals with the setup and realization of a cooper-ation network with multiple partners for a long-term strategic collaboration span-ning across several projects with different clients. Furthermore, none of the existing models focuses on a sustainable life-cycle optimization by interlinking the relevant key capabilities, using modern industrialization technologies for off-site production, and securing the design results with function and cost guarantees for the operation phase.

9.2   Objectives of the new business model

This chapter tackles this research gap by developing a new cooperative business model to overcome the linear, highly fragmented design and construction processes and to enable construction companies to master the required cross-linking of capabilities

Figure 9.2   Fully coordinated prefabrication for wall frame construction.

and know-how using modern industrialization technologies for off-site production. Figures 9.2 and 9.3 show two examples of positive and negative utilization of the benefits of off-site prefabrication. Figure 9.2 shows fully coordinated construction of wall panels for multi-storey apartment buildings, using mechanical connections for speed and structural integrity, whilst Figure 9.3 shows the reliance of time consuming cast insitu infill in order to achieve the same goal.

The essential innovative elements of the new cooperative business model are:

- A holistic sustainable project-specific optimization of the building as overall system with regard to life-cycle costs supplemented with guarantees covering the operating phase
- Through the creation of synergies between different interdependent subsystems (such as carcass, building envelope, HVAC, electrical energy) using modern industrialization technologies for off-site production and
- Repeated collaboration between the companies contributing key capabilities, which drives continuous learning and improvement and can be used as incentive for cooperation compliant behaviour.

The new business models strives to secure the best possible economic position of the enterprises providing the new service offer as members of the cooperation network by identifying service goals with consideration of their social and environmental value and transforming them into service results, which generate the required client value advantage. An interdisciplinary cooperation across trades enables the combination of knowledge of all project participants in a synergetic way. A system integrator

Figure 9.3   Cast insitu concrete infill has a negative impact on off-site prefabrication.

develops sustainably optimized buildings as system service provider in cooperation with various specialist contractors, the system suppliers (e.g., heating, ventilation, air conditioning (HVAC), water, electrical energy, facility management).

9.3   Modelling

The modelling process comprises two steps, the formal structuring and the subsequent contextual configuration.

9.3.1   Formal structuring

The formal structuring of the model is based on the business model framework by Girmscheid (2010), which comprises three strategic elements as depicted in Figure 9.4. The first two strategic elements (i) purpose and value of the business idea and (ii) competitive strategy, determine the service offer. With the new business model the client does not only receive a building according to his specifications but receives additional cost and function guarantees for the operation phase. These elements are modelled in the so-called *outside view* of the business model, which describes the interaction of the business model with its environment (economy, society, ecology, technology). The final element (iii) realization of the value creation process is modelled in the *inside view*. The objective on this level is to optimize the

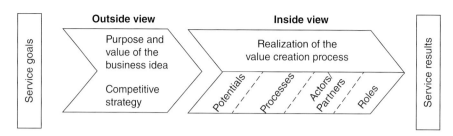

Figure 9.4   Theoretical framework for business models (adapted from Girmscheid, 2010).

cooperative value creation in order to maximize the economic value for the partners involved in the cooperation network and to optimize the social and environmental impact.

9.3.2   Contextual configuration of the outside view: development of the new service offer

The objective on the level of the *outside view* is to develop a service offer with regard to the structural constraints in relation to economy, society, ecology, and technology, which generates the required client value advantage compared to competitors. The new business model pursues the use of sustainable, renewable or renewed building materials, an energetic optimization resulting in reduced energy consumption and CO_2 emissions as well as the increased use of regenerative sources of energy and innovative forms of energy supply and the provision of flexible usability.

The new service offer is gradually expanded starting with a basic service offer, which covers design and construction of building where investment cost is restricted through a guaranteed maximum price contract and life-cycle optimization is substantiated with predictions of the future energy consumption and model calculations using prior specified objective criteria to demonstrate the amortisation of increased investment cost over the building's life cycle. In order to keep ahead of competitors, and to sustain client satisfaction by generating outstanding client value, life-cycle offer I and II are suggested as expansions for the new service offer (Table 9.1).

9.3.3   Contextual configuration of the inside view: Realization of the value creation process

The level of the *inside view* deals with the realization of the value creation process. First, the organization model and the key roles with the corresponding capabilities are discussed. Then, the three sub-models of the *inside view*, which refer to different levels of action and chronological phases, are described:

*   initiation process model
*   project delivery process model, and
*   cooperation steering model.

Table 9.1 Scope and step-wise expansion of the service offer.

Basic offer	
Scope of services	Supplementation of a design-build service offer with the service component energy supply to create an integral life-cycle service offer
Scope of guarantees	Guaranteed maximum price for investment cost Limited guarantee of operation costs for 5 years
Service objective	Optimization of the HVAC-system with regard to renewable sources of energy and innovative forms of energy supply under consideration of the interdependence between the building envelope and the HVAC-system
Key roles	System integrator System supplier Building envelope and Carcass System supplier HVAC System supplier Facility management
Life cycle offer I	
Scope of services	Optimization of the entire building according to the client requirements in terms of function, utility, aesthetics, quality, yield and value conservation with the relatively lowest life cycle costs
Scope of guarantees	Guaranteed maximum price for investment cost Guarantee of operation costs for 10 years
Service objective	Holistic system optimization under consideration of the building envelope, energetically passive components, HVAC-system using renewables sources of energy and innovative forms of energy supply, water supply and disposal, electrical energy as well as systems for building automation and control
Key roles	System integrator System supplier Building envelope and Carcass System supplier HVAC System supplier Electrical energy System supplier Water System supplier Facility management
Life cycle offer II	
Scope of services	Design, construction and operation of client-specifically optimized useful areas with finishing standards in specified quality for a fixed price over a certain period
Scope of guarantees	Guaranteed maximum price for investment cost Guarantee of operation costs for 25 years
Service objective	Technological and administrative optimization of the building's entire life cycle with regard to design, construction, operation, alteration and demolition if applicable
Key roles	System integrator System supplier Building envelope and Carcass System supplier HVAC System supplier Electrical energy System supplier Water System supplier Finishing System supplier Facility management

9.3.3.1   Organisation concept, key roles, and capabilities

The essential components for the creation of client-specifically optimized buildings are the key capabilities which contribute to the network by the system suppliers and are project-specifically integrated by the system integrator. The remaining, complementary services are conventionally passed to subcontractors. For the basic service

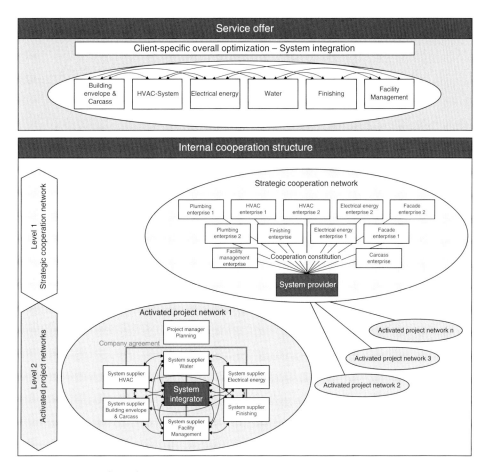

Figure 9.5   Service offer and organization concept.

offer, for example, the system suppliers cover the building envelope and framework, HVAC, and facility management. Later also electrical energy, water, and finishes are included as system suppliers. A two-tier network organisation model (Figure 9.5), consisting of a cooperation network as basis for the setup of an activated project network, is chosen (Mack, 2003; Camarinha-Matos *et al.*, 2005).

The cooperation network (CN) constitutes a pool of potential partners, who all sign the cooperation constitution thereby agreeing to common values and norms. It is founded independently of any particular project and serves as the basis for the project-specific collaboration which is taking place in the activated project network (APN). The CN is governed by the cooperation board, which is composed of representatives of the participating enterprises. When a new project is acquired, the client requirements are evaluated by the system integrator and an APN is founded by signing a company agreement with the enterprises possessing the required key capabilities. The APN is at the level of value creation. Here the contributions of all system suppliers are integrated to create a holistically optimized sustainable building.

The key role for the success of the new service offer is that of the system integrator, who is the leading actor in the configuration and realisation of the APN. While

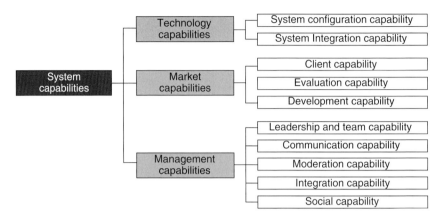

Figure 9.6   System capabilities for the role of the system integrator.

the system suppliers contribute the component knowledge and technological capabilities concerning the subsystems, such as carcass, building envelope, HVAC, electrical energy, the system integrator is responsible for the integration of the subsystems and optimization of the building as overall system. It is the system integrator's task to connect the interfaces between the interactive subsystems in order to obtain a client-specific integrative system optimization. The system capabilities required to successfully fulfil the tasks of the system integrator include technology, market, and management capabilities (Figure 9.6). In larger projects a project manager supports the system integrator by dealing with organizational issues and steering the project in terms of cost, time and quality.

The system suppliers are responsible for the life-cycle optimization of the subsystems with regard to interfaces to other subsystems during design and construction, and for the construction of the subsystems in compliance with cost, time, and quality objectives. Therefore they need excellent technological capabilities particularly with regard to life-cycle costs of the subsystems, function-oriented optimization, and consideration of interfaces in terms of an integrated project delivery as well as outstanding knowledge concerning modern industrialization technologies for off-site production. Each system supplier should nominate a representative who is the designated contact person for all activities and inquiries concerning the cooperation network and member of the cooperation board. The system suppliers excel due to the depth of their know-how relating to the subsystems, whereas the system integrator features an enormous width of know-how covering all the aspects which constitute a sustainable, optimized building. The following example, of a multi-storey building in mixed precast-*insitu* concrete and steelwork (Figure 9.7), shows the role of the specialist "system suppliers" together with the task of the "system integrators". The suppliers are specialist in the design and construction of (a) prestressed concrete hollow-core slabs, (b) concrete filled tubular steel columns, (c) prefabricated steel box beams and (d) roof finishes. The system integrator, that is, the consultant for the entire project has wide knowledge of the practical aspects of these subassemblies and in particular the structural integration of them, for example, composite action between (a) and (c), and the horizontal stability between (a) and (b).

Figure 9.7 Mixed precast-insitu concrete with steelwork illustrates the role of system suppliers and system integrators.

9.3.3.2   Initiation process model: setting up a cooperation network

The initiation process model's objective is the setup of a cooperation network for the development of sustainable life-cycle service offers with extensive cost guarantees. It enables companies to generate a resource- and operation-oriented building optimization through the creation of synergies of collaboration across trades and stages of the building's life cycle. The initiation process model defines the normative elements of the new business model and describes the strategic processes for the initiation of the cooperation network in the development phase thereby creating the constitutive basis, which enables and supports the development of the cooperation network (Bleicher, 2011). This sub-model concerns the main process initiation, which comprises the three module processes: analysis, recruiting, and constitution. These are detailed further on the level of elementary processes. The elementary processes are assigned to actors via roles. To set up the cooperation network the initiating enterprise builds a founding team, which is responsible for the recruiting of cooperation partners as system suppliers like the general contractor, HVAC-contractor, electrical contractor and others.

The initiation process model comprises the conduction of an environment and potential analysis to evaluate the current market situation as well as future

Table 9.2 Requirements for cooperation partners.

Type of field	Field of potential		Field of social relations
Type of fit	Strategic fit	Organizational fit	Relational fit
Type of selection criteria	Task-related criteria	Partner-related criteria	Relation-related criteria
Type of capabilities	Technological (epistemic and heuristic) capabilities	Organizational capabilities and situational/structural characteristics	Relational and reputational capabilities, past experience
Type of connection	Complementarity	Compatibility	Communality
Type of competence	Core competence	Complementary competence	Complementary competence
Type of condition	Necessary condition for ability to cooperate	Sufficient condition for ability to cooperate	Sufficient condition for ability to cooperate

developments. Next, a value creation and implementation analysis is undertaken to identify an appropriate competitive strategy and develop the service offer. Following this, the cooperation partners are recruited using a standardized partner selection process to support the decision (Selberherr and Girmscheid, 2013).

For the development of the partner selection process the CN is described as a function of the field of potential and the field of social relations (Mack, 2003):

$$Cooperation\ network = f\ (field\ of\ potential;\ field\ of\ social\ relations)$$

The objective, rational characteristics, like resources, competencies, and capacities, which qualify an enterprise as a potential partner form the *field of potential*. The social factors and relational aspects, which make an enterprise a potential partner, constitute the *field of social relations*. These social and relational aspects can affect the personal level, the level of the enterprise, or the institutional level. From these fields the selection process for cooperation partners is developed (Table 9.2). Strategic and organizational fit are vital for companies to belong to the field of potential. Strategic fit (Dyer and Singh, 1998) concerns task-related criteria (Geringer, 1991) and aims at the complementarity of technological capabilities in order to successfully create synergies. Companies as potential partners have to possess excellent technological capabilities which constitute the core competencies of the system suppliers. The integration of these competencies is the core competence of the system integrator. Organizational fit (Dyer and Singh, 1998) concerns partner-related criteria (Geringer, 1991) and subsumes organizational capabilities as well as the compatibility of situational and structural characteristics of the enterprise. Cooperative arrangements are more successful, if the partners are compatible in terms of characteristics like corporate culture, corporate strategy, process execution, firm size, and nationality (Müller, 2005). The field of social relations aims at generating relational fit to build a basis of trust and is therefore connected to relation-related criteria like past experience and relational and reputational capabilities. In the assessment of a partner's ability to cooperate, strategic fit constitutes a necessary condition, while organizational fit and relational fit are regarded as sufficient conditions (Mack, 2003).

Based on these requirements the hierarchical structure of the partner selection processes is developed. The criteria in Figure 9.8 are logically deductively derived.

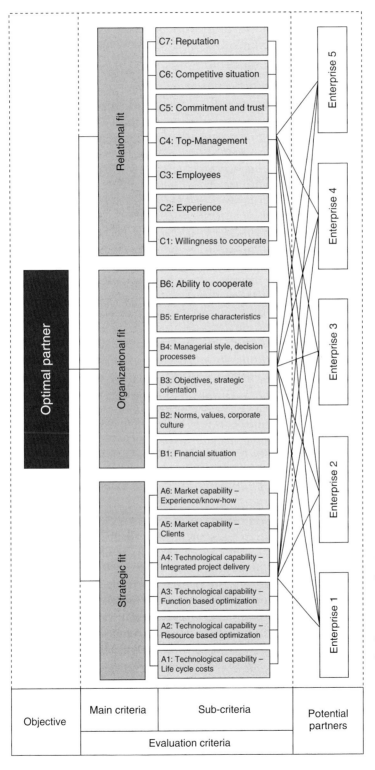

Figure 9.8   Criteria for the partner selection process.

The definite criteria for the qualification of partners have to be situationally adapted to the specific constraints depending on type and size of projects. Intuitive unstructured pondering and intuitive structured brainstorming are possible methods for the identification of alternative or additional requirements.

Enterprises with whom prior business experiences already exist are especially qualified as cooperation partners because they share a basis of trust and mutual understanding which is an important prerequisite for fair and non-opportunistic behaviour in the cooperation network (Girmscheid and Brockmann, 2010). They are contacted directly by the founding team in order to evaluate the strategic, organizational, and relational fit. Additionally, the initiation of cooperation network is publicly announced and interested enterprises are invited to apply as potential partners. If the applicants appear to be qualified they are invited for an interview.

The assessment of potential partners is conducted in two stages. Before a detailed assessment is undertaken, cooperation-specific No-Go-criteria are determined. These lead to the immediate exclusion of the applicant. For the evaluation of the remaining candidates the Analytic Hierarchy Process method (AHP method) (Saaty, 2001) is used.

The applicants with the highest degree of target achievement are then selected as partners in the cooperation network. As soon as the composition of the cooperation network is fixed, the negotiation of a cooperation constitution is initiated in order to agree on common values and to set rules for cooperation based on mutual trust and commitment. Furthermore, it is essential to discuss and resolve concerns raised by potential partners concerning the cooperation network, for example financial risk or knowledge drain. Subsequently, the cooperation constitution as well as the common values and rules have to be communicated to the employees to create a shared understanding as basis for trust. The cooperation network constitutes a pool of partners who are ready to undertake the cooperative service provision and integral optimization in activated project networks.

9.3.3.3   Project delivery process model: integrating interdependent key capabilities

The project delivery process model describes the strategic planning of tasks as well as the operational execution for the provision of innovative, sustainable life-cycle service offers using modern industrialization technologies for off-site production (Figures 9.9 and 9.10).

Figure 9.11 shows an example of where project delivery process models have been used in coordinating the construction of mixed precast and *insitu* concrete, steelwork, glue-laminated timber and structural glazing, of which about 95% of the construction involves off-site production. Figure 9.12 shows a completed building involving similar strategic planning using prefabricated concrete, timber and steelwork.

On the strategic level the model focuses on the planning of tasks and the coordination of the project participants to enable the cooperative life cycle building optimization subject to the cost, schedule, and quality specifications. On the operational level, this sub-model aims at an efficient task execution by all project participants. In the course of the contextual configuration of this sub-model processes are also hierarchically structured in main, module, and elementary processes, which are connected to actors via roles. The processes comprise the description of service provision,

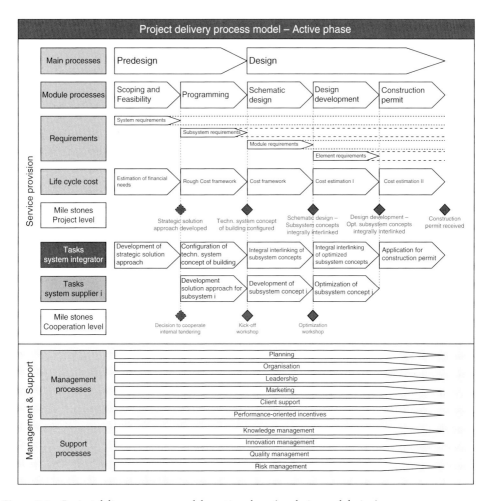

Figure 9.9   Project delivery process model – active phase (predesign and design).

management, and support processes for the realization of the value creation in activated project networks.

Two phases of the project delivery process are distinguished as "active" and "consulting". The key tasks of the system integrator in the course of the technical system integration for a client-specific overall building optimization are allocated in the "active phase", comprising the main processes Predesign and Design. In this phase, the system integrator plays a key role in regard to the system configuration and the technical system integration (Figure 9.5). In the subsequent "consulting phase", encompassing the main processes Construction and Operation, the system integrator takes on a consulting function (Figure 9.6). The system integrator regularly receives feedback from the construction site as well as from the subsequent operation phase concerning the latest results and suboptimal processes. These insights from the construction and operation phase are seized by the system integrator as lessons learned in the "active phase" of future projects to generate a continuous integrative learning and improvement process.

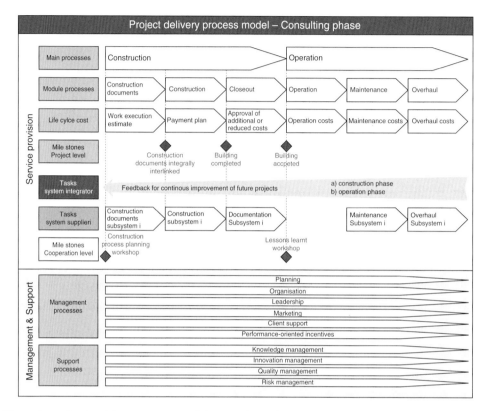

Figure 9.10   Project delivery process model – consulting phase (construction and operation).

Figure 9.11   Strategic planning for the construction of Oslo International Airport using prefabricated concrete, steelwork and timber elements.

Figure 9.12   Coordination of prefabricated concrete, steelwork and timber elements.

In order to generate a holistic optimization of the building each system supplier develops a solution for his subsystem. The system integrator is then responsible for the integral interlinking across the different subsystems. Therefore, he may conduct workshops for the identification of synergy potentials between different sub-systems with regard to resource-oriented as well as function-oriented optimization generating a holistically optimized life cycle building. These workshops are organized with the system suppliers in each design phase with phase-wise increasing level of detail. Based on the workshop result each system supplier adapts and improves his subsystem accordingly. This feedback process is undertaken iteratively generating a holistically optimized building through cooperation.

This process is supported with the extensive use of modern industrialization technologies. Building information modelling enables an efficient coordination of the diverse project participants in the design phase thereby reducing re-work and other non-value creating activities. Closing the digital chain from design to production using computer integrated manufacturing an optimized construction process in terms of time, cost, and quality can be generated. The physically segregated off-site production processes can be temporally parallelized and only the on-site assembly processes have to be coordinated in space and time. Furthermore the controlled conditions for production off-site generate a high degree of quality and cost certainty.

9.3.3.4   Cooperation steering model: developing trust and commitment

Besides the technical integration, which is targeted by the project delivery process model, the motivations for cooperation-compliant behaviour and incentives to reduce opportunism have to be investigated. Two types of corresponding risks, competence risk and relational risk (Das and Teng, 2001) can be identified in the context of collaboration. Competence risk is considered in the above presented project delivery process model, which focuses on the technical coordination and integration of the system suppliers. For the management of relational risk, the cooperation network steering model is developed. It provides the cooperation partners with steering instruments for the cooperative collaboration in the course of the service provision on the strategic and on the operational level. The following steering instruments are employed:

- information system
- conflict resolution mechanism
- team-building activities
- continuous improvement process and monitoring
- social activities
- incentive system

9.3.4   Overview

The cooperation steering model serves as guideline for establishing a cooperative attitude and increasing the efficiency of the collaboration in particular with regard to building mutual trust, reducing opportunistic behaviour and creating the foundations for efficient collaboration.

The new business model consisting of three sub-models describes the required processes to achieve efficient collaboration in the field of concrete prefabrication. The initiation process model enables contractors to found a cooperation network and to agree on a common code of conduct. The project delivery process model then shows the steps necessary in the design and construction of a building to attain an integration of capabilities of all the contractors involved. As efficient collaboration is not only a matter of formal organization and process design but even more of informal, relational aspects, the cooperation steering model equips the cooperation partners with instruments for aligning interests and creating common incentives.

9.4   Conclusion

This chapter has described the development of a cooperative business model for sustainable, life-cycle–oriented building optimization. The modelling process comprises the formal structuring and the contextual configuration. In the course of the formal structuring the *outside view*, which describes the interaction with the environment, and the *inside view*, which deals with the cooperative value creation, are analytically separated. In the subsequent contextual configuration of the *outside view* of the business model its purpose and use are specified and differentiation as competitive strategy is chosen. Then the service offer is developed. This service offer fulfils the client requirements through an innovative, integral solution in a better and more sustainable way and thereby generates a discernable surplus value for the client. Starting from a basic offer a stepwise expansion to the life cycle offer I and II is suggested. This gradual expansion allows for the continuous maintenance of a competitive advantage.

The *inside view* focuses on the realization of the cooperative value creation with the objective of generating an economic advantage for the cooperating enterprises as service provider. On the level of the *inside view* three sub-models are developed.

The initiation process model describes the setup of the cooperation network as pool of potential system suppliers. As the selection of appropriate partners is crucial to the success of the new business model a standardized partner selection process is developed. With the selected partners a cooperation constitution is negotiated. By signing this document they officially state their willingness to contribute to life cycle optimization as system suppliers in an activated project network as soon as a project is acquired.

The project delivery process model describes the realization of the value creation process in an activated project network. It aims at a project-specific, integral optimization using customized industrial production technologies. Each system supplier pursues the optimization of his subsystem and the system integrator is responsible for the identification and seizing of cross-system synergies. Therefore he conducts several optimization workshops with ever increasing level of detail. The workshop results are then fed back to the system suppliers for further improving their respective subsystem. Furthermore the system integrator subserves the chronological coordination of the different system suppliers with particular consideration of potentials for increasing efficiency through parallelization of processes using modern industrialization technologies in order reduce the project delivery time.

The cooperation steering model concerns the relational aspects and enables an efficient collaboration while excluding opportunistic behaviour to a large extent. It contains several steering instruments, which influence the strategic and operational level of service provision.

In conclusion the new business model enables especially small and medium-sized enterprises to pursue a differentiation strategy by contributing their excellent technological capabilities in a cooperative network with the objective of offering sustainably optimized life-cycle buildings that are client specifically created through the integration of interdependent subsystems.

References

Abudayyeh, O. (1994). Partnering: A Team Building Approach to Quality Construction Management. *Journal of Management in Engineering* **10**(6): 29–29.

Bleicher, K. (2011). Das Konzept Integriertes Management (Integrated Management Concet), Frankfurt, Campus.

Camarinha-Matos, L., Afsarmanesh, H., *et al.* (2005). Ecolead: A Holistic Approach to Creation and Management of Dynamic Virtual Organizations. Collaborative Networks and Their Breeding Environments. L. Camarinha-Matos, H. Afsarmanesh and A. Ortiz, Springer Boston. **186**: 3–16.

Cheng, E. W. L., & Li, H. (2004). Development of a Practical Model of Partnering for Construction Projects. *Journal of Construction Engineering and Management* **130**(6): 790–798.

Cho, K., Hyun, C., *et al.* (2010). Partnering Process Model for Public-Sector Fast-Track Design-Build Projects in Korea. *Journal of Management in Engineering* **26**(1): 19–29.

Crane, T. G., Felder, J. P., *et al.* (1997). Partnering Process Model. *Journal of Management in Engineering* **13**(3): 57–63.

Das, T. K. & Teng, B.-S. (2001). Trust, Control, and Risk in Strategic Alliances: An Integrated Framework. *Organization Studies* **22**(2): 251–283.

Dreyer, J. (2008). *Prozessmodell zur Gestaltung einer Public Private Partnership für den kommunalen Strassenunterhalt in der Schweiz (Process Model for the Desing of a Public Privat Partnership for Municipal Street Maintenance in Switzerland)*, Dissertation, ETH.

Dyer, J. H. & Singh, H. (1998). The Relational View. Cooperative Strategy and Sources of Interorganizational Competitive Advantage. *The Academy of Management Review* **23**(4): 660–679.

Gadde, L.-E. & Dubois, A. (2010). Partnering in the Construction Industry. *Problems and Opportunities. Journal of Purchasing and Supply Management* **16**: 254–263.

Geringer, M. (1991). Strategic Determinants of Partner Selection Criteria in International Joint Ventures. *Journal of International Business Studies* **22**(1): 41–62.

Girmscheid, G. (2010). Potentials of Computer Aided Construction. New Perspective in Industrialisation in Construction. A State-of-the-Art Report. G. Girmscheid and F. Scheublin. Zürich, Eigenverlag des IBB an der ETH ZUrich: 29–35.

Girmscheid, G. (2010). Strategisches Bauunternehmensmanagement. Prozessorientiertes integriertes Management für Unternehmen in der Bauwirtschaft (Strategic Construction Enterprise Management. Process-oriented Integrated Management for Enterprises in the Construction Industry), Berlin Heidelberg, Springer Verlag.

Girmscheid, G. & Brockmann, C. (2010). Inter- and Intraorganizational Trust in International Construction Joint Ventures. *Journal of Construction Engineering and Management* **136**(3): 353–360.

Khanzode, A., Fischer, M., & Reed, D. (2008). Benefits and Lessons Learned of Implementing Building Virtual Design and Construction (VDC) Technologiew for Coordination of Mechanical, Electrical, and Plumbing (MEP) Systems on a Large Healthcare Project. *ITcon* **13**: 324–342.

Mack, O. (2003). Konfiguration und Koordination von Unternehmungsnetzwerken. Ein allgemeines Netzwerkmodell (Configuration and Coordination of Corporate Networkd. A General Network Model), Wiesbaden, Deutscher Universitäts-Verlag.

Müller, N. (2005). *Die Wirkung innovationsorientierter Kooperationsnetzwerke auf den Innovationserfolg. Eine empirische Untersuchung auf Basis des Competence-Based View und des Relational View*. Dissertation, Universität Bremen.

Pacheco Torgal, F. & Jalali, S. (2011). Eco-efficient Construction and Buidling Materials. Heidelberg Dordrecht London New York, Springer Verlag.

Rinas, T. (2011). Kooperationen und innovative Vertriebskonzepte im individuellen Fertigteilbau. *Entwicklung eines Geschäftsmodells (Cooperations and Innovative Sales Concepts in the Individual Prefabrication, Development of a Businsee Model)*, Dissertation, ETH.

Saaty, T. L. (2001). Decision Making for Leaders - The Analytic Hierarchy Process for Decisions in a Complex World. Pittsburg, RWS Publishing.

Selberherr, J. (2015). Sustainable Life Cycle Offers Through Cooperation. Smart and Sustainable Built Environment 4(1).

Selberherr, J. (2014). Geschäftsmodell für kooperative Lebenszyklusangebote (Business Model for Cooperative Life Cycle Service Offers), Dissertation, ETH Zürich.

Selberherr, J. & Girmscheid, G. (2014). Sustainable Business Management. New Challenge for the Constructin Industry. *Journal of Management Science and Practice* **2** (1): 1–7.

Selberherr, J. & Girmscheid, G. (2012). A Business Model for Life Cycle Service Provision through Cooperation. Sustainable Business Management. *Joint CIB International Symposium of W055, W065, W089, W118, TG76, TG78, TG81 and TG84. CIB. Montreal.* **2**: 571–582.

Selberherr, J. & Girmscheid, G. (2013). A Partner Selection Process Model for the Provision of Life Cycle Service Offers through Cooperation. 8° Simpósio Brasileiro de Gestão e Economia da Construção, Inovação e Sustentabilidade (SIBRAGEC 2013), Salvador, Brazilian Association of Technology for the Built Environment (ANTAC): 1–12.

United Nations Environment Programme. (2009). Buildings and Climate Change. Summary for Decision Makers.

Chapter 10

Retrospective View and Future Initiatives in Industrialised Building Systems (IBS) and Modernisation, Mechanisation and Industrialisation (MMI)

Zuhairi Abd. Hamid, Foo Chee Hung and Ahmad Hazim Abdul Rahim
Construction Research Institute of Malaysia (CREAM), Kuala Lumpur, Malaysia

Prefabrication is a technology that emerges out of the necessity and desires of the societies. The success of prefabrication in the U.S., Scandinavia and Japan is mainly due to the more homogenous cultures in these countries that have uniquely developed the technology which eventually serves the society in fulfilling both the wants and needs during the industrial revolution. By taking Malaysia as a case study, the context of prefabrication is discussed, in an attempt to outline the development and social acceptability of prefabrication and standardisation in the developing country. Following this, the potential of prefabrication in the housing industry market is discussed.

10.1   Industrialisation of the construction industry

As like many other industries, the construction industry is under constant pressure to improve productivity, reduce cost, and minimize waste in its operations. However, productivity in the construction industry over the last century has stayed flat or even worsened – turned negative, as compared to the manufacturing industry, which has improved its productivity four fold (about 400%) (Chapman & Butry, 2008) from which Figure 10.1 is extracted. The reason behind this obvious difference in productivity is mainly due to the fact that the construction industry has yet to

Modernisation, Mechanisation and Industrialisation of Concrete Structures, First Edition.
Edited by Kim S. Elliott and Zuhairi Abd. Hamid.
© 2017 John Wiley & Sons Ltd. Published 2017 by John Wiley & Sons Ltd.

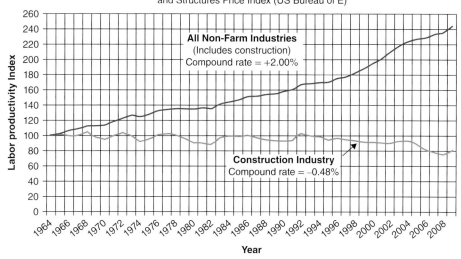

Figure 10.1   Productivity of the U.S. construction industry (Chapman & Butry, 2008).

undergo a complete phase of industrialisation, in spite of the best attempts to do so in Northern and Eastern Europe and Scandinavia in the past 40 years. Based on the report published by the Economist Intelligence Unit (2015), high-productivity firms are likely to be early adopters of productivity-enhancing technologies as technology helps firms to support greater collaboration and information-sharing, thereby facilitating the boost of its productivity and the adaptation to the global business trend.

Modernising the construction industry through industrialisation has become a worldwide agenda (Nawi *et al.*, 2014). This is evident by looking the market share, where industrialisation of construction industry has indicated a great potential of progress to the industry in the developed countries. In the UK, for example, both the Latham Report (Latham, 1994) and the Egan Report (Egan, 1998) emphasized the advantages of standardisation and pre-assembly, and stressed the importance and capability of industrialized systems in reducing the costs, improve the quality, and make complex products available to the vast majority of people. The Egan report also called for 20% improvement in predictability, that is, no more surprises on site, and called for the standardisation of building products and processes. For the concrete industry this clearly spelled out "pre-fabrication", and to this end precast concrete elements and building systems have formed the central core to this book.

Industrialisation clearly reveals the potential for better work preparation, logistics, optimization and continuous improvements which have a major impact on the cost structure of modern projects (CIB, 2010). For instance, the cost saving that could be achieved by optimizing construction logistics is more than 20% of the total labour costs. It is also able to optimize construction supervision by up to 19% by moving works away from the construction site to the manufacturing floor (Figure 10.2).

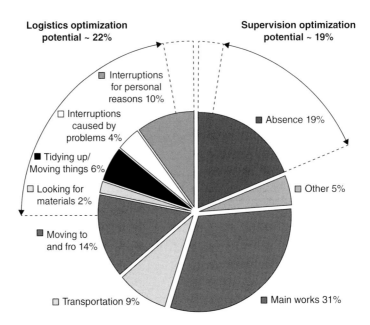

Figure 10.2   Potential cost reduction of industrialized construction (CIB, 2010).

10.2   Overview on global housing prefabrication

As part of the industrialisation process, prefabricated construction has *stormed* into the twenty-first century as a plausible solution to improve current construction performance and image (Kamar *et al.,* 2009), which has long been characterized as labour-intensive and surrounded by significant risks associated with market, site and weather conditions, and lower productivity relative to other industries. Prefabricated construction refers to those buildings where the structural components are standardized and produced in plants located away from the building, and then transported to the site for assembly. The distance to site can of course be just a few hundred metres in the case of site-precast (e.g., stack casting wall panels and slabs), and up to hundreds of kilometers for high-value products, (e.g., façade panels manufactured in Kentucky, USA destined for the UK).

Housing constitutes the most explored building type of prefabrication experimentation (Davies, 2005). The term "prefabricated housing" is a general term that may indicate manufactured housing, modular housing, and production housing, depending on different categorization system created by the authorities from different countries. It is not a new method in the housing industry, as the logic of prefabricated housings had been fully formulated by the end of the 1940s in the UK (Glendinning & Muthesius, 1994), France (Sebestyen, 1998) and the USA (Abel, 2004). Prefabrication allows for a higher in-factory completion of housing components and takes advantage of mass production with its promise of global markets. Due to its dual benefits of economy as well as speed in construction, prefabrication is the obvious choice to meet the ever-increasing demand for housing in the countries. In a time of economic and environmental challenges, particularly, prefabrication holds the potential to aid in fostering a paradigm shift in the construction industry in the USA and other nations. Among the successful prefabricated housing in the world are Sekisui Home (Japan), Living Solution (United Kingdom), Open House (Sweden), and Wenswonen (Netherlands) (Oostra & Joonson, 2007).

Prefabrication is a technology that emerges out of the necessity and desires of the societies. It relies upon the social and cultural context: labour, factory ability, knowledge base, and especially the market to determine what is developed. The lessons from the countries that have become proficient at prefabrication for both mass housing and customised buildings alike is that in order for innovation to occur it must be socially driven. The greatest developments in prefabrication in the USA housing industry, for example, can be attributed to events that have spurred this on including the California Gold Rush, Tennessee Housing Authority, war time and post war housing, and most recently the energy and housing crisis of the late twentieth and early twenty-first century (Smith, 2009). The developments of the USA steel frame technology, the precast industry, and the proliferation of the balloon frame wall throughout the world are of no less impacting on the prefabrication across the globe.

Prefabrication in Scandinavia follows a similar pattern to the USA: the industrial revolution, war time and post war housing crisis, and the resurgence of interest in utilising CAD/CAM technology, but with little differences. This culture has taken the traditions of shipbuilding and its greatest natural resource – timber, to explore the various methods of housing construction. The society of Scandinavia does not shirk at prefabrication but in effect sees it as a way to produce affordable quality buildings. As a result, the Scandinavia prefabricated housing is more affordable than the traditional onsite counterparts, and the market has made not only an ideological shift, but an economic and market driven shift toward prefabrication (Smith, 2009). In the case of Japan, it is the culture of collaboration, integration, and perseverance that has advanced its capacity to prefabricate in housing effectively. Besides, the legal flexibility and the continual emphasis on craft and pride in ones work also play an important role in motivating this unique culture to prefabrication, bettering the construction process, and delivering quality architecture (Smith, 2009).

10.3   Housing prefabrication in Malaysia – the industrialisation building system (IBS)

The potential for prefabrication if delivered correctly is to allow more client choice and involvement, particularly in the case of housing where a variety of different features and systems can be realised from manufacturers. Whilst prefabricated housing has been taking place throughout the world, nowhere has the contrast between traditional cast *insitu* concrete and the attempted move towards prefabrication (in the widest sense, not just precast concrete elements) been more focused than in Malaysia.

The Malaysian government's vision to be a developed nation by 2020 has pushed forward the use of innovative technologies in most sectors and industries. With the implementation of various government projects under the Entry Points Projects (EPPs) through the Economic Transformation Programme (ETP), a platform has been established where the increasing use of mechanized and enhanced automation in the construction industry is highly encouraged. Among the construction technologies that are much favoured by the Malaysian government is the use of Industrialized Building System (IBS). The main drivers for this have been:

i.   control of labour, particularly in relation to unskilled migrant labour
ii.   quality of concrete production
iii. accuracy and safety of construction

all of which have been a major force behind the implementation of this book.

IBS is the term used to represent the prefabrication concept – it should not be confused with the post WWII industrialisation of concrete buildings, for example, tower blocks and apartments (Figures 11.2 and 11.3). IBS is a construction philosophy as well as a technique, where components are manufactured in a controlled environment on-site or off-site, transported, positioned, and assembled into a structure with minimal additional site work, and has been pioneered by the Construction Industry development Board (CIDB, 2003). Other definitions of IBS compiled from different studies are given in Table 10.1.

The fundamental idea of IBS is to move *some* effort away from the construction site to the controlled environment of the factory, as explained in Chapters 1 to 3 of this book. As such, IBS exploits the huge benefits of BIM (Chapter 5 of this book), it takes advantage of mass production and mechanisation of slabs and walls and so

Table 10.1 Various definitions of IBS.

References	IBS Definition
Fathi *et al.*, 2012	A construction technique which components are manufactured in a controlled environment (on or off site), transported, positioned and assembled into a structure with minimal additional site works contributing to less wastage.
Kamar *et al.*, 2011	As prefabrication process and construction industrialisation concept.
Abdullah & Egbu, 2009	A method of construction established based on innovation and on rethinking the various techniques of construction.
Chung & Kadir, 2007	Mass production of building components in factory (offsite) or at site (onsite).
Rahman & Omar, 2006	A construction system that is built using pre-fabricated components.
Marsono *et al.*, 2006 Haron *et al.*, 2005	A construction method through the use of best construction machineries, equipment, materials and extensive planning of the construction projects.
Lessing *et al.*, 2005	An integrated manufacturing and construction process with well organization and activities management.
Gibb, 1999	The process of preassembly, organization and completion of final project assembly before installation.
Sarja, 1998	A set of interrelated elements that act together to enable designated performance of building which includes several procedures (managerial and technological) for the production and installation of these elements.
Jaafar *et al.*, 2003	An industrialised system of components production or building assembly or both.
Junid, 1986	An integrated system including software and hardware which building components are planned, fabricated, transported and assembled at site.

Table 10.2   Categorization of IBS components.

Period	IBS Component		Reference
Early 1960s	i. ii. iii.	Frame system Panel system Box system	Badir *et al.*, 2009
Early 1990s	i. ii. iii. iv.	Precast concrete framing, panel and box systems Load bearing block Sandwich panel Steel frame	Badir & Razali, 1998
2003	i. ii. iii. iv. v.	Precast concrete framing, panel and box systems Formworks systems Steel framing system Prefabricated timber framing systems Block work system	CIDB, 2003

Source: Adapted from Azman *et al.*, 2009.

on (Chapters 6 of this book), with its promise of creating synergy among industry players by undergoing a paradigm shift from using conventional technology to a more systematic and mechanized system that adopts efficient product management process to provide more products of better quality in less time (a key feature in Chapters 7 to 9 of this book). IBS can be further divided into categories and has undergone changes following the new trend of technology (Table 10.2). At present, it is categorized into precast concrete component systems, fabricated steel structures, innovative formwork systems, modular block/masonry systems and prefabricated timber structures.

The construction industry in Malaysia has started to embrace IBS as a method of attaining better construction quality and productivity. The use of IBS assures valuable advantages such as reduction of unskilled workers, less wastage, less volume of building materials, increased environmental and construction site cleanliness. It also offers benefits concerning cost and time certainty, reducing risk related to occupational safety and health, alleviating issues skilled workers and dependency on manual foreign labour, and achieving the ultimate goal of reducing the overall cost of construction. Towards the end, IBS is believed to be able to stimulate a better performance through the development of an industrialized construction sector and achieving open building.

10.3.1   *Chronology of IBS development in Malaysia*

The IBS was first introduced to the Malaysian construction industry in 1966. It was engaged in the construction of low-cost high-rise residential building as to overcome the increasing demand for housing needs. Two pilot projects for the precast housing were launched, involving the construction of Tunku Abdul Rahman Flats in Kuala Lumpur (similar to Figure 10.3) and the Rifle Range Road Flat in Penang (similar to Figure 10.4), by utilising the large panel system that required large concrete panel cast in the factory and transported to site on trailers for assembly. While the speed of construction is much faster the building design was very basic and did not consider the aspect of serviceability such as the local needs to have wet toilet and bathroom areas (Rahman & Omar, 2006). The industrialisation of construction was never sustained

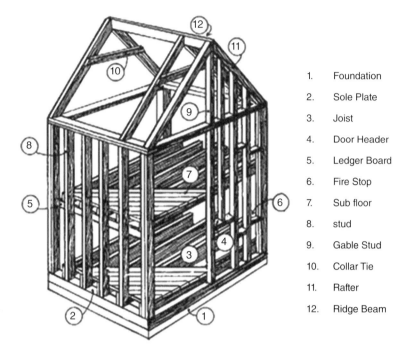

1.   Foundation

2.   Sole Plate

3.   Joist

4.   Door Header

5.   Ledger Board

6.   Fire Stop

7.   Sub floor

8.   stud

9.   Gable Stud

10. Collar Tie

11. Rafter

12. Ridge Beam

Figure 10.3   Balloon framing.

Figure 10.4   The Tunku Abdul Rahman Flats in Kuala Lumpur.

Figure 10.5   The Rifle Range Road Flat in Penang.

at this period because the conventional construction – consisting of a reinforced concrete frame and brick, beam, column, wall, and roof, which was cast insitu using timber framework while steel reinforcement was fabricated off-site – was still preferable in the industry. Moreover, the introduced foreign precast systems were also found to be not suitable with Malaysia climate and social practices.

The use of structural steel components took place during the period of early 1980s up to 1990s, particularly in high-rise buildings in Kuala Lumpur and the construction of 36-storey Dayabumi complex (Figure 10.5) in 1984 by Takenaka Corporation of Japan. The use of precast concrete systems in residential buildings was also increasing with the growing demand for the new township in the state of Selangor. For example, the system from Praton Haus International was used to construct around 52,000 housing units in Shah Alam; the system from Taisei Marubumi was used for the construction of 1,237 housing units and 11 shop lots at PJS; 3,222 flat units and 1,112 housing units at Bandar Baru Bangi were built by using the Hazama system; the Ingeback system, a Swedish method using large panels in vertical battery mould and tilt-up table mould, was used to construct 3,694 flat units. All these projects were constructed by local contractor with international technical support from established international firms in joint venture partnership (Hassim *et al.*, 2009; Sarja, 1998). Although the system was originated overseas, the local contractors made modification to suit local requirement. Instead of steel, high quality film coated plywood shuttering was used in an innovative mould system. The form could easily be dismantled and handled by a small crane and adjusted to suit the architectural requirement (Sarja, 1998).

The development of IBS in the period of 1990s and 2000s saw the application of hybrid IBS in many national iconic landmarks, such as the Kuala Lumpur Convention Centre – using the steel beam and roof trusses and precast concrete slab from the Victor Buyck Steel Construction; Lightweight Railway Train (LRT) – using a steel roof

Figure 10.6   The Dayabumi complex in Kuala Lumpur.

and precast hollow core slabs from the RSPA – Bovis; KL Tower – using steel beams and columns from the Wayss and Freytag; the Kuala Lumpur International Airport using a steel roof from Eversendai; and the Petronas Twin Towers – using steel beams and steel decking for the floor system from the Mayjus JV and SKJ JV (Figure 10.6). Apart from teaming up with foreign companies for technology transfer, the Malaysian companies also developed their own IBS technologies which were widely used for private residential projects, government's schools and teachers housing complexes, hospitals, colleges and universities, custom and immigration complexes, private buildings as well as police quarters. Among these local IBS technologies are such as the Zenbes, CSR, IJM Formwork, Pryda, Baktian, and HC Precast. The country's IBS agenda was further boosted in 2008 with the formulation of regulations on the use of IBS in the construction of public buildings. The Treasury Malaysia issued a Treasury Circular Letter (now referred to as SPP 7/2008) to all Malaysian government agencies to incorporate IBS as part of their contract document for tender , thus increasing the IBS contents of their building development projects to a level not less than 70 points of the IBS score (Hamid *et al.*, 2008).

The circular letter took effect immediately and the Implementation and Coordination Unit (ICU) of the Prime Minister's Department was given the task of monitoring the level of compliance to this directive by the respective agencies. The decision was to create sufficient momentum for the demand of IBS components and to create a spill-out effect throughout the nation. The government also established the National IBS Secretariat to monitor the IBS implementation. Such secretariat involves the coordination between inter-ministry levels to make sure the policy is successfully implemented. During a period from October 2008 and May 2010, about 331 projects under 17 ministries were awarded and constructed using IBS. The majority of these projects were public schools, hospitals, higher learning institutions and government offices. The total cost of the projects was about RM9.6 billion (US$2.4 billion).

10.3.2 IBS roadmap

As a statutory body under the Ministry of Works, the CIDB has been actively promoting the use of IBS in the local construction industry. Its commitment can be seen with the development of the IBS Roadmap 2003–2010 and its successor IBS Roadmap 2011–2015, both of which aim to provide guidelines towards the establishment of an industrialized construction sector, as well as to steer the right direction for fundamental changes in the construction industry. The progress of IBS implementation was generally on track since the launching of the IBS Roadmap. This can be seen by assessing several quantitatively measurable Key Performance Indicators (KPIs) set in the IBS Roadmap:

i.   IBS score
ii.   number of IBS certified contractors
iii. number of ISO certified IBS Manufacturer

10.3.2.1 IBS score

The IBS score is a physical measure of the IBS usage in a construction by calculation based on a predetermined formula covering structural systems, walls, and peripherals of a building, as defined in the CIDB: CIS 18:2101, Manual for IBS Content Scoring System (IBS Score). It has been set as one of the targets to be achieved in the IBS Roadmap 2011–2015:

i.   to sustain the existing momentum of 70% IBS content for public sector building projects through to 2015
ii.   to increase the existing IBS content to 50% for private sector building projects by 2015.

Figure 10.7 National iconic landmarks in Malaysia with hybrid IBS application.

Through a questionnaire survey carried out by the authors in 2014, the mean IBS score achieved by building type of residential, non-residential, and social amenities in the government project were 70, 80, and 76, respectively; while in the case of private project, the average IBS scores were 65, 63, and 64, respectively (Figure 10.7), which indicate that the targets set by the Roadmap (2011–2015) have been achieved.

10.3.2.2 ISO certified IBS manufacturer

As stated in the Roadmap, the targeted number of ISO certified IBS manufacturer should reach 10% of the registered IBS manufacturers in that particular year. As of November 2014, there were 189 IBS manufacturers in Malaysia producing 381 IBS components (Figure 10.8). Out of these 189 IBS manufacturers: 20 achieved class A IBS manufacturers (for those who are ISO certified; possess product testing report; and with QA/QC); 69 achieved class B manufacturers (for those who possess product testing report; and with QA/QC); and 12 achieved class C manufacturers (for those who only possess QA/QC) (Figure 10.9). If the number of class A, B, and C IBS manufacturer was taken as point of measurement, one may find that there were 53% of the registered IBS manufacturers in 2014.

10.3.2.3 IBS certified contractor

The number of IBS certified contractors is obtained by referring to the number of contractors attending IBS courses organized by CIDB. These courses are classified based on the IBS skill specifications: precast concrete; steel frames; formwork systems;

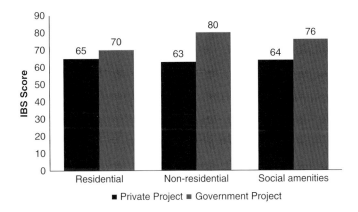

Figure 10.8   Average IBS Score by project category.

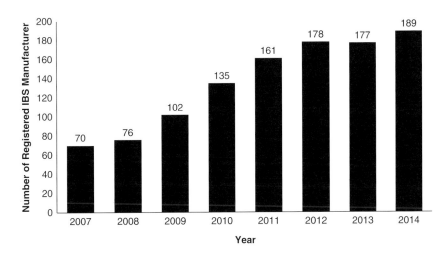

Figure 10.9   Number of IBS manufacturers registered with CIDB (2007–2014) (IBS Centre, IBS Digest, 2014).

block system; and timber frames. A contractor has to go through these courses in order to obtain the certificate of recognition for the use of particular IBS system. As of November 2014, there were 9,423 contractors (approximately xx% of the total number in Malaysia) attending IBS courses organized by CIDB, which far exceeds the target of 1,000 set in the Roadmap. Figure 10.10 shows the number of contractors who attended IBS courses by states in 2014.

10.3.3   IBS adoption level in Malaysia

At first glance, the above stated KPIs – IBS score, number of IBS certified contractors and number of ISO certified IBS Manufacturer – indicate that the country's IBS implementation was in line with the targets set in the IBS Roadmap. However, the efficiency of IBS in reducing the number of foreign labourers, particularly in relation

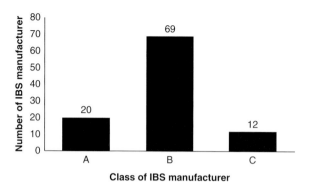

Figure 10.10   Number of IBS manufacturer with different classes (IBS Centre, 2014).

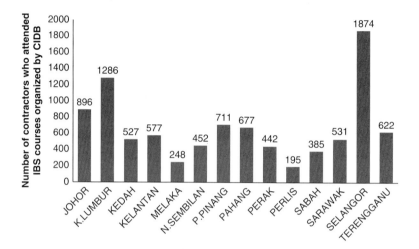

Figure 10.11   Number of contractors who attended IBS courses by state in 2014 (IBS Centre).

to unskilled migrant labour, is still a question mark. As shown in Figure 10.11, the recruitment of foreign workers was expected to decline from 285,000 in 2011 to 260,000 in 2013 and further to 225,000 in 2015 (21% in total), while in actual fact, the number increased from 276,325 in 2011 to 307,500 in 2013, as swing of 32%.

Labour usage is paramount in the Malaysian construction industry, where the foreign labours are imported to fill the shortage of local labour. The total foreign workers in the construction sector in 2013 accounts for 2.3% of total employment (Table 10.3). While these figures may seem less compared to manufacturing and agriculture, the induced impact cannot be overlooked as the amount of foreign workers in construction industry could be more because the statistic did not capture illegal foreign workers. It is estimated that there could be 600,000 foreign workers on construction sites (CREAM, 2011). The biggest block of foreign workers in construction sector in 2010 was Indonesia (80.6%), followed by Myanmar (6.5%), Pakistan (3.3%), Philippines (1.8%), Nepal (1.6%) and Bangladesh (1.6%) (Figure 10.12).

Table 10.3   Distribution of foreign labours in manufacturing, construction, and agriculture.

Sector	2006	2007	2008	2009	2010	2011	2012	2013
Manufacturing	628k	766k	737k	356k	539k	422k	389k	409k
Construction	273k	298k	286k	204k	188k	276k	266k	307k
Agriculture	162k	162k	220k	116k	151k	371k	355k	373k

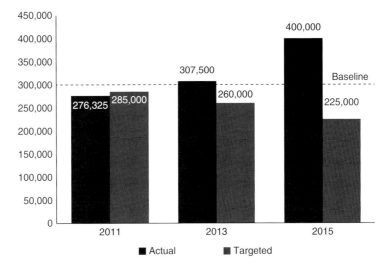

Figure 10.12   Observed and targeted number of foreign workers in construction industry (CIDB, 2010).

The increasing number of foreign workforce in construction sector carries an important message: the country's construction sector still rely heavily on foreign manpower rather than on mechanisation and industrialisation. The industry player's response to IBS has been lukewarm because the majority of them have been exposed and trained in the conventional building system for decades. This is evidenced through the IBS surveys carried out from 2003 to 2010. The first IBS survey (in 2003) as to studied the awareness and usage of IBS in construction from the viewpoint of contractors. The second IBS survey (in 2005) studied the architects' opinion and acceptance on IBS. The third IBS survey (in 2008) measured the acceptance of construction industry towards IBS system and the ranking of IBS benefits. The fourth IBS survey (in 2010) focused on measuring the drivers, barriers, and the critical success factors of contractors in adopting IBS construction. In 2007, the IBS Roadmap's mid-term review was also conducted to gauge the state of IBS implementation and progress on the roadmap's recommendations.

Findings from these studies point out that the initial take-up of IBS was not as high as it was anticipated, particularly from the private sector. The limited take-up of IBS is mainly due to the cost factor and budget constraint along with availability of cheap foreign labour. Since moving towards mechanized and industrialized systems involves high capital investment on heavy equipment and mechanized construction facility, the contractors are unlikely to switch to an unfamiliar system in order to secure their projects, particularly for those small contractors who involved in the small scale development. To note, the usage of IBS is reported to result in an additional 10% of construction cost as compared to the conventional construction method (cast *insitu*) (CIDB, 2007), whilst the levy exemption given to those who adopt IBS as an incentive is only 0.125% of the project value, which is unable to cover the additional cost incurred. Besides, the negative perception on IBS among industry players also affects IBS adoption in the industry. Since the quality of the overall finished work with IBS construction still left much to be desired, industry players tended to perceive that IBS may lead to less productive, more costly, and cause the delay of project. As such, the industry has become reluctant in accepting IBS.

The latest IBS survey, conducted by the authors in 2014, also indicates a low IBS adoption level, where the overall IBS adoption in 2013 was 15.3%; with the adoption level of 61% and 14% in both government and private projects, respectively (Figure 10.13). Residential appeared to be the building type that adopted IBS the most (i.e. 42% of government IBS project; 57% of private IBS project), particularly for projects covering the construction of houses for farmers/labours, and apartments (Figure 10.14). With such low adoption rate, the influence of IBS in reducing foreign workers in construction industry, especially the country basis, is not obviously shown, unless more buy-in is received from the private sector.

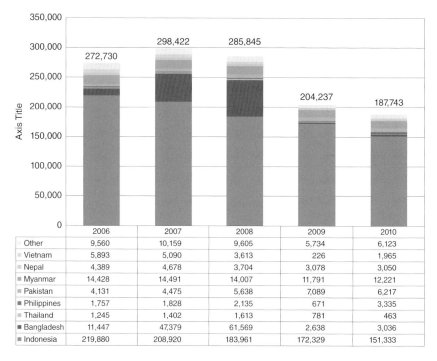

	2006	2007	2008	2009	2010
Other	9,560	10,159	9,605	5,734	6,123
Vietnam	5,893	5,090	3,613	226	1,965
Nepal	4,389	4,678	3,704	3,078	3,050
Myanmar	14,428	14,491	14,007	11,791	12,221
Pakistan	4,131	4,475	5,638	7,089	6,217
Philippines	1,757	1,828	2,135	671	3,335
Thailand	1,245	1,402	1,613	781	463
Bangladesh	11,447	47,379	61,569	2,638	3,036
Indonesia	219,880	208,920	183,961	172,329	151,333

Figure 10.13   Foreign labours in construction sector by country of origin (Department of Statistics Malaysia, 2015).

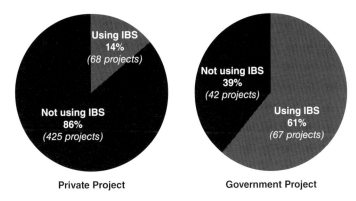

Private Project   **Government Project**

Figure 10.14   Breakdown of projects with value RM10 million and above, using/not using IBS (Foo *et al.*, 2015).

10.4   Social acceptability of IBS in relation to housing

The housing industry does not exist in a vacuum but is a product of structural influences such as economic conditions, the regulatory framework and cultural factors (Barlow and Ozaki, 2005). Thus, considering social acceptability and how it may act on the construction industry is a core part of understanding the drivers and barriers influencing the uptake of prefabricated housing (Goulding *et al.*, 2012). If the potential benefits of IBS are to be realised, there is a need to overcome people's resistance to prefabrication and standardisation in housing.

Various perceptions, opinions and images spring to mind in the popular imagination when considering the concept of prefabrication and standardisation in housing. This is mainly due to the experiences that a number of buildings constructed in the past making use of prefabrication were judged to be of poor quality. Studies show that residents of mass housing in Malaysia are generally not satisfied with their housing conditions (Karim, 2012; Isnin *et al.*, 2012), where a presumed need for flexibility in design, room size and fittings has precluded use. According to Sahabuddin and Gonzalez-Longo (2015), criticism has been made with regard to the People's Housing Project Scheme (PHP) – an initiative by the government to solve the problem of existence slums and squatter areas – on the architectural design including the lack of storage area, small size and deep location of the kitchen, minimum external wall area, complicated partitions, less cross ventilation etc. Most of mass housing dwellers end up renovating houses to tailor-suit their needs before occupancy (Rostam *et al.*, 2012; Nurdalila, 2012; Erdayu *et al.*, 2010). Figure 10.15 shows the layout plan of a PHP unit.

While some of the problems to IBS housing may not now be an issue with the application of new technology, such as using advanced water-proofing or innovative jointing method for solving the leakage problem, the image of IBS is still strongly tarnished by the general perception that the lifespan of such housing is likely to be less than that of "traditional" built housing. The programmes of poor quality and poorly designed "system building" gave rise to a notion that the process of prefabrication per se, rather than particular products, was at fault. Most modern construction, which uses non-masonry materials, also tends to have predicted a lifespan less than those of

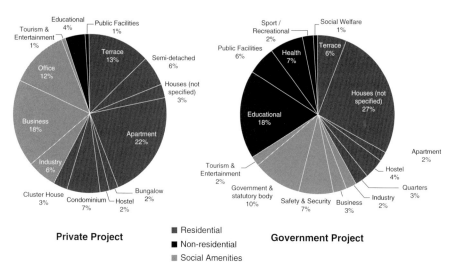

Figure 10.15   Breakdown of IBS projects by project category (Foo *et al.*, 2015).

masonry built housing. Any perceptual links between the ideas of prefabrication, standardisation and non-permanence may therefore prove vital in the acceptance of new approaches to construction. To allow successful application of prefabrication these past mistakes need to be recognised and in particular the following addressed:

- Quality – previous lack of quality has led to the perceived reduction of value associated with IBS housing;
- Attention to detail – poor detailing in the past has led to technical problems and a general perception of poor value;
- Life-cycle performance – failure to consider the practicalities and costs associated with maintaining these buildings had led to some IBS housing becoming difficult to maintain.

Developers, among the wide range of industry players, hold a very important position to ensure the success of IBS adoption. It is because the adoption of IBS is hugely dependent on readiness and maturity of developers to move from existing contracting role into IBS system integrator (see also Chapter 7). Based on the questionnaire survey conducted by the authors, *"client's requirement to use IBS"* has been the main driving factor for IBS adoption in both private and government sectors (Figure 10.16). If more of the developers, especially from the private sector, can be convinced to adopt IBS, an overall higher level of industrialized construction industry can be achieved. Private developers were found to be aware of the importance of IBS as an implication for business in future construction project. However, conventional *insitu* construction is still preferable over IBS due mainly to the sheer cost of investment and the inadequacy of market size. As shown in the questionnaire survey conducted by the authors, 33% of the private project respondents were not using IBS due mainly to the high cost of construction resulted from IBS construction (Figure 10.17). IBS is still not viewed as cost-effective because of the existing closed system in IBS supply chain that may cause an increase in the price of components and tender pricing (see Section 2.2.3).

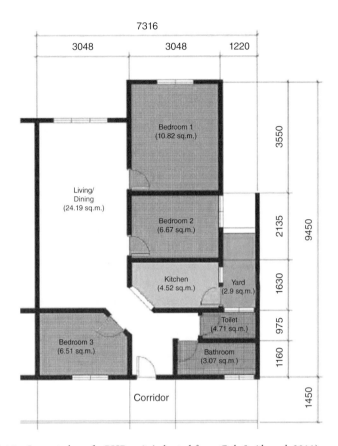

Figure 10.16   Layout plan of a PHP unit (adapted from Goh & Ahmad, 2011).

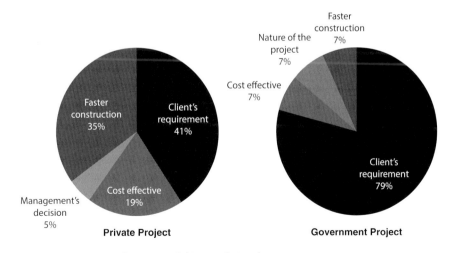

Figure 10.17   Reason for using IBS (Foo *et al.*, 2015).

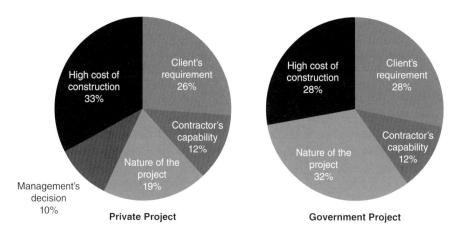

Figure 10.18   Reason for not using IBS.

In addition, IBS and conventional construction are totally different in various aspects, be it the idealism, processes, construction method (see Figure 10.21 later), management, or even the skill sets. Figure 10.18 compares the housing construction process flow between conventional and IBS approaches. While the two processes are almost similar, IBS approach requires high level of collaboration among project parties to account for major constraints in the design with respect to transportation of components, installation logistics, permits and inspection schedules. In short, it requires fundamental structural change to the industry, both in the design and construction stage. During the design stage, IBS approach requires the integration of the architectural plan, civil and structural (C&S), mechanical and electrical (M&E), and workshop drawings, to ensure the effectiveness of design coordination; while during the construction stage, IBS components are manufactured in a factory and then delivered on-site for assembly and erection. Overall, IBS requires a different strategy for the supply chain, planning, scheduling, handling, as well as purchasing of materials, which lent the adopter to serious rethinking about how construction projects are planned and executed (see also Chapter 7). With this regard, a new business approach, investment, and financial planning including the effective combination of cost control and selection of projects that give enough volume to justify the investment is a must in IBS construction.

The limited take up of IBS in private projects is also affected by the different payment system required when using IBS. From the developer's point of view, the costs of material, labour, and machinery in the IBS system are not deemed as a good business investment compared with the conventional system. In current practice, before a construction begins, the client pays between 10 and 25% of the total amount of the contract value as an initial payment. However, in an IBS project, initial spending has to be made to the manufacturers before any progress in the payment is made. IBS manufacturers are normally required to advance approximately 75% of the capital to manufacture the IBS components before delivering these components to construction sites. Without sufficient financial backup, the developers are hardly convinced to use IBS in their projects. Moreover, the adoption of IBS mainly depends on the readiness and maturity of the developers/constructors in terms of know-how and expertise. In several cases, the use of IBS by the developers/contractors has not led to total satisfaction, and actually has been less productive, lacking in quality, and is more costly than the conventional method. Sometimes certain building projects were awarded and constructed using IBS system but suffered project delays and bad qualities. This condition has left the industry with noticeable difficulties when using IBS. Consequently, the industry has become reluctant in accepting IBS, except when it is required by the clients.

At present, almost all projects in Malaysia are conceived, and most are designed in a conventional way. If IBS is to be adopted in these projects, they have to be first converted into IBS, namely, standardisation and modularisation, grid patterns and floor layouts. As such, the developers may face many difficulties to accommodate the changes in the design phase as well as getting approval from the authorities thereafter, since the conventional design did not consider the modularity and buildability of components. In order to ease the process during the tendering stage and also to reduce cost amount, conventional system that can guarantee a rather smooth path is highly preferable.

The issues of expertise and technology cannot be avoided in ensuring the successful IBS implementation. IBS labour competency and skills are still a critical area to be developed. Currently, the country remains lacking in professional skilled workers to lead the implementation of IBS on site, production, moulding, and fabrication, as well as lacking of land surveyor for guideline panel. Poor human capital development on IBS will affect not only the contractors but also the entire supply chain. The skill level of IBS workers is more demanding compared with the skills demanded of workers under conventional construction methods. The system demands more machine-oriented skills, thus requiring the reorientation in terms of the training and education of human resources in a given organisation. Finally, technology transfer is also one of the barriers in implementing IBS system. Currently, local developers depend on foreign expertise and technology. Since most of the machines and materials used in the design of IBS components are imported from developed countries, the cost of producing IBS components and their installation is not competitive. Given that profit is the main motivation, most of the developers will put a second thought in attempting to apply IBS system in their construction projects.

10.5   IBS in future – opportunity for wider IBS adoption

The introduction of IBS – embedded within the CIMP, championed by the CIDB, and promoted by the government – is paving the way for a radical change in the Malaysian construction industry towards the globalization era where an increase in productivity, quality and safety is a must. While IBS has admittedly been less prevalent in privately owned housing, the future development of IBS in Malaysia is encouraging. Considering that labour cost has dramatically increased due to the rising of living standard and the levy imposed by the government on foreign workers, it is important for the builders to explore other building system that requires the least labour input, like IBS, as shorter construction time may eventually imply lower site staff overhead and cost saving on equipment rental. Besides, following the commencement of various mega projects as well as the increasing demand for affordable housing around the country, greater adoption of IBS is assured in order to meet the time requirement with a cost-effective measure.

10.5.1   *Greater Kuala Lumpur*

Kuala Lumpur has been the pillar of the country that drives the national economy. This is evidenced from its share in contributing to the country's gross domestic product (GDP), which rose from 13.4% to 14.8% within the period of 2006 and 2009. Meanwhile, Selangor has been the biggest contributor all the time, with a share as high as 22.1% of the country's GDP. If consideration is being made in the urban or metropolitan scale, Klang Valley – consisting of Kuala Lumpur and the four districts of the state of Selangor (Figure 10.19) – is far outperforming than the other metropolitan towns, with a contribution of GDP eight times more than the others (Figure 10.20).

According to the Performance Management and Delivery Unit (PEMANDU) (2010), Klang Valley with an approximate 6 million population – equal to 20% of the national population – is contributing 30% of the nation's economy. It is expected that when more intense urbanization takes place, the economic activities will become increasingly concentrated spatially, so as the infrastructure development, talent, and productivity growth. As such, Klang Valley has been identified as one

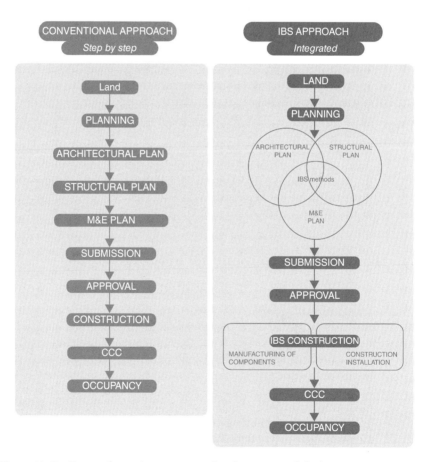

Figure 10.19   Process flow using conventional and IBS approach for housing construction by developer (adapted from CIDB, 2014).

Figure 10.20   Map of Peninsular Malaysia, state of Selangor, and Klang Valley region.

of the twelve National Key Economic Areas (NKEA) in the country's Economic Transformation Program (ETP) to accelerate the country's economy in the near future. Following the launching of EPT, IBS is predicted to play a significant role in building construction. Most of the Entry Point Projects (EPPs) need to be completed on time for fast utilization to create business and values. As a method of construction, IBS has been proven to be quicker compared to conventional construction projects due to the usage of standardized components and a simplified construction process. The use of precast concrete framing, panel and box systems, steel framework systems, prefabricated timber framing systems, steel framing systems, and blockwork systems can ensure faster delivery with lesser wastage. Figure 10.21 demonstrates this point at a multi-storey supermarket, where the traditional concrete construction method has been swiftly over taken by IBS (on the left of photo).

The application of IBS will promote efficient assembly layout and process and accurate resource and material allocation. Its repetitive production method means that costs for construction will be lowered, higher quality is guaranteed and delivery time is shortened. Besides, IBS contributes to the reduction of foreign labour. Currently, the construction industry has taken the cheaper path by utilizing a cheap supply of foreign labour. The government has been looking for ways to reduce the number of foreign workers. The reduced number of foreign workers could hamper construction activities at site. As a result, some of the larger projects will face possible delays due to limited available workforces, without compromising quality. IBS offers improvements in quality, productivity, and efficiency from the use of factory-made products, thus reducing the possibilities of poor workmanship and lack of quality control (for example see Figures 2.1a and b).

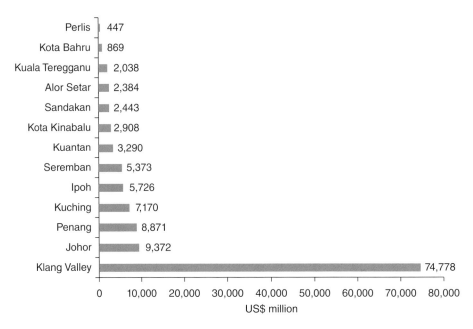

Figure 10.21   Urban Gross Domestic Product (GDP) contributions in 2010 (10th Malaysian Plan, 2011–2015).

10.5.2   *Affordable housing*

IBS has been associated with a cost increase of 10% compared to conventional construction methods, and plagued with bad publicity such as water leakage, inflexible for repair and maintenance, etc. Perceptions of IBS buildings also resemble low cost housing projects which are basic and pay minimal attention to the aesthetics of the building (Figures 10.3 and 10.4). However, one should not confuse "industrialized" with "simplification of design" and "low quality". IBS is capable of facilitating the mass production of houses for mass population with an affordable price in a shorter period of time, especially for countries that are facing high degree of urbanization during the period of transition from the status of developing to a developed country.

The increasing population has prompted the country to put emphasis on housing, particularly the low cost housing. This is evidenced in the periodic five-year Malaysia Plan, where the number of low cost housing to be constructed were increasing in every plan. For example, during the Third Malaysia Plan (1976–1980) and the Fourth Malaysia Plan (1981–1985), a total of 500,000 units and 923,000 units of various categories of houses were planned (i.e. flats, terrace, detached houses), respectively, with the built-up area of 550 to 600 square feet, including two bedrooms, one living room, one kitchen, and one bathroom. The concern for the low-cost housing provision continued to be seen in the Ninth Malaysia Plan (2006–2010), where the allocation for low cost housing has increased 110% as compared to the Eighth Malaysia Plan (2001–2005).

With the announcement of the Tenth Malaysia Plan (2011–2015) the country embarks on developing affordable medium-cost houses, in order to cope with the affordability problem that is widespread in most of the major cities and towns

Figure 10.22   Comparison of traditional cast insitu (right) and prefabricated (left) concrete construction.

in Malaysia. In the Malaysian context, the affordable housing for middle income household is defined as a housing in which the house payment no greater than 33% of the gross household income. The explosion of new growth centres such as new townships, commercial hubs, industrial parks and offices complexes inevitably result in an increase in the working population, leading further to the shortage of affordable housing in the urban centres. According to the study carried out by the Khazanah Research Institute (2015), the median price for the Malaysian housing market exceeds the three times median annual household income threshold for affordability. In 2014, it stood at 4.4 times, and has consistently exceeded 4.0 times from 2002 to 2014, signalling a seriously unaffordable housing market in the country (Figure 10.22). In certain areas, the housing markets even stand out as severely unaffordable such as Kuala Lumpur and Pulau Pinang (Table 10.4).

This trend suggests that both the bottom 40% and the middle 40% of household income earners are likely to end up in some form of social housing if the relevant interventions are not made urgently. As such, the Perbadanan PR1MA Malaysia (in short PR1MA) was put forward to assist individuals and families in the middle income bracket. PR1MA originated as a result of the PR1MA Act 2012 where the government will endeavour to plan, develop, construct and maintain affordable housing for these middle-income people in key urban areas. The main thrust of the programme is to build 500,000 affordable homes by 2018 (Waskett, 2001). Although first-time home owners will be favoured under these schemes, the government plans to offer PR1MA homes in areas where the demand by first-time home owners is low to existing homeowners. However, a question arises concerning how far this program has succeeded in providing physical and social wellbeing of the population, considering that at least 100,000 affordable housing units needed to be built each year, which is equivalent to 274 units per day. Besides, according to the Selangor Structural Plan

Table 10.4 Comparison of housing affordability based on annual household income and median all-house prices across states in Malaysia, 2014

Area	Annual median income	Median All-House Price	Median Multiple Affordability	Affordability
Terengganu	45,324	250,000	5.5	5.1 & Over (Severely unaffordable)
Kuala Lumpur	91,440	490,000	5.4	
Pulau Pinang	56,424	295,000	5.2	
Sabah	44,940	230,000	5.1	
Pahang	40,668	200,000	4.9	4.1 to 5.0 (Seriously unaffordable)
Kelantan	32,592	157,740	4.8	
Malaysia	55,020	242,000	4.4	
Perak	41,412	180,000	4.3	
Perlis	42,000	181,000	4.3	
Johor	62,364	260,000	4.2	
Selangor	74,568	300,000	4.0	3.1 to 4.0 (Moderately unaffordable)
Negeri Sembilan	49,536	188,888	3.8	
Sarawak	45,336	164,667	3.6	
Kedah	41,412	140,000	3.4	
Melaka	60,348	180,000	3.0	3.0 & Under (Affordable)

Source: Khazanah Research Institute, 2015.

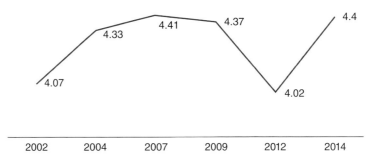

Figure 10.23   Median multiple affordability (Khazanah Research Institute, 2015).

2020, 21,414 hectares of land will be used for housing development purposes in the year 2020 (Figure 10.23), to cater for increasing housing needs which is projected to increase by 93% from 2005 to 2020 (Figures 10.24 and 10.25).

To support this initiative, IBS can improve the building rate of housing schemes dramatically by increasing the number of houses completed over a period of time. The conventional construction system which is presently being used by the construction industry is unable to cope with the demand in a stipulated period because it is labour intensive and relies heavily on foreign workers. Thus, productivity research attention shall be devised toward IBS which employs the philosophy of assembly activity. There is an immense potential for productivity improvement in the building industry from craft activity to assembly activity. Albeit, the statistics presented in Table 10.5 are cited from the Singapore construction industry, the paradigm is similar to Malaysia construction industry as the workers in both countries are shying away from the construction industry. This will eventually help developers to meet demands in housing

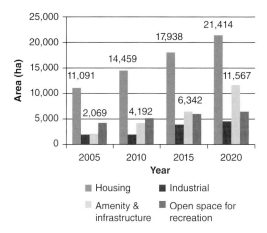

Figure 10.24   Projection of land required for development in Selangor, Malaysia, 2005–2020 (Selangor Structural Plan 2020).

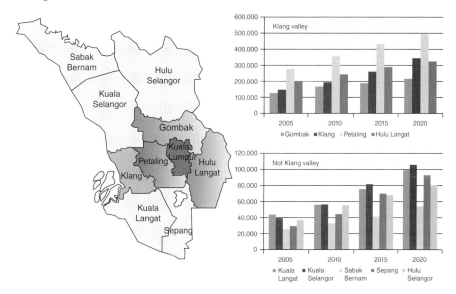

Figure 10.25   Projected number of housing units needed in every district of Selangor state (State Economic Planning Unit, Selangor, Malaysia).

Table 10.5   Usage of workers and potential for productivity improvement in building work

Work type	Usage of workers (%)	Usage of foreign workers (%)	Potential for productivity improvement	Skill	Replaceable
Structural	50	80 – 85	High	Craft	Assembly
Finishing	30 – 35	50 – 60	Medium	More craft and less assembly	Less craft and more assembly
Mechanical and electrical	15 – 20	30	Low	Assembly	Assembly

Source: Anon, 1992.

and contribute to the government's aim to provide a sufficient supply of affordable housing. The rise of housing developments also means the need for new business parks and shopping malls. With IBS, construction time will be shortened; lowering interest and capital outlays for both buyers and developers. This in return will make the development projects more attractive to both buyers and investors.

In summary, achieving cost and speed improvements to deliver affordable housing on a large scale will require, above all, a shift in mind sets. The affordable housing challenge has been complicated by the limitations of the construction industry. Simply put, it costs too much and takes too long to build housing, restricting supply of affordable housing and contributing to the growing affordable housing gap. Even if land is available and proper incentives and financing mechanisms are in place, the affordable housing gap cannot be filled unless inefficiencies in construction can be addressed. Thus, utilising the already established and well developed system of IBS construction in manufactured housing for the future housing in Malaysia has real potential to delivery more affordable and higher quality building.

10.6   Conclusion

The prefabricated housing industry continues to grow. From the perspective of Malaysia, it is the more homogenous cultures during the industrial revolution in which prefabricated housing as a method of practice was developed that ensures the success of prefabrication in the US, Scandinavia, and Japan. It is the cultures that have uniquely developed the technology which eventually serves the society in fulfilling both the wants and needs. In contrast, prefabrication in Malaysia has been seen as a threat to conventional construction, especially in concrete. The failure of IBS to penetrate the market is due to the misconception that IBS will replace the conventional construction method, which is still considered to be more cost competitive according to most of the stakeholders. In reality, many IBS technologies like block works exist together with conventional practices, where they can be applied in tandem to promote best practice in construction. Using what we learn from the U.S., Scandinavia, and Japan – that ideology, products and processes must be the hallmarks of a prefabrication society, Malaysia may yet be able to hone its present interest and capacity for modular construction in order to deliver higher-quality housing.

References

Abdullah, M.R. & Egbu, C. (2009). *IBS in Malaysia: Issues for Research in a Changing Financial and Property Market*, BuHu 9th International Postgraduate Research Conference (IPGRC), Salford, United Kingdom, 15–25.

Abel, C. (2004). *Architecture, technology and process*. London: Architectural Press.

Anon. (1992). Raising Singapore's construction productivity. Singapore, CIDB Construction Productivity Taskforce Report.

Azman, M. N. A., Ahmad, M. S. S., & Hanafi, M. H. (2009). Conceptual spatial site selection for one stop centre for industrialized building system manufacturing plant in Pulau Pining. National Conference AWAM 2009, 70pp.

Badir, Y. F., Kadir, M. R. A., & Hashim, A. H. (2009). Industrialized building systems construction in Malaysia. *J. Architec. Eng.*, **8**(1), 19–23.

Badir, Y. & Razali, A. (1998). *Theory of classification: its application and Badir-Razali building systems classification. Journal Institute Engineering Malaysia*, October.

Barlow, J. & Ozaki, R. (2005). Building mass customised housing through innovation in the production system: lessons from Japan. *Environment and Planning A*, **37**(1), 9–20.

Chapman, R.E. & Butry, D.T. (2008). Measuring and improving the productivity of the U.S. construction industry: Issues, Challenges, and Opportunities. The National Institute of Standards and Technology (NIST).

Chung, L.P. & Kadir, A.M. (2007). Implementation Strategy for Industrialized Building System, PhD thesis, Universiti Teknologi Malaysia (UTM), Johor Bahru.

CIB. (2010). Industrialized construction: state of the art report, TG 57 Publication, International Council for Research and Innovation in Buildings and Construction.

Construction Industry Development Board (CIDB) Malaysia. (2007). Construction Industry Master Plan (CIMP 2006–2015), CIDB, Kuala Lumpur, Malaysia.

Construction Industry Development Board (CIDB) Malaysia. (2003). IBS Roadmap 2003–2010, Kuala Lumpur, Malaysia.

Construction Research Institute of Malaysia (CREAM). (2011). Report on The Current State of Local and Foreign Workforce in Malaysian Construction Site: Building Projects.

Davies, C. (2005). *Prefabricated Home*. London: Reaktion Books.

Egan, J. (1998). *Rethinking construction, report of the construction task force on the scope for improving the quality and efficiency of UK construction industry*, Department of the Environment, Transport and the Regions, London.

Omar, E.O., Endut, E., & Saruwono, M. (2010). Adapting by altering: spatial modifications of terraced houses in the Klang Valley Area. *Asian Journal of environment-Behaviours Studies*, **1**(3), 1–10.

Fathi, M.S., Abedi, M., & Mirasa A.K. (2012). *Construction Industry Experience of Industralised Building System in Malaysia*, 9th International Congress on Civil Engineering (9ICCE), Isfahan University of Technology (IUT), Isfahan, Iran, May 8-10.

Gibb, A. (1999). Offsite Fabrication, Whittles Publishing, Scotland, UK.

Glendinning M. & Muthesius, S. (1994). Tower Block: Modern Public Housing in England, Scotland, Wales and Northern Ireland. Yale University Press.

Goulding, J., Rahimian, F.P., Arif, M., & Sharp, M. (2012). Offsite Construction: Strategic Priorities for Shaping the Future Research Agenda. *Architectoni. ca*, **1**, 62–73.

Hamid, Z., Kamar, K.A.M., Zain, M., Ghani, K., & Rahim, A.H.A. (2008). Industrialized Building System (IBS) in Malaysia: The Current State and R&D Initiatives", *Malaysia Construction Research Journal*, **2**(1), 1–13.

Haron, N.A., Hassim, S., Kadir, M.R.A., & Jaafar, M.S. (2005). Building Cost Comparison Between Conventional and Formwork System: A Case Study on Four-Story School Buildings in Malaysia, *American Journal of Applied Sciences*, **2** (4), 819–823.

Hassim, S., Jaafar, M.S., & Sazali, S.A.A.H. (2009). The Contractor Perception Towards Industrialised Building System Risk in Construction Projects in Malaysia, *American Journal of Applied Sciences*, **6**(5), 937–942.

Isnin, Z., Ramli, R., Hashim, A.E., & Ali, I.M. (2012). Sustainable issues in low cost housing alteration projects. *Procedia – Social and behavioural Sciences*, **36**, 292–401.

Jaafar, S., *et al.* (2003). *Global Trends in Research, Development and Construction*, Proceeding of the International conference On Industrialised Building System (IBS 2003), CIDB (1997).

Junid, S. (1986). *Industrilised Building System*, Proceeding of UNESCO/FEISEAP, Regional Workshop.

Kamar, K.A.M., Hamid, Z.A., Azman, M.N.A., & Ahmad, M.S.S. (2011). Industrialized Building System (IBS): Revisiting Issues of Definition and Classification. *International Journal of Emerging Sciences*, **1**(2), 120–132.

Kamar, K. A. M., Alshawi, M., & Hamid, Z. A. (2009). *Barriers to industrialized building systems: The case of Malaysia*. Paper proceedings in BuHu 9th International Postgraduate Research Conference (IPGRC 2009), The University of Salford, 29th–30th January 2009, Salford, United Kingdom.

Karim, H.A. (2012). Low cost housing environment: Compromising quality of life? *Procedia – Social and Behavioral Science*, **35**, 44–53.

Khazanah Research Institute. (2015). Making Housing Affordable.

Latham, M. (1994). Constructing the team, final report on joint review of procurement and contractual agreements in the UK construction industry. HMSO, London.

Lessing, J., Ekholm, A., & Stehn, L. (2005). *Industrialized Housing – Definition and Categorization of the Concept*. 13th International Group for Lean Construction, Australia, Sydney.

Nawi, M. N. M., Hanifa, F. A. A., Kamar, K. A. M., Lee, A., & Azhari Azman, M. N. (2014). Modern method of construction: An experience from UK construction industry. *Australian Journal of Basic and Applied Science*, **8**(5), 527–532.

Nurdalila, S. (2012). A review of Malaysian terraced house design and the tendency of changing. *Journal of Sustainable Development*, **5**(5), 140–149.

Oostra, M. & Joonson, C. C. (2007) Best practices: Lesson Learned on Building Concept (edited by) Kazi, A. S., Hannus, M., Boudjabeur, S., Malone, A. (2007), *Open Building Manufacturing – Core Concept and Industrial Requirement'*, Manubuild Consortium and VTT Finland Publication, Finland.

Marsono, A.K., Tap, M.M., Ching, N.S., & Mokhtar, A.M. (2006). *Simulation of Industrialized Building System (IBS) Components Production*, Proceedings of the 6th Asia-Pacific Structural Engineering and Construction Conference (APSEC 2006), Kuala Lumpur, Malaysia.

Rahman, A.B.A. & Wahid Omar. (2006). *Issues and Challenge in the Implementation of IBS in Malaysia*, Proceeding of the 6th Asia Pacific Structural Engineering and Construction Conference (ASPEC), Kuala Lumpur, Malaysia, 5 & 6 September.

Rostam, Y., Hamimah, A., Mohd Reza Esa, & Norishahaini, M.I. (2012). Redesigning a design as a case of mass housing in Malaysia. *ARPN Journal of Engineering and Applied Sciences*, **7**(12), 1652–1657.

Sahabuddin, M.F. M. & Gonzalez-Longo, C. (2015). Traditional values and their adaptation in social housing design: Towards a new typology and establishment of 'air house' standard in Malaysia. *International Journal of Architectural Research*, **9**(2), 31–44.

Sarja, A. (1998). *Open and Industrialised Building*, International Council for Building Research, E&FN Spoon, London.

Sebestyen, G. (1998). *Construction-Craft to Industry*. E & FN Spon, London.

Smith, R.E. (2009). *History of Prefabrication: A Cultural Survey*. Proceedings of the Third International Congress on Construction History, Cottbus, May.

The Economist Intelligence Unit (EIU). (2015). *Rethinking productivity across the construction industry: The challenge of change*. The Economist Intelligence Unit Limited.

Waskett, P. (2001). *Current Practice and Potential Uses of Prefabrication*. DTI Construction Industry Directorate Project.

Chapter 11

Affordable and Quality Housing Through Mechanization, Modernization and Mass Customisation

Zuhairi Abd. Hamid[1], Foo Chee Hung[1] and Gan Hock Beng[2]

[1]*Construction Research Institute of Malaysia (CREAM), Kuala Lumpur, Malaysia*
[2]*G&A Architect, Malaysia*

Affordable housing in the past has never been designed to last as it was aimed to meet the urgent housing demand in the shortest possible time. However, changes in the demographic make-up due to the diversity of family typologies and household arrangements have generated a need for housing that can adapt to different privacy, space, use requirements, and life styles. These trends, overlain with the demands for better environmental performance, suggest the need for affordable housing that is intrinsically sustainable – reducing construction cost and life-cycle costs (i.e. building operations, long-term maintenance, refurbishment) while maintaining liveability. Inspired by the vernacular architecture – Malay *Kampung* House, this chapter presents a new typology of affordable housing design, which is not only able to adapt to the changes of a household over its life cycle, but also enable retrofit and reconfiguration to be made quickly, economically, and repeatedly without involving excessive site labour, time, and cost. Most importantly, the proposed typology is able to facilitate the delivery of the customised products with near mass production efficiency based on the existing technology and performance of the housing industry.

11.1 Introduction

Reducing construction costs while meeting consumers' various needs for housing quality are the foremost challenges faced by today's house builders. As shown in Figure 11.1 – a schematic illustration of the common trade-off between cost, square area and quality – given a fixed contest (the triangle), cost reduction seems to be

Modernisation, Mechanisation and Industrialisation of Concrete Structures, First Edition.
Edited by Kim S. Elliott and Zuhairi Abd. Hamid.
© 2017 John Wiley & Sons Ltd. Published 2017 by John Wiley & Sons Ltd.

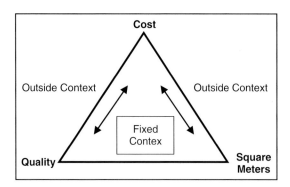

Figure 11.1   Trade-off between cost, square area and quality (adapted from Dluhosch, 2006).

achievable only by reducing either quality or area (m^2); conversely, increase in floor area will have impact on both cost and/or quality. This well explains the problem of affordable housing development as its primary determinant is cost. Realizing that land acquisition and construction costs are not likely to reduce in the future, while regulations restricting affordable housing development are not likely to become less burdensome, quality is often one of the few components to be traded off for cost-effective strategy in affordable housing development. Consequently, most of the affordable housing projects – as a reflection of government's effort in providing adequate and affordable housing that meet the minimal shelter requirements – have brought with them a set of quality deficiencies. These projects fall short of the well-planned, well-designed, and well landscaped environments that often lead to the question whether mass population is being accommodated sustainably.

While moving towards a higher level of urbanization, the developing countries of South East Asia are of no exception in facing mass housing problems that require a solution with the integration of the dual aims of the principles of affordability and sustainability. Keen architects, urban planners, and engineers might have achieved advances in generating single meaningful buildings, but most of them fail to come up with effective sustainable mass housing design when dealing with concentrated urban populations (Gan, 2013; Gan et al., 2013a). Major mass housing developments of the past 50 years, for example as shown in Figures 11.2 and 11.3, has only served to highlight the paradox between the structural and social aspects in building design, a pattern also repeated in Europe and South America.

Prefabrication is, no doubt, an obvious choice for quick, efficient, and inexpensive housing construction and delivery. The use of industrialized building systems (IBS) – as discussed in Chapter 10 – has the potential to eliminate building site inconveniences, reducing the lapsed time and cost of construction, and contributing to an end product that conforms to the same standards and codes as site-built housing (Hullibarger, 2001). In some affordable housing developments, industrial approaches to construction coupled with value engineering, were found to help reduce cost by about 30% and delivery times by up to 50% (MGI, 2014). However, prefabricated houses may have a rigid structure, an interlocking plan, and predetermined functions, in which very few of them are sufficiently open plan to enable retrofitting and reconfigurations to be made quickly, economically, and repeatedly.

Figure 11.2   Factory produced precast concrete housing in the 1970s is now due for demolition.

Figure 11.3   Serviceability and social problems, rather than structural design, have condemned precast concrete buildings to demolition after less than 40 years.

Changes in the demographic make-up due to the diversity of family typologies and household arrangements have generated a need for housing that can adapt to different privacy, space, use requirements, and life styles. Questions arise whether the mass population is being accommodated in suitable dwellings, and are homes now being developed capable of adapting to the occupant's ever-changing requirements. Studies show that residents of mass housing in Malaysia are generally not satisfied with their housing conditions, in terms of construction activities, materials used, aesthetic value, amenities, and so on (Karim, 2012; Isnin *et al.*, 2012). Most of them end up renovating houses to tailor-suit their needs before occupancy (Rostam *et al.*, 2012; Nurdalila, 2012; Erdayu *et al.*, 2010).

This is largely due to the nature of the current mass housing architectural strategy, namely the convergent design system (Figure 11.4), which is a "one fits all" design initiative where housing is likely to be designed around the capability of a given product, instead of around the end-user (Zuhairi *et al.*, 2015). Thus, houses designed for the average family are deficient in meeting the mass housing sustainability objectives as they are leading to further compromise the occupant's needs. Besides, the convergent design system implies extreme compartmentalization and dissociation of internal elements, where service spaces such as the kitchen, bathroom, and so on, are built internally by interlocking with space, making the service spaces difficult to interchange (Zuhairi *et al.*, 2015). Houses designed and built with this system are solely based on the economic concepts of housing that only measure affordability, ignoring the potential of sustainable housing design that offers social and environmental

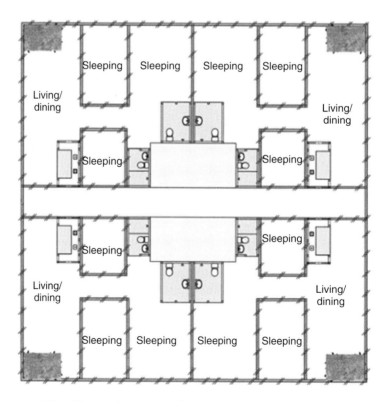

Figure 11.4   Typical layout of convergent design system.

benefits. Even though the convergent dwelling design offers a reasonable alternative of housing needs for the general population, it is found to consider more on the physical development of housing rather than on the sustainable inhabitation.

Meanwhile, as the era of sustainability is taking its stand, unsustainable construction practices in the building industry are getting increasing attention from the public (Gan, 2013). Issues such as excessive energy and electricity consumption, consuming more raw materials than other industrial sectors, generating large amounts of wastes, using heavy materials subject to wear and tear, disruption to nature, etc. have become the main topics of public concern. Hence, house builders should be aware of the rising demands for better environmental performance, so as to take advantage of technology in the housing production process, either by reducing initial and lifecycle cost, or making affordable housing an attractive and sustainable option for the mass population.

In this chapter, the potential of applying flexibility – a design approach which has been widely adopted in the vernacular architecture – is proposed as an inherent design strategy for the modern urban mass housing. The aim is to explore how vernacular architecture grant insights to produce flexible mass housing design that will respond to the changing demands of users during occupancy. Within this framework, the focus is on the following topics of discussion:

i.   what are the features of flexibility offered by the vernacular architecture
ii. what is the scope of flexible housing
iii. does the proposed design system address the issues of affordability and sustainability within the context of modern mass housing?

11.2 Design for flexibility – insight from the vernacular architecture

Flexibility refers to the idea of accommodating changes over time (Siddharth & Ashok, 2012). It is an innovative approach to architectural design that enables facilities to be retrofitted quickly, economically, and repeatedly. The concept of flexibility has long been a hallmark of the office and commercial spaces design. In-line with the new trend of residential housing design, the concept of flexibility is further intensified in the design of housing, as housing in the present day is required to consider not only about housing one family or group of occupants throughout their lifecycle, but also allowing new residents to adapt the dwelling to their needs, or to allow a suitable mix of dwellings existing in an ever-changing environment. Flexibility, in the context of housing, represents a comprehensive research on cases in the European context beginning from the early twentieth century (Siddharth & Ashok, 2012). According to (Schneider & Till, 2007), who introduce "flexible housing" by providing a criticism on the current condition of housing in the UK, housing flexibility addresses a number of issues related to the current and future needs of the users as it:

i.   offers variety in the architectural layout of the units
ii. includes adjustability and adaptability of housing units over time
iii. allows buildings to accommodate new functions.

In Malaysia, the idea of flexibility has long been the key feature of design in the vernacular architecture. This is well reflected in the traditional Malay house, where the residential environment is not only designed to respond to the occupant's living demands, but also fully integrated with the tropical climate and the use of local resource (Figures 11.5 and 11.6). The traditional Malay house is constructed by employing sophisticated architectural processes that has been proven to be in harmony and successfully maintaining the capacity of the rainforest ecosystem (Che Amat *et al.*, 2009; Nordin *et al.*, 2005). Its design approach is ideally established based on the accumulated local knowledge, way of life, culture, and the deep understanding of natural environment. Apart from being the richest component of the country's cultural heritage, this type of house is also recognized as the most sustainable building in the past, even until today, due to its design and construction process that takes into account energy efficiency, indoor environmental qualities, sustainable site planning, and the uses of local materials and resources. Typical example can be seen from the orientation of the house, where a high-pitched roof not only encourages stack effect function but also acts as solar shading devices (Ramli, 2012). Building on stilts allows cross-ventilating breezes beneath the dwelling to cool the house whilst also mitigating the effects of occasional flood as well as ensuring safety from possible attack by wild animals (Amad *et al.*, 2007). Plenty of windows and openings allow more natural lighting while capturing high-velocity of air movement. All these features have been proven to be the most effective passive design for a tropical building, as it increases the overall building's thermal comfort and energy efficiency. The application of Malay houses' features into the design of low energy building can be seen from the Ministry of Energy, Green Technology & Water (MEGTW)

Figure 11.5   Traditional Malay and Indonesian house with natural ventilation (in Sarawak, Borneo).

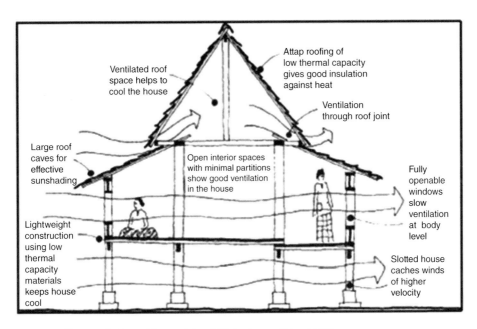

Figure 11.6   Climatic design of the traditional Malay house (Lim, 1991).

buildings, where among the features to be re-adapted are such as the consideration of building orientation, fenestration design, the application of natural lighting system, natural air ventilation system, and the arrangement of interior spaces (Ramli, 2012).

In terms of construction, the Malay traditional house utilizes the technique of housing components by prefabrication, which has permitted flexibility in the development of the house form. All the components that make up the structural framework, roof trusses, stairs, wall and floor panels, and roof surfaces, are made before erection of the house begins (Hassan, 2001). The structural framework, such as columns (*tiang*), beams (*rasuk*), secondary beam (*gelegar*), primary roof beam (*alang panjang*) and secondary roof beam (*alang pendek*) are assembled on site to form the primary framework. Through this flexible housing construction approach, the traditional Malay house is capable of expanding from time to time, based on the occupant's affordability level and demographic concerns. With this the basic type (i.e. *rumah ibu*) forms the "core house", the house size is changeable along with the occupant's future requirement. Whenever additional finance is available or as the family grows larger, the house is capable to evolve, not only in terms of configuration and appearance, but also in respond to the explicit need through accommodating a wide diversity of users and household types (Figure 11.7).

11.3   Scope of flexibility in residential housing

Today's residential housing in developing countries does not incorporate the concept of flexibility in its design (Gan *et al.*, 2015). The present mass housing design is said to have lost its build form identity in terms of rainforest and tropical landscape. This is largely due to planning patterns and construction systems, which by adopting

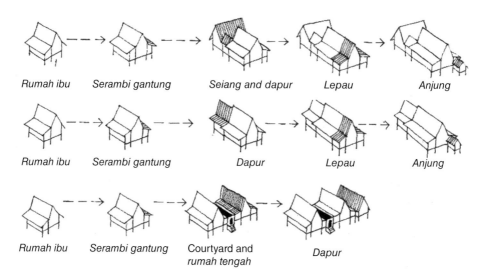

Rumah ibu   *Serambi gantung*   *Seiang and dapur*   *Lepau*   *Anjung*

Rumah ibu   *Serambi gantung*   *Dapur*   *Lepau*   *Anjung*

Rumah ibu   *Serambi gantung*   Courtyard and   *Dapur*
                                       rumah tengah

Figure 11.7   Possibility of expansion – the flexibility of traditional Malay house (Lim, 1991).

planning laws, building codes, and regulations borrowed from the West that promote heavy weight construction using bricks and reinforced concrete as the main materials has forbidden housing development based on traditional concepts (Hassan, 2001). Besides, the practicing housing construction method which is derived from systems used since the 1800s (the period of Industrial Revolution in Europe) is outdated. It is unlikely taking environmental concern as a primary consideration. What makes the condition even worse is that almost all housing programmes that initially aimed to tackle housing shortages is simply to augment the housing supply with minimum concern for social tradition issues (Hassan, 2001). For example, the housing industry is required to construct at least 100,000 affordable housing units per year (or 274 units per day) following the launching of "1 Malaysia People's Housing Scheme" (PR1MA). The desirable building rate in most European countries is about 2500 housing units per 1 million population. The question arises concerning how far these programmes have succeeded in providing physical and social wellbeing of the population, considering the country's current housing design strategies are unable to cope with the incredible demands on such massive scale in sustainable manner.

For mass housing to be an attractive option for the average family, the provision of architectural flexibility is essential (Singh *et al.*, 1999). Since each dwelling unit is a primary structure that would contribute to the quality of life through its flexible organization, and the root causes leading to housing quality problem are identified as issues related to housing layout and design, surrounding environment, maintenance, location, amenities, and building material (Živković & Jovanović, 2012), flexibility should be reflected, as much as possible, within all aspects of the housing type. According to (Friedman & Krawitz, 1998) elements to be considered for a flexible housing should include:

i.   the composition of the varied households within the single structure
ii.   the choice of components that are available
iii. the ability to make future modifications with minimal inconvenience.

Therefore, each dwelling unit should be designed in such a way that it is economically and easily adjustable, whilst it adheres to the context of contemporary technology, tropical adaptation, and cultural responses. The key design element is the realization that lifestyle – as one of the defining characteristics of peoples' lives as citizens, consumers, and householders – is a feature that shifts in accordance with a dynamic lifecycle process. A home that can be altered with minimum effort and expense at a time of change in the lives of its owners, whether through such a minor intervention as the rearrangement of furniture in a non-restrictive space or through more vigorous modification such as the relocation of living or storage spaces, is a home that evolves with the lifecycles of its household rather than becoming rigidly obsolete in the conventional manner (Friedman & Krawitz, 1998).

11.4   Divergent Dwelling Design (D3) – proposed mass housing system for today and tomorrow

The proposed mass housing design system – Divergent Dwelling Design (D3) – is an inherent design strategy of sustainable development that fully utilizes the idea

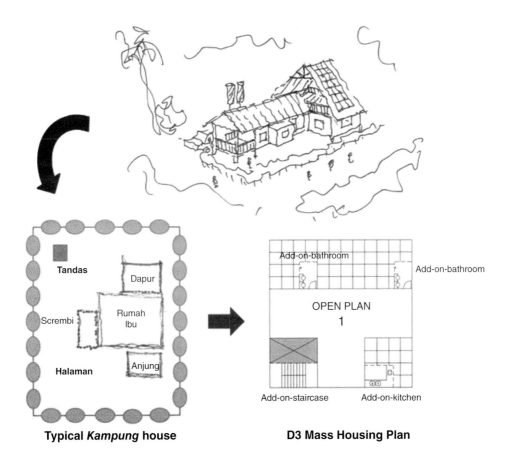

Figure 11.8   D3 design process.

of flexibility. It is inspired by the traditional Malay house design approach that accommodates freedom for change of preference even after the structure is built. Each function unit (*rumah ibu, dapur, serambi, anjung,* etc.) combines divergently to reduce the immense intricacy of architectural phenomenon to simple constant units and brings about an effective formation of a flexible dwelling system (Figure 11.8). By having the same models, structures, and constructions, D3 can produce millions of combinations, each of which is of a high level of flexible form and function in architectural organization with sustainable manner – capable of continuous modification, renewal, and redesign (Figure 11.9).

In layman terms, D3 is an architecture where service spaces are attached externally to form dwelling of which the service spaces can be changed to accommodate different functions with minimum or no disturbance to the core structure (referred to here as COR) as shown in Figure 11.10. The concept is similar to the other existing facilities in modern industry (automotive innovations, electrical appliances, furniture, etc.) where each individual functional unit is freely bonded with the COR to serve different occupants' requirement. The COR, as a module, houses every independent functional unit of the dwelling and also forms the space for further expansion.

The sustainability issues focus on providing spaces to be used for a variety of purposes over time, be it the changes of household demography or the changes of resident's living satisfaction. Since this kind of functional change can be done by merely switching independent units within the configuration through a simple process, the

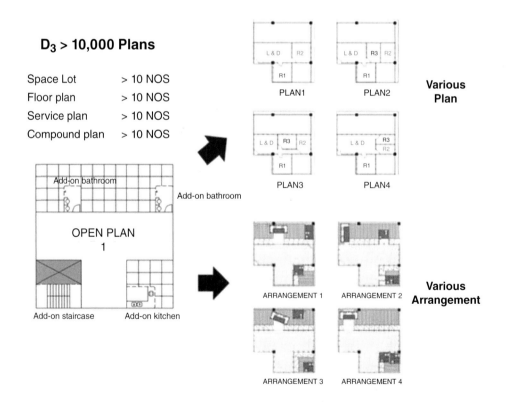

Figure 11.9   Possible combination of D3.

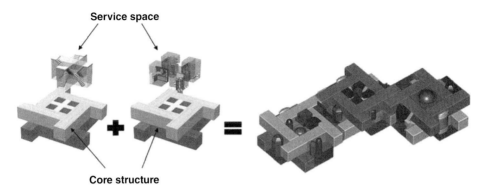

Figure 11.10   The concept of Divergent Dwelling Design (D3).

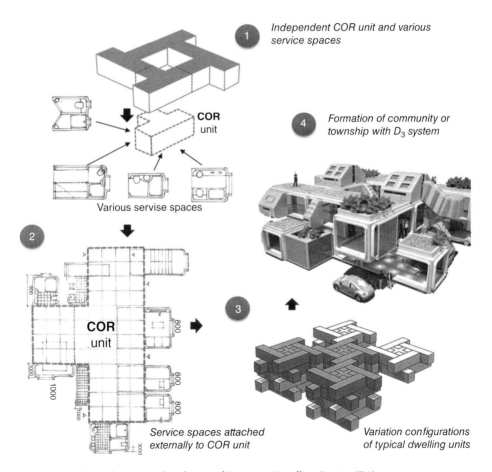

Figure 11.11   The application and evolution of Divergent Dwelling Design (D3) system.

function of the dwelling unit can be cultivated and adapted to the occupant's need whenever it is required. For example, a residential space can be converted into a café by just incorporating a larger kitchen and more toilets; a laboratory or playroom or computer room when added with a unit space for teaching can be used as an educational institution. Similarly the kind of unit space or constant space can change its function from residential to commercial, without ever needing to change the basic unit. As such, D3 systems makes possible the creation of dwellings which may grow old yet without becoming obsolete; incorporating the latest design ideas and technologies, yet have a sense of history on the Malaysian housing design (the *rumah kampung* design), allowing the communities to live for generations, yet incorporating the potential of adaptation.

Another feature of D3 is that the system utilizes amply science, technology and industrialization in the formation of a unit. For example, bathroom and kitchen dimensions are fixed for mass production. The occupant has wide spectrum of choice with regards to products in the market. Since each unit is independently constructed by machine production, the development entails the use of technology and innovation, without the involvement of excessive site labour, time, and cost. In this sense, divergent design concept comprehends the advances of science and technology over time, thereby resulting in faster production at economical rates. More crucially, it helps to boost a greater productivity, better quality, and an assurance of a growing and interested housing market in the twenty-first century. Once the design system is in tandem with serial production and standardization, there will be no bounds for the development of sustainable community, as outlined in Figure 11.11. It is because every detail can be perfected – just like the automobile and telecommunications industry, has seen continued advancement in technological innovations that have benefited consumers in the long run.

11.5   Design principles of D3

D3 directly responds to the fundamental demographic and economic pressure that have recently heightened the need for a new housing alternative that appropriately integrates affordability and sustainability. To ensure mass housing populations could enjoy eco-housing with affordable price, affordability is designed in at the beginning by adopting a simple design layout. This is to avoid the use of unnecessary technology which is costly to buy or repair, so that the house is easy for maintenance in the future. Besides, defining affordability in relation to housing within the ambit of sustainability broadens its scope by involving the social and environmental perspectives. The element of sustainability in D3 is ensured through a flexible design prototype that suits climatic conditions using environmental friendly methods, contributes to the sustainable development of the construction industry, offers what people demand from a house and that they can live how they want to within it, by taking into account of:

i.   the spatial and functional arrangement
ii.   the potential to expand spaces for increased occupant's usage
iii.   maximizing natural lighting and ventilation
iv.   continuity of the traditional housing concepts into a modern contemporary residential development.

Figure 11.12   The design unit plan of the D3 system.

In general, D3 systems utilizes four level of housing design principles:

i.   design of the unit plan
ii.   design of unit configurations
iii.   design of sustainable strategies
iv.   design of structure and construction.

11.5.1   The design of the unit plan

In the D3 system, the unit plan is designed to respond to the demand for more space and the changing circumstances of the occupants. It allows the flexibility in forming the basic unit configuration which then responds to the nature of the site conditions as well as the improvement of living environment quality (Figure 11.12). Internal space is adjustable according to the requirement of the user through adopting an open plan design. The dwellers are able to choose the interior components to tailor the design to their individual lifestyles and budgets, and can easily modify these initial parameters as the need arises. By enabling the floor plan to be adapted to the future users and the changing needs of families, the D3 unit plan is also said to take into account different family types:

i.   dynamic family which is likely to have more children in future, and is thus requires a high degree of space flexibility to cater for continuously changing and increasing needs

   ii.   stable family which is not going to have any more children (either the children have left home or are too small to leave home) and thus requires a relatively lower degree of space flexibility

   iii.   stagnant family which is expected to live in the same dwelling for a long time and thus has sufficient opportunity to benefit from flexible building elements, which provides for lower life-cycle cost of such elements.

With the dwelling unit built according to these flexible design principles, users are not only given the chance to choose the floor plan they want to live before moving in, but are also able to achieve harmony between the basic structure and the various sizes of dwellings in the long term, in accordance with rising space standards and the possibility of new family members in the future.

11.5.2   Unit configurations design

According to the Construction Industry Master Plan (CIMP 2006–2015), the quality in the construction industry encompasses more than contractors alone. Architects and engineers may have to be involved as well since they are the contributing factor to quality failures (material faults, construction faults, and design faults), where 50% of failures can be attributed to design faults, while the remaining 40% and 10% are due to the construction and material faults, respectively. Since the formulation of the National Urban Policy is to the provision of adequate, affordable and quality housing, it is essential to include environmental clauses in every project development so as to strengthen environmental control and preventive measures. A flexible design can respond to events, even when they unfold in unpredictable ways. On a much larger scale, there is current interest in the adaptation of the built environment for climate change. Because the rate and severity of future climate change is unpredictable, design for adaptation is best achieved by providing lifecycle options that will allow future decision makers to respond appropriately to the trajectory of climate change that actually occurs. D3 will surely place sustainable architecture along the curve followed by science, technology and industrialization, facilitating a shift towards higher quality housing development that will eventually create a sustainable community for everyone anywhere in the country.

Figure 11.13 shows some possible formations that can be achieved with the D3 system. By combining each and every sustainable individual dwelling, a greater sustainable community or township is formed. No part of it need ever be obsolete. One does not have to determine in advance the overall makeup of the group of units, for the whole community may just be cultivated or generated in a naturally evolving manner. This programmable dwelling pattern will enable a variety of dwellings to be processed, to constantly renew themselves owing to industrialization and with rational rearrangement of all available habitual spaces and incorporation of machine production resulting in a community that has no slum, and where no redevelopment is necessary. Fundamentally, the evolutionary nature of D3 – the notion that housing is to be designed to evolve not only in configuration and appearance but also in use – responds to an explicit need to accommodate a wide diversity of users and household types. Figure 11.14 shows the example of a finished D3 building, while Figure 11.15 shows the application of D3 design during construction and on completion.

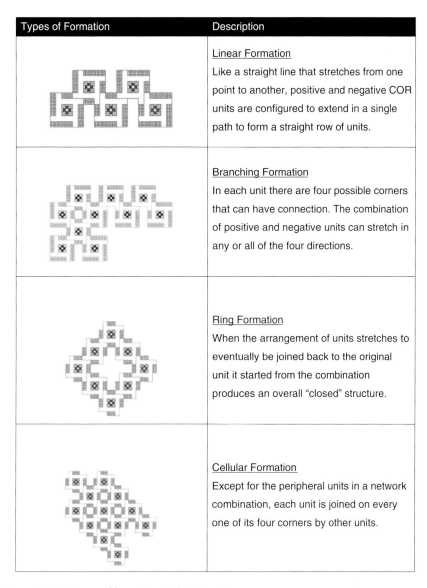

Types of Formation	Description
	Linear Formation Like a straight line that stretches from one point to another, positive and negative COR units are configured to extend in a single path to form a straight row of units.
	Branching Formation In each unit there are four possible corners that can have connection. The combination of positive and negative units can stretch in any or all of the four directions.
	Ring Formation When the arrangement of units stretches to eventually be joined back to the original unit it started from the combination produces an overall "closed" structure.
	Cellular Formation Except for the peripheral units in a network combination, each unit is joined on every one of its four corners by other units.

Figure 11.13   Types of formation in the D3 system.

11.5.3   *Sustainable strategies design*

Flexibility and adaptability are key strategies for sustainable design of the D3 system. The use of concrete construction ensures many of the qualities that aid flexibility in housing design. Concrete is an inert material, with no harmful off-gassing emissions. Coupled with its structural form, concrete is the construction material most commonly associated with building designed with enhanced natural ventilation and daylighting. Concrete construction presents great opportunities to meet the needs of the users by helping to improve the function, value, and whole life performance of the

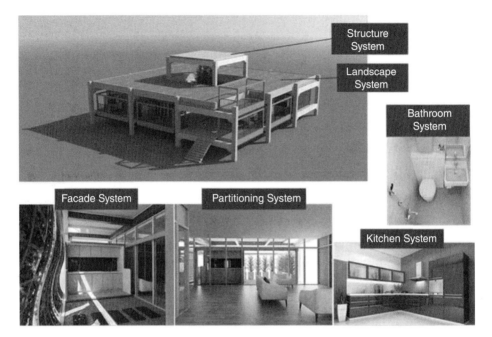

Figure 11.14   Finished D3 building with different views.

house. With its high thermal mass, concrete construction can reduce energy requirements during the operation of the building by moderating energy demands in cooling buildings, thereby adding to the value in use of residential building. Besides, concrete contributes to the range of other inherent benefits at no extra cost, such as its proven integral fire resistance, high levels of sound insulation, and robust finishes. Through its very nature, concrete provides robust surfaces for walls, partitions, columns, soffits and cladding that are easily sealed and free of ledges or joint details. All these may finally lead to the lower maintenance costs of the building while set in motion an efficient, cost effective and practical method for solving housing needs and overcrowding concerns in urban areas.

The use of a reinforced concrete skeleton structure which allows the design of floor plans that are variable to accommodate different family structures, coupled with the constant improvement in structural design and technology supported by the incorporation of lightweight, durable, smooth edged, space efficient and universally adopted specifications ensure that mass housings remain affordable and sustainable for the long term. The overall result/outcome of the sustainable strategies design can be seen by observing how the dwelling units encourage maximum cross ventilation, minimizing heat island effect, facilitating new technology installation, and promoting green architecture (Figure 11.16).

11.5.4   Structure and construction design

There is usually a cost with the provision of a structure that allows the most flexibility and adaptability. However, with early consideration of the benefits of using concrete

Figure 11.15   Construction flow of D3 building.

during the design phase, flexibility of the dwelling can be optimized at little or no extra expenses. It is because the use of concrete is compatible with fast housing construction, in part due to its easy mobilisation at the start of a project. In a departure from the conventional mass housing design, the D3 system generates a better and cheaper habitat option through the application of existing science, technology and machine production capability. By incorporating IBS and industrialization into the construction process, a compressed construction schedule is assured, leading to not only cost-saving but also enables the dwelling to get into productive use sooner, as well as reducing finance periods. It is because the use of modern methods of construction, such as sophisticated formwork systems, post-tensioning, prefabricated roll out reinforcement, and precast elements can all shorten the construction time.

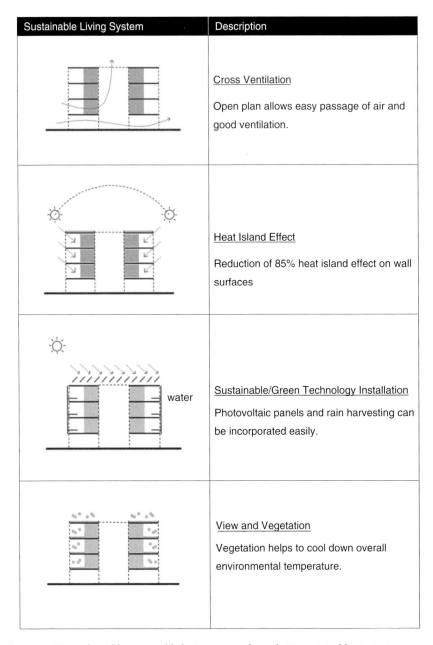

Sustainable Living System	Description
	Cross Ventilation Open plan allows easy passage of air and good ventilation.
	Heat Island Effect Reduction of 85% heat island effect on wall surfaces
	Sustainable/Green Technology Installation Photovoltaic panels and rain harvesting can be incorporated easily.
	View and Vegetation Vegetation helps to cool down overall environmental temperature.

Figure 11.16   Achievable sustainable living system through D3 sustainable strategies.

In the D3 system, modular construction concept is widely adopted, where prefabricated elements or existing structures are used as the basic structure while additional natural materials are used to add insulation or to adjust the aesthetics. This allows for the construction of various building types, be it the low-rise low-density configuration, vertical tower configuration, low-rise high density pyramid configuration, or the high-rise high density configuration (Figure 11.17).

Figure 11.17   Different building types with D3 design.

Short-term flexibility can be achieved by movable partitions. Since common walls between the dwelling units are non-load-bearing walls, the dwelling unit floor areas can be arranged independent from boundaries, thereby providing an entirely flexible arrangement of living room and wet room locations in the plan layouts. Partition walls in the D3 system that enable flexible planning will encompass the following features:

i.   easily applicable
ii.   produced in standardized dimensions and not requiring a base coat
iii.   easy to remove any traces left by demounted partition elements on the adjoining elements, such as floors, ceiling, face walls, and fixed partition walls
iv.   possibilities to coat them with different materials and to change the colour and texture of their surfaces in accordance with the requirements of the space and the individual taste of the users.

In the longer term, adaptability is needed over the life of the dwelling unit to allow internal walls to be moved, to change the size or use of spaces or suites of spaces. As such, flexibility is achieved through both the arrangement of columns and load-bearing walls and the possible clear span. Steel or pre-stressed reinforced concrete floor systems and components are used to obtain maximum clear space in the plan layout. The ceiling surfaces are clear, and beams are hidden in the exterior wall axes or fixed infill wall axes. All structural elements are located at the exterior of the layout to allow for unlimited unobstructed clear spaces that can be freely arranged over the life of the dwelling unit.

There is also the bonus of a gain in flexibility in the process of designing the primary-use spaces. Architects and planners have more time to work with clients, consultants, and others to plan these spaces more effectively. Additional costs may be saved by increasing the number of contractors who can bid on a single job that can now be divided into several jobs. Smaller, more competitive firms, whose bond limitations might have precluded their undertaking a project as costly as an entire hospital, can bid on individual aspects of the project – say, the M/E core alone.

11.6   Conclusion

Affordable housing in the past has never been designed to last. It was aimed to provide a short-term solution – maximum number of houses in the shortest possible time (Figures 11.2 and 11.3) – to meet the urgent housing demand as if poverty and lack of affordable housing is a short-term problem. Although it is a government effort in providing adequate and affordable housing for the general population, the new contemporary household with its diversity of interior design needs in their consideration of future housing prototypes can no longer be ignored. Moreover, in the realm of current architecture and urban planning, affordable housing development is acquired to last into the future without becoming obsolete, as sustainability is increasingly embedded into building regulations and is no longer perceived as a novel idea pertinent to certain locations, populations, or building typologies. Realizing that past trends of energy and resource consumption for constructing and operating the built environment are no longer feasible to continue environmentally or economically, how architects design affordable housing that is intrinsically sustainable becomes the most essential. Many people think that sustainability is about planting more trees. Yet even more people believe that sustainability is about producing green energy with more solar panels. But very few realize that sustainability is all about ensuring the dynamic balance of the environment through the regulatory mechanism of divergent design processes.

This chapter suggests that the incorporation of flexibility in architecture is essential for the design of affordable housing that is environmentally, economically, and socially sustainable. D3 – a new design approach to the sustainable living system that derived from the tropical vernacular architecture – has lots to offer towards sustainable construction solutions in the development of affordable housing, as it is not only able to reduce both the initial build cost and running expenses of housing construction, but also balance affordability, durability, and adaptability in designing sustainable solutions that are resistant to obsolescence. Due to its consideration in addressing the shortcomings in the current housing development, as well as taking into account the beginning of the design process for the dwelling unit, the construction process, and the flexibility in future development, D3 is capable of redefining tropical housing as adaptable and resilient, and shows promise for the creation of more environmentally and economically sustainable architecture and infrastructure.

Demographics in developing countries are changing rapidly, with average households becoming smaller as an increasing number of people live independently in their later years. Prefabrication and modular construction are believed to be the solution for constructing houses that meet the vast number of demands in urban areas. Yet, as a housing strategy it is still considered unresponsive to local climates and conditions with low acceptance rate. One of the problems that Malaysian prefabricated housing industry does not perform as well as other countries (i.e. U.S., Japan, Korea) may likely be due to the lack of variability and an individual identified design. At this juncture, how prefabricated housing design can evolve from mass repetitive production to mass customization that accounts for flexibility and variability is the primary issue to be explored. The D3 system is deemed to bring improvement to the country's prefabricated housing industry with respect to time, cost, and quality, through the design of adaptability of individual residential units, the buildings that contain those units, and the surrounding site. On a much larger scale, there is current interest in the adaptation of the built environment for climate change. Because the rate and severity of future climate change is unpredictable, design for adaptation is best achieved by providing lifecycle options that will allow future decision makers to respond appropriately to the trajectory of climate change that actually occurs. D3 will surely place sustainable architecture along the curve followed by science, technology and industrialization, facilitating a shift towards higher quality housing development that eventually creates sustainable dwellings for everyone anywhere in the country.

References

Amad, A. M., Sujud, A., & Hasan, H. Z. (2007). Proxemics and its relationship with Malay Architecture. *Human Communication*, **10**(3), 275–288.

Che Amat, S. & Rashid, M.S. (2009). An Analysis of the Traditional Malay Architecture as Indicators for Sustainability: An Introduction to its Genius Loci. Proceedings of Arte-Polis 3rd International Conference on Creative Collaboration and the Making of Place, Bandung, Indonesia.

Erdayu, O. O., Esmawee, E., & Masran, S. (2010). Adapting by altering: spatial modifications of terraced houses in the Klang Valley Area. *Asian Journal of environment-Behaviours Studies*, **1**(3), 1–10.

Friedman, A. & Krawitz, D. (1998). The Next Home: Affordability through flexibility and choice. *Housing and Society*, **25**(1 & 2), 103–116.

Gan, H. B., Zuhairi, A. H., & Foo, C. H. (2015). Unleashing the potential of traditional construction technique in the development of modern urban mass housing. *Malaysian Construction Research Journal*, **16**(1), 59–75.

Gan, H. B. (2013). *Molecular Architecture*, Things To Come, GMA Publication.

Gan, H. B., Zuhairi, A. H., Foo, C. H., Mohd Khairolden, G., Maria Zura, M. Z., & Ong, K. T. (2013a). Rethinking the affordable housing in Klang Valley Region – An introduction to the divergent dwelling design (D3) Concept. Master Builders Journal, 3rd Quarter, 71-74.

Hassan, A. S. (2001). Towards sustainable housing construction in Southeast Asia. Agenda 21 for Sustainable Construction in Developing Countries, Asia Position Paper.

Hullibarger, S. (2001). Developing with manufactured homes. Arlington: Manufactured housing institute.

Isnin, Z., Ramli, R., Hashim, A. E., & Ali, I. M. (2012). Sustainable issues in low cost housing alteration projects. *Procedia – Social and behavioural Sciences*, **36**, 292–401.

Karim, H. A. (2012). Low cost housing environment: Compromising quality of life? *Procedia – Social and Behavioral Science*, **35**, 44–53.

Lim, J. Y. (1991). The Malay house: rediscovering Malaysia's indigenous shelter system, Institute Masyarakat/Central Books, Malaysia.

McKinsey Global Institute (MGI). (2014). A blueprint for addressing the global affordable housing challenge. McKinsey & Company.

Nordin, T. E., Husin, H. N., & Kamal, K. S. (2005). Climatic Design Feature in the Traditional Malay House for Ventilation Purpose. Proceedings of International Seminar Malay Architecture as Lingua Franca, 22-23 June 2005, Jakarta, Indonesia, 41–48.

Nurdalila, S. (2012). A review of Malaysian terraced house design and the tendency of changing. *Journal of Sustainable Development*, **5**(5), 140–149.

Ramli, N. H. (2012). Re-adaptation of Malay house thermal comfort design elements into modern building elements – Case study of Selangor traditional Malay house & low energy building in Malaysia. *Iranica Journal of Energy & Environment*, **3**, 19–23.

Rostam, Y., Hamimah, A., Mohd R. E., & Norishahaini, M. I. (2012). Redesigning a design as a case of mass housing in Malaysia. *ARPN Journal of Engineering and Applied Sciences*, **7**(12), 1652–1657.

Schneider, T. & Till, J. (2007). *Flexible Housing*. Oxford, United Kingdom: Architectural Press.

Siddharth, I. & Ashok, K. (2012). *Flexibility Concept in Design and Construction for domestic Transformation*. 7th International Conference on Innovation in Architecture, Engineering & Construction, 15th-17th August 2012, Sao Paulo, Brazil.

Singh, A., Barnes, R., & Yousefpour, A. (1999). High-turnaround and flexibility in design and construction of mass housing. Proceedings IGLC-7, 26th-28th July 1999, University of California, Berkeley, CA, USA.

Živković, M. & Jovanović, G. (2012). A method for evaluating the degree of housing unit flexibility in multi-family housing. *Architecture and Civil Engineering*, **10**(1), 17–32.

Zuhairi, A. H. Foo, C. H., & Gan, H. B. (2015). Crossing the unlimited Architecture possibilities: Advancing the art of design and aesthetic using precast innovation. *Concrete Plant International*, **3**, 168–171.

Index

Modernisation, Mechanisation and Industrialisation of Concrete Structures, First Edition.
Edited by Kim S. Elliott and Zuhairi Abd. Hamid.
© 2017 John Wiley & Sons Ltd. Published 2017 by John Wiley & Sons Ltd.